A Covenantal Imagination

A Covenantal Imagination

Selected Essays in Christian Social Ethics

WILLIAM JOHNSON EVERETT

RESOURCE *Publications* • Eugene, Oregon

A COVENANTAL IMAGINATION
Selected Essays in Christian Social Ethics

Copyright © 2021 William Johnson Everett. All rights reserved. Except for brief quotations in critical publications or reviews, no part of this book may be reproduced in any manner without prior written permission from the publisher. Write: Permissions, Wipf and Stock Publishers, 199 W. 8th Ave., Suite 3, Eugene, OR 97401.

Resource Publications
An Imprint of Wipf and Stock Publishers
199 W. 8th Ave., Suite 3
Eugene, OR 97401

www.wipfandstock.com

PAPERBACK ISBN: 978-1-6667-3154-5
HARDCOVER ISBN: 978-1-6667-2418-9
EBOOK ISBN: 978-1-6667-2419-6

.   NOVEMBER 10, 2021 10:49 AM

Contents

Acknowledgements | vii

Introduction | xi

1	CYBERNETICS AND THE SYMBOLIC BODY MODEL	1
2	LITURGY AND AMERICAN SOCIETY *An Invocation for Ethical Analysis*	13
3	ECCLESIOLOGY AND POLITICAL AUTHORITY *A Dialogue with Hannah Arendt*	32
4	VOCATION AND LOCATION AN EXPLORATION IN THE ETHICS OF ETHICS	44
5	LAND ETHICS: *Toward a Covenantal Model*	68
6	STEWARDSHIP THROUGH TRUST AND COOPERATION	89
7	SHARED PARENTHOOD IN DIVORCE *The Parental Covenant and Custody Law*	97
8	OIKOS: *Convergence in Business Ethics*	111
9	TRANSFORMATION AT WORK	131
10	SUNDAY MONARCHISTS AND MONDAY CITIZENS?	151
11	COUPLES AT WORK *A Study of Patterns of Work, Family, and Faith*	157

| 12 | HUMAN RIGHTS IN THE CHURCH | 182

| 13 | CONSTITUTIONAL ORDER IN UNITED METHODISM AND AMERICAN CULTURE | 203

| 14 | SEALS AND SPRINGBOKS
Theological Reflections on Constitutionalism and South African Culture | 242

| 15 | RECONCILIATION AS NEW COVENANT, NEW PUBLIC | 255

| 16 | SERVING THE CHURCH AND FACING THE LAW
Virtues for Committee Members Evaluating a Pastor | 267

| 17 | PUBLIC WORKS
Bridging the Gap between Theology and Public Ethics | 277

| 18 | RECONCILIATION BETWEEN HOMECOMING AND FUTURE
A Case Study from America's Struggle with the Vietnam War | 294

| 19 | JOURNEY IMAGES AND THE SEARCH FOR RECONCILIATION | 308

Bibliography | 335

Acknowledgements

"Cybernetics and the Symbolic Body Model" was originally published in *Zygon* 7:2 (June 1972) 98–109. It is reprinted here with permission from John Wiley & Sons.

"Liturgy and American Society: An Invocation to Ethical Analysis" was originally published in *Anglican Theological Review* 56:1 (January 1974) 16–34.

"Ecclesiology and Political Authority: A Dialogue with Hannah Arendt" was originally published in *Encounter* (Indianapolis) 36:1 (Winter 1975) 26–36, and is reprinted here with permission from the ATLA Index.

"Vocation and Location: An Exploration in the Ethics of Ethics" was originally published in *Journal of Religious Ethics* 5:1 (Spring 1977) 91–112. It is reprinted here with permission from John Wiley & Sons.

"Land Ethics: Toward a Covenantal Model" was originally published in *American Society of Christian Ethics, 1979: Selected Papers from the Twentieth Annual Meeting* (Waterloo, Ontario: Council on the Study of Religion, 1979) 45–74.

"Stewardship Through Trust and Cooperation" was originally published in *Stewardship Papers* ("Spirituality Series, No. 3"; Chicago: Pax Christi USA, 1980) 1–4.

"Shared Parenthood in Divorce: The Parental Covenant and Custody Law" was originally published in *Journal of Law and Religion* 1984 (2:1) 85-99. It is reprinted here with permission from Cambridge University Press.

"OIKOS: Convergence in Business Ethics" was originally published in *Journal of Business Ethics* 5 (1986) 313–25. It is reprinted here with permission from Springer Nature.

Acknowledgements

"Transformation at Work" was originally published in *Religious Education as Social Transformation*, edited by Allen Moore, 153-76. Birmingham: Religious Education Press, 1989.

"Sunday Monarchists and Monday Citizens?" was originally published in *The Christian Century* 106:16 (May 10, 1989) 503-5.

"Couples at Work: A Study in Patterns of Work, Family and Faith" (with Sylvia Johnson Everett) was originally published in *Work, Family, and Religion in Contemporary Society*, edited by Nancy Tatom Ammerman and Wade Clark Roof, 305-29. New York: Routledge, 1995. It is reprinted here with permission from Taylor and Francis Group LLC.

"Human Rights in the Church" was originally published in *Religious Human Rights in Global Perspective: Religious Perspectives*, edited by John Witte Jr. and Johann van der Vyver, 121-42. Dordrecht, London, Boston: Martinus Nijhoff, 1995. It is reprinted here with permission from Brill Publishers.

"Constitutional Order in United Methodism and American Culture," co-authored with Thomas E. Frank, was originally published in *Connectionalism: Ecclesiology, Mission, and Identity*, edited by Russell E. Richey and Dennis Campbell, 41-73. Nashville: Abingdon Press, 1997. It is reprinted here with permission from United Methodist Publishing House.

"Seals and Springboks: Theological Reflections on Constitutionalism and South African Culture" was originally published in *Journal of Theology for Southern Africa* 101 (July 1998) 71-82. Republished with permission from the JTSA.

"Reconciliation's Public Face" was originally published in *New Routes: A Journal of Peace and Action* (Uppsala) 1998 (3:4) 29-33. It also appeared as "Reconciliation as New Covenant, New Public," *Agenda for Peace: Reconciliation*, ("Loccumer Protocolle," 55/98; Rehburg-Loccum, 1999) 140-51.

"Serving the Church and Facing the Law: Virtues for Committee Members Evaluating a Pastor" was originally published in *Practice What You Preach: Virtues, Ethics, and Power in the Lives of Pastoral Ministers and Their Congregations*, edited by James F. Keenan and Joseph Kotva Jr., 268-79. Franklin, WI: Sheed and Ward, 1999. It is reprinted here with permission from Rowman and Littlefield.

"Public Works: Bridging the Gap Between Theology and Public Ethics" was originally published in *Theological Literacy for the Twenty-First Century*, edited by Rodney L. Petersen, 150-65. Grand Rapids: William B.

Eerdmans Publishing Company, 2002. It is reprinted here with permission from William. B. Eerdmans Publishing Co.

"Reconciliation between Homecoming and the Future: A Case Study from the Vietnam War" is a translation by the author of "Versöhnung zwischen Heimkehr und Zukunft: Eine Fallstudie aus dem amerikanischen Vietnam-Krieg," which was originally published in *Politik der Versöhnung*, edited by Gerhard Beestermöller und Hans-Richard Reuter, 169-80. Stuttgart: Verlag W. Kohlhammer, 2002.

"Journey Images and the Search for Reconciliation" was originally published in *Journal of the Society of Christian Ethics* 23:2 (Fall/Winter 2003) 155-78.

Introduction

THESE ESSAYS FROM OVER thirty years of research and teaching in the field of Christian social ethics display the development of several enduring themes that have guided my work. As I have reviewed these essays, whether about the public life of the church, the challenge of ecological responsibility, or the meaning of reconciliation, what continually comes to the fore knitting these thoughts together is the overarching theme of covenant. Only dimly present in my early work on body metaphors in social thought, an understanding of covenant has emerged to guide many facets of my work, providing an overarching framework of interpretation and constructive ethical vision.

It has performed this work with a team of other concepts: publicity, vocation, federalism (itself a covenant term), reconciliation, and the concept of the *oikos* knitting together work, family, faith, and the land. Together they have woven the tapestry of thought laid out in this series of essays. As you read through them you will see that in some way I am ringing contemporary changes on this ancient concept rooted in Biblical religion and the history of European and American constitutionalism. You will also see ways this concept is being deployed as a lens to investigate analogous images of human relationship beyond these cultures. Other times it appears as a tincture that brings out other colors and dimensions of its companion concepts. At all times, it works to serve an ethical task of imagining future action that better embodies the values, virtues, and goods that underlie life's possible flourishing on this planet. In this introduction I want to walk you through this journey as it unfolded.[1] While I have lightly edited the essays for consistency of format and

---

1. I have laid out the wider context of this journey in the companion to this volume, Everett, *Making My Way*.

clarity, I have left other aspects, like gendered language in my early work, intact for the historical record.

"Cybernetics and the Symbolic Body Model," published in *Zygon: A Journal of Religion and Science* in 1972, lifted up the central theme of my doctoral dissertation, "Body Thinking in Ecclesiology and Cybernetics." The centrality of powerful symbols like the body in shaping our ethical life, both personally and institutionally, has remained a core awareness in my writing and research. In the dissertation I tried to show how the "body thinking" accompanying this symbol in Western social theory and ecclesiology could be reworked using cybernetic social theory in order to avoid the authoritarianism or social or religious domination often expressed through the classical use of this symbol. At the time, I was more concerned with this work of understanding and revision than with the development of a covenantal vision as an alternative. Though this work was already present in my study at Harvard with James Luther Adams as well as my reading of Michael Walzer's *The Revolution of the Saints*, I was not ready to develop this programmatic possibility.

"Liturgy and American Society: An Invocation to Ethical Analysis" was presented at the 1972 Annual Meeting of the American Society of Christian Ethics and subsequently published in the *Anglican Theological Review* in 1974. Drawing on the sociological perspectives I had absorbed through my work at Harvard with Robert Bellah and Talcott Parsons, it extended symbolic analysis into a wider cultural context and began my exploration of the way symbols shape institutional ethics through the activity of worship. Going beyond the sense of society as a single organism that is implicit in foundational work by Emile Durkheim and others, this essay takes on the task of trying to understand how these liturgical dynamics play out in a pluralistic and more conflicted society.

Hannah Arendt first entered my awareness when she was in residence in my senior year at Wesleyan University in 1961–62, a period when she was finishing up the work published as *On Revolution* in 1963. While I was not in her seminar, her thought left a lasting impression on me and has shaped much of my thinking about the life of action at the core of public life. I subsequently was involved in a study group formed in the American Society of Christian Ethics to study her significance for Christian ethics. This resulted in an invitation to her to address the Society at its annual meeting in January 1973. I was privileged to present a paper on the possible ecclesiological significance of her thought at that meeting. While she wrote to me later that her understanding of

Christianity did not support the line of thought I was taking in this paper, the perspective opened up in it continued to develop in my later work in what I and others have found to be very fruitful ways. A further draft of this paper was published in *Encounter* (Christian Theological Seminary) as "Ecclesiology and Political Authority: A Dialogue with Hannah Arendt" in 1975. Here the concepts of action, of public assembly, and the more dynamic political meaning of my covenantal imagination began to take shape.

The social and political turmoil of the late 1960s and early 1970s forced many scholars and teachers in ethics to confront the tension between active advocacy in public life and the work of careful reflection and study characteristic of academic professions. In "Vocation and Location: An Exploration in the Ethics of Ethics" I introduced the typology of systemic, pluralist, and dualist social theories to trace out the impact of ethical study in the wider culture and society. These were components of the methodological work I was doing with my late colleague T. J. Bachmeyer, which emerged in 1979 as *Disciplines in Transformation: A Guide to Theology and the Behavioral Sciences*. After pointing out how the social location of most ethicists shapes their work and their impact, I turned to the way the very requirements of being a profession, and not merely an academic discipline, should lead them to promote and defend a range of societal conditions, such as freedom of information, relative equality of access to public life, and maintaining space for a plurality of opinions. This article attuned me more deeply to the demands of the professional work I was taking on in the midst of many commitments in church and local political life that were emerging for me in Milwaukee, Wisconsin, in those years.

It was the ecological question of land that first brought out an explicit covenantal focus in my writing and research. This was driven in part by my own family's struggle to preserve a farm operation in Virginia in the face of overwhelming forces of urbanization and agricultural policy. "Land Ethics: Toward a Covenantal Model" was presented to the Annual Meeting of the American Society of Christian Ethics in 1979. It displays the careful methodological framework (one reader spoke of my "Everettian grids") that shaped my work at the time, but directs it to the very practical problems of land tenure, use, alteration, and transfer. But with this paper methodological questions began to fade backstage in service of the production of substantive proposals for ethical action. While it didn't save the family farm, it began to open up the wider ecological work that emerged five years later in the OIKOS Project on Work, Family, and Faith. "Stewardship Through Trust and Cooperation" is a short paper based on these ideas for a conference sponsored by Pax Christi in 1980 on the idea of stewardship.

Personal experiences also motivated the turn to family law expressed in the article "Shared Parenthood in Divorce: The Parental Covenant and Custody Law," which appeared in the *Journal of Law and Religion* in 1984. Here the covenantal perspective begins to be fleshed out in terms of family ethics and public policy, mirroring some of the work I was doing in land ethics. In both cases the covenantal perspective seeks to honor the exercise of personal freedom with the deep responsibilities of our human condition, as Arendt would formulate it. Many of these policy directions have been introduced into American family law in the intervening years, but the questions of how we nurture the covenantal bonds required in the exercise of greater personal freedom remain to be addressed by religious and cultural institutions.

The OIKOS Project on Work, Family, and Faith, which my wife Sylvia and I began in 1984, brought together these ecological and familial concerns around the practical problems of developing more sustainable social relationships informed by classical faith commitments. It led to numerous workshops, presentations, research, and writing over the next dozen years. "OIKOS: Convergence in Business Ethics" appeared in the *Journal of Business Ethics* in 1986. It laid out the basic concepts and images that shaped my analysis of the differentiation of work, family, and faith in the wake of the industrial revolution.

Though its title did not include land, the project gave considerable attention to the wider ecology of these developments. The Greek word "*oikos*" was a way of capturing the essential unity of all these components, still contained in our English words economics, ecumenics, and ecology. At that time the field of business ethics was beginning to struggle with issues of work and family in response to the sea change of women's participation in the paid work force. This article attempted to expand the conversation and continued to influence my teaching, research, and writing in subsequent years. The article also further developed the significance of deep symbols of trusted relationships, such as body or *oikos*, for concepts of leadership and management, an emphasis which continued to inform my work in preparing people for formal ministry in the church, especially their work of institutional leadership.

"Transformation at Work," which appeared in *Religious Education as Social Transformation*, edited by Prof. Allen J. Moore at Claremont School of Theology, carried the concerns of this project into the field of religious education. It focused especially on the way religious education in America for the past two centuries has responded to changes in the structure and content of work. It also began to develop the significance of the concept of "covenant publicity" (or "covenantal publicity") as a bridge for formulating the mission of religious education in light of radical changes in the workplace. This concept would be greatly expanded in my subsequent work.

Covenant publicity was drawn from my 1987 book *God's Federal Republic: Transforming our Governing Symbol*, which has oriented much of my work to this day. Written as a companion to my *Blessed Be the Bond*, which focused on marriage and family, it shaped the wider context for concerns of the OIKOS Project as well as the ongoing exploration of the relationship between key governing symbols and public life, including the church.

The short article "Sunday Monarchists and Monday Citizens?" that appeared in *The Christian Century* in 1989 took up the struggle between monarchy and federal republicanism that emerged in *God's Federal Republic*. Here I explored the contradiction in much Christian worship between our deepest commitments to democratic participation in constitutional republics and the symbols of monarchy and patriarchy that infuse Christian tradition. In this little piece you can see elements of my concern for the connection of worship, ethics, and political theory that were expanded greatly in subsequent books and other writings.

The article "Couples at Work" emerged from a research project that Sylvia and I conducted in collaboration with Nancy Ammerman and Scott Thumma at Emory University in the early 1990s. It was motivated in part by our own search for a pattern of work collaboration that would best express our deeper bond in marriage. Working out of the framework of the OIKOS Project, it identified the types of relationships and organizational patterns generated by couples and identified where such collaboration could be fostered and how it could be thwarted. Picking up from leads in the article "Transformation at Work," it identified the impact of these work and marriage patterns on relationships to religion and traditional religious values.

My constructive work in *God's Federal Republic* led to an extended period of research into the relationship of churches to constitutional republics in various cultural settings, culminating in 1997 with *Religion, Federalism, and the Struggle for Public Life: Cases from Germany, India, and America*. In the subsequent article, "Human Rights in the Church," I explored the various relationships between churches with very different ecclesiologies and the rise of human rights doctrines and practices in the twentieth century. Presented at a conference convened by John Witte Jr., Director of Emory University's program on Religion and Law, it dealt with issues of freedom of speech in the church, confidentiality, church autonomy, marriage law, and church administration as they are affected by human rights laws in their respective political environments.

The importance of ecclesiology in understanding religious issues in public life received additional expression in "Constitutional Order in United Methodist and American Culture." This article, written with Thomas E. Frank, my colleague at Candler School of Theology, Emory University, appeared in slightly briefer form in 1997 in the collection *Connectionalism: Ecclesiology, Mission and Identity*, edited by Russell Richey and others. It brings together my own interest in the relation of ecclesiology to the constitutional principles of public assembly and federalism with Thomas Frank's deep knowledge of United Methodist polity and history. Since it was written, several changes in the United Methodist *Book of Discipline* have been made that are not explored in this text. Moreover, deep conflicts over issues of sexuality and scriptural interpretation have brought about a seemingly inevitable division and restructuring of the United Methodist Church, partly along the lines we were discussing in this document.

In particular that article lifted up the way in which covenant and federalist theory might help Methodists move beyond their historic ties to episcopal order and corporate-bureaucratic institutions to organizational forms that can deal more adequately with vastly different cultural contexts as well as people's needs to express their baptismal citizenship in appropriate church publics. Once again, concepts of covenant publicity and public assembly played an important role in the vision we lifted up there.

My exploration of the relation of religion to the development of federal republics focused on Germany, India, and America around the time of the collapse of the old Soviet Union in 1989 to 1992. It was not until 1998 that I was able to spend time in South Africa examining the role of religion in their own transition from apartheid to democracy. Over the next fifteen years the South African experience lay at the center of much of my research and writing.

"Seals and Springboks" refocused my long interest in governing symbols and covenantal publicity on two symbols of South Africa's transition—the transformation of the prison on Robben ("Seal") Island into a museum and the transformation of South Africa's sports culture, specifically their revered rugby team, the Springboks. I concluded with the image of the sheltering tree of the village council ground, a symbol that gained new centrality in my life as I turned to woodworking for inspiration in my transition to retirement from teaching.

My first visits to South Africa focused on the work of the Truth and Reconciliation Commission, anchoring that work in the core of my developing understanding of covenantal publicity. "Reconciliation as New Covenant, New Public" was presented at a conference at the Evangelische Akademie in Loccum, Germany. Here I elaborated the way reconciliation processes underlie the envisioning of a new future essential to the forging of a new covenant for public life. I also linked the covenants necessary for public life to a renewed covenant with the earth, that is, to the ecological dimension of our common life, symbolized by the ancient practice of sabbath. Finally, in brief compass, I tied in the work of confession and truth-telling essential to the life of a public assembly, something even more compelling twenty years later.

In 1995 I moved to Andover Newton Theological School to become the Herbert Gezork Professor of Christian Social Ethics. This put me in close contact with the graduate theological program of the faculty at Boston College as well as other schools in the Boston Theological Institute. In this context I could renew my conversation with Roman Catholic ethicists sensitive to the relation of ethics to ecclesiology.

James Keenan SJ invited me to write the essay on "Serving the Church and Facing the Law," which appeared in a collection on virtue ethics and pastoral ministry. In it I rehearsed some of the issues in the tension between ecclesial autonomy and civil law, probing the importance of the character and moral compass of actors within the church, acting often in the gray areas beyond the bright lines of legal prescription. This need for professional autonomy tests the meaning and limits of publicity and formal processes of adjudication. It heightens the need to cultivate virtues that elicit trust through humility, truth-seeking, and capacity for communicative interaction. The case study that frames this article helped me add some finer lines to the way I have come to understand the work of covenantal publicity.

In the same period, Rodney Petersen, Director of the Boston Theological Institute, asked me to contribute my thoughts to a wide-ranging collection of essays on what he called "theological literacy." My own contribution focused on the need for bridge language between the discourse of theology and that of the general publics in which theologians are speaking. "Public Works" explores the familiar landscape of separation between government and religion in the United States in order to examine how theologians can find bridge concepts to cross the classic wall of separation to serve both their religious mission and the needs of the wider common good.

Reaching back to the methodological values of my early work, I sought to respect the integrity of the two realms of discourse and find bridge concepts that might enable them to interact with one another effectively. To this purpose I elaborated on the concepts of covenant, public assembly, household (the *oikos*), and nature as bridges, given their widespread use in civil as well as religious discourse over many generations.

My work in South Africa brought the theme of reconciliation more clearly into the center of my work, leading to participation in a conference convened by my German colleague Hans-Richard Reuter and his colleague Gerhard Beestermöller in Heidelberg, Germany, where I presented the piece I have now translated into English as "Reconciliation between Homecoming and Future." In presenting this to a German audience I was also bringing together the experiences with efforts at reconciliation in two national settings that had been estranged in war and united in common peaceful purpose. My own piece is a detailed examination of the motivations, meanings, and impact of the memorial to what Americans call the Vietnam War, designed by Maya Lin in 1980 and dedicated in November 1982. The memorial serves a complex purpose, because, as I point out here, it brings together not only the tragedy of the war itself but also America's history of slavery and genocide as well as its exploitation and destruction of the earth. Thus, it embraces the dimensions of the classical covenant, both its violation and its repair. As such, the memorial creates a new public of shared grief as well as new resolve, linking the processes of covenant, reconciliation, and publicity.

My research on the American and South African struggle for greater reconciliation between settler and native peoples led me to write a historical novel, *Red Clay, Blood River*, that dramatized the intricate links

Introduction

between America's own "Trail of Tears" of Indian Removal and South Africa's "Great Trek" of Afrikaner consolidation. Both of these were journeys that became highly mythologized and laden with bitterly contested memories needing deliberate practices of reconciliation to overcome their ruinous consequences. As I prepared for this new departure in my writing I presented my reflections on the significance of journey stories for the work of reconciliation at the Annual Meeting of the Society of Christian Ethics. I added the story of Mao's "Long March" to the material from which I drew out eight types of stories, such as those of exodus, exile, exploration, and pilgrimage to explore the different modes of reconciliation they entail. Since journey stories always implicate a geography, I then related them to stories of place before ending with some reflections on the process of re-covenanting they can nurture.

This collection is in itself a kind of journey, one with enduring interests as well as developing understandings and approaches to the challenges they present. While these essays are expressions of only one uniquely situated observer they also clearly reflect much wider concerns of the global era emerging from the wreckage of the Great Depression and World War II. As the structures that were erected in that period are being attacked from many sides, this collection offers a perspective for claiming an ancient longing for trusted relationships among human beings and with the other creatures sharing this earth. Rich covenants upheld within public assemblies of equally respected persons remains one of the beacons to steer by in this storm.

1

CYBERNETICS
and the SYMBOLIC BODY MODEL

THE HUMAN BODY PLAYS a leading role in social and religious symbolism. Terms like Corpus Christi, body politic, corporation, body of knowledge, and corpus juris fill the speech of academics and laymen alike. Our perceptions of social and material reality are deeply affected by this metaphorical use of the word "body." Exhortation to the loftiest tasks of self-sacrifice or salvation appeal frequently to body symbols. Calls to civic action often depend on an analysis of the "cancerous growth" or "sickness" in the social body. Men are even willing to die for a body politic which is their Mother or Bride. A people's sense of the historical fittingness of acts can be greatly determined by the view that a society's history is the growth of a body to maturity and senescence. Similarly, the human body can be used to depict the whole universe and cosmic history as an eternal or near-immortal body. The symbolic use of the body conditions much human thought and action. In this regard it is a basic component of ethical reflection and morally purposeful action. In this essay I shall develop the concept of the symbolic body model and indicate how cybernetics may affect its implications for contemporary society.[1]

I have spoken of the body as a symbol. By a symbol I mean a representation (usually an image), rich with associations and extrapolations, which is strongly tied to basic human purposes. Paul Tillich's notion of

1. I have developed the ideas presented here more expansively in Everett, "Body Thinking."

a "religious symbol" and Susanne Langer's idea of a "charged symbol" are very close to this view.[2] In Freudian terms, the symbol is an object of cathexis. It elicits deep and often pre-rational response and is therefore a primary aspect of human motivation. The body symbol has frequently been bound to the deepest kind of drive for self-perfection and survival. In religious contexts it plays a fundamental role in formulating and expressing the drive for salvation.

Symbols can undergo refinement and inner differentiation in order to make a precise impact on our more purposive actions and thought. They become models for us in rational life. A model is a means for depicting some unfamiliar process or object in terms of one that is familiar. usually by using the familiar one to draw out the basic structural properties of the other. A symbolic model is a symbol that has found modular elaboration. Some basic symbols, such as fire, water, or earth, have relatively limited rational refinement. The body symbol, however, has adopted and refined some of the most sophisticated models in human history. The progression from symbol to symbolic model usually demands the incorporation of other models from different contexts. Thus, the body symbol has drawn upon the machine, plant, lower organisms, architecture, and shipbuilding to gain full modular precision.

Once a symbolic model has become refined, it can slip from reference point to reference point. With regard to human organization, it can refer to a small group. a large organization, the nation, and humanity, not to mention the whole cosmos. In Buddhism and Christianity it has been applied to the religious community. Through this referential slippage it can direct human loyalties to many different areas. The sphere to which the symbol is attached becomes the center of value for the self. That referent now bears the hopes of that self for perfection. survival, and salvation. The referent of the symbolic body model becomes some kind of "pure," "subtle," or "mystical" body, to which one must adhere to overcome the fragilities of the individual self. The symbolic body model has thus functioned as a basis for evoking deep loyalty to social and cosmic "bodies" and has bound these rather inchoate loyalties to sophisticated schemes for defining human action and organization. The relation of "head" to "members" and the relations of the members to each other can be spelled out in considerable detail. Whether this social body refers to a specific

---

2. Tillich, "The Religious Symbol," and Langer, *Philosophy in a New Key*.

organization or to a future transcendent body makes enormous differences for ethics and action.

CYBERNETIC MODELS

We have seen recently the emergence of yet a new symbolic body model which draws upon cybernetics for its rational elaboration. What is this cybernetic body model? What are some of its implications? Will it really gain importance in human affairs? In the next paragraphs I shall expose the cybernetic anthropology and sociology which is typified in the work of Karl Deutsch.

The cybernetic view of the body has emerged from the work of men such as Norbert Wiener, Karl Deutsch, Arturo Rosenblueth, Anatol Rapoport, and W. R. Ashby, as well as the efforts in general systems theory pursued by Ludwig Bertalanffy, Kenneth Boulding, and others.[3] Wiener defined cybernetics as the study of "communication and control in animals and machines." Cybernetics maintains that control is ultimately a matter of communication. Moreover, this communication can be understood in such a way—namely, mathematically—that the same rules pertaining to machines can also apply to men, whether they be individuals or organizations. At first blush we seem to have returned to the world of mechanics as the model for understanding. Mechanical procedures are used to describe the transfer of patterns of information. However, through immense complexification of this mechanical interaction, phenomena emerge which hitherto could only be explained in terms of "spirit" or "organism." How is this so?

The basic unit of communication is a "bit" of information conveying the signal "yes" or "no" (i.e., "on" or "off"). With enough yes–no indicators one can detect grades of intensity as well as handle mathematical problems. The machine of yes-no devices thus processes incoming signals in accordance with certain preconditions in its program. We have now reached the point in computer research that even these conditions can be changed with regard to more general "purposes" present in the program

---

3. Deutsch, *Nerves of Government*, presents the furthest-ranging exploration of the implications of cybernetics for political science. I have drawn on his work extensively in formulating my own position. See also Wiener, *Cybernetics*, and Wiener, *Human use of Human Beings*. For Ludwig Bertalanffy, see Bertalanffy, *Robots, Men, and Minds*, as well as the journal *General Systems*. For Kenneth Boulding, see Boulding, *Organizational Revolution*.

at a higher level. By extrapolating this property of computers we come to an explanation of mind itself.

Every discrete aspect of information processing, or "mind," is thoroughly rational, quantifiable, and calculable. There are no wispy spirits or transcendent incursions involved. All novelty, purposefulness, and memory are a matter of complexification of the basic units of operation. Thus, it is claimed that we have a thoroughly rational theory of mind without sacrificing explanation of matters hitherto relegated to mystery.

Hence, the self appears under two aspects. On the one hand, it is totally mind, in the sense that it is an elaborate system of communication. The self is fundamentally a particular form of organization of information—from its genes and chromosomes to its cerebrum. On the other hand, it is entirely body, or material. All information processing is the operation of complex material mechanisms. We thus have a comprehensive model for the bodily self.

This self is characterized by homeostatic mechanisms, memory, and hierarchy. It is a system of interlocking structures which preserves its unity by tending toward some kind of equilibrium or homeostasis. It tries to achieve some kind of balance between demands from the outside environment and demands from the inside environment. If it exceeds certain critical margins, it will destroy itself. Achieving this equilibrium is not only the simple matter of obtaining food, water, warmth, and safety. The complex processes of the mind also spin out very general purposes and values according to which action is determined. In some cases, to forsake these deeply held purposes, even at the risk of other deprivations, is a mortal threat to the balance of the self. These values need not arise out of the peculiar operations of each self. They may be the product of the public mind, whose processes span many centuries. Therefore, we cannot really say cybernetics is "materialistic," for it deals with the transfer of *patterns* of information. These patterns are desired states described as principles of arrangement, which in themselves are quite ephemeral indeed.

The purposive activity of the self can be explained in terms of the survival necessity of adjusting to internal and external environments as well as to the established values of the self's information-processing mechanisms. What we call purpose is the action of tending toward a given state of equilibrium among the many forces constituting the self. These margins of survival, of course, will differ from self to self and culture to culture. Detection of the gap between actual behavior and these margins occurs through feedback mechanisms which monitor the effects of actions. Since feedback has become a very popular concept, I shall not develop it here. It must be noted, however, that it refers to the monitoring of both the internal and external conditions of the body system and of the impact made on those conditions by the self.

If the mind is fundamentally an enormously complex computer, then its primary characteristic is memory. The mind is a process of memory. It not only recalls information from feedback sources but also contains the established goals, purposes, and values according to which the self selects various alternatives presented by the environment. The memory breaks down complex inputs into their separate components and then can imagine a great variety of possible novel recombinations of these units. Thus arise proposals for new kinds of actions, responses, and goals. Memory is not a graveyard of the past but a process of assembly—a beehive of continuously interacting information units.

Finally, the processing of information is a hierarchical process. The homeostatic demands of the self require decisions among alternative courses of action. The self is a decision center. Decisions can be made only with regard to some hierarchical criterion of importance. Not only are the various values and goals hierarchically arranged, so are the very structures of feedback. Some signals can be processed at very simple levels, such as instinctive blinking or subconscious reactions to stimuli. Others must reach higher levels to receive adequate treatment. There is a relative decentralization of mind according to the character of recurring needs, variety of possible actions open to the self, and simplicity of response necessary. In every case, the mind is hierarchical because it is a means for deciding upon responses to environmental conditions.

In all three of these aspects mind emerges here as not a receiver and transmitter of some static images, which are then conveyed to the "lower members," as in the classical Platonic view of the relation of mind to body. But mind is merely the activity of the body as it responds to changing environments. Any generalized images of God or the cosmos are the results of the complexification of human mind in making these responses.

The self and human culture generally are therefore complex constructions. Any mysteries, spirit, and non-empirical realities they contain are sheerly the result of immense complexification. Such a view is quite congenial with the modern scientific temper but clashes harshly with traditional realistic philosophies. The dispute does not involve the rejection or acceptance of certain experiences (especially religious), but the means for interpreting them. Some cybernetic theories, such as Karl Deutsch's, contain an ethical methodology similar to that of natural law theories. They move from a statement of actual tendencies, such as the drive for dignified survival, to a set of prescriptions for behavior, such as openness, flexibility, adaptation.

Cybernetic approaches, sketched here only broadly, are already influential in theories of cognitive development, neurology, anthropology, and certain forms of psychotherapy.[4] But what are their implications for the use of the symbolic body model in human affairs, especially at the societal level? Let us deal with these implications in two stages: first, those which are unequivocal and, second, those which are equivocal and ambivalent.

UNEQUIVOCAL IMPLICATIONS FOR HUMAN AFFAIRS

Certainly, a cybernetic model locates the sources of control in the information centers of an organization or society. Even with regard to American society, we see that the locations of power in data banks, files, and intelligence agencies require new interpretations of the Constitution and Bill of Rights. Power in a highly cybernated society assumes forms unknown to the Founding Fathers. A cybernetic model helps us understand the actual configuration of power in the society.

Second. this kind of power is implicitly public because it is rational and quantifiable. Its secrecy is only a policy decision. The power of information does not arise from an arbitrary will of persons or their strength but from the capacity of information to reveal to us the objective conditions under which we must operate in order to achieve goals. Arbitrary action only leads to lack of coordination and powerlessness over complex environments.

Third, a cybernetic model emphasizes the necessity for clear separation between the functions of feedback and those of execution. If the hierarchy of monitoring and that of execution are confused, then the whole system loses accurate control over its effects. It becomes ignorant of what it is doing or loses sense of its goals by adapting execution orders merely to accord with previous responses. In the case of American

---

4. I have in mind the work of John Dollard and Jerome Bruner in psychology, Warren McCullough in neurology, and Claude Levi-Strauss in anthropology. Cybernetics has even appeared as the basis for self-help psychotherapy of the positive-thinking variety in Maxwell Maltz, *Psychocybernetics*.

government this implies an organizational expansion in the legislative sphere to monitor the activities of the executive branch. The idea that Congress should only legislate is a pre-cybernetic conception of governance. Similar changes have already occurred in industry with the rise of elaborate hierarchies of quality control.

Separation of channels is just one way an information system overcomes entropy, that is, the deterioration of messages. Cybernetics leads us to improve social communication in an effort to overcome damaging conflicts. However, such increased clarity may actually sharpen conflicts in the short run by bringing groups to self-consciousness about their own interests. Moreover, truly rational decision making in a cybernetic scheme requires that all messages be translated into one quantifiable spectrum in order to weigh one against the other. But how are subjectively oriented demands to be compared? Cybernetics can direct us to calculations of the consequences of various alternatives, usually in terms of money cost, but this has great limitations for policy making. However, the goal of rational decision making is held out to us. Whether the world and human affairs are ultimately rational is an eschatological and theological question.

Finally, cybernetics emphasizes the organizational aspects of any large or complex grouping. Human affairs are to be understood in terms of the problems of adapting societal systems to environments. This requires policy decisions based on adequate information, clear goals, and effective execution. Of secondary analytic importance is legitimation of accommodation to a plurality of conflicting, fairly independent groups within the social arena. Survival demands tighter coordination. In times of acute social sensitivity to survival needs we would expect increased rhetorical use of symbolic body models. This would also be true of the cybernetic body model, with the proviso that "unity" demands a fairly free circulation of objective information about the environment. Moreover, these messages must not be contaminated by preexisting decisions. In broader respects, however, the cybernetic model falls in the tradition of organic social theory, which emphasizes the needs of the total social system, rather than nominalistic or conflict theories, which emphasize the needs of persons or small groups. This is only another indication of the basic compatibility between the cybernetic model and other symbolic body models.

One can still question whether a jaundiced view of the gaps and conflicts among participating groups in a society is really bound to the cybernetic body model. Could it not be that the cybernetic system does not refer to the overall society but to the conflicting groups within it? Yes, of course, though if the small groups were tightly organized, the society would be seen as an open arena or theater of history. With this observation we can move to the second stage of implications, in which our choice of referent makes a great difference to our description and ethical prescription.

AMBIVALENT IMPLICATIONS FOR HUMAN AFFAIRS

The capacity to slip from one to another among various referents makes it possible for a symbolic model to be powerful in human affairs, for it can transfer loyalties from smaller and more familiar groupings to much more impersonal, distant, and comprehensive realities. According to the reference point taken, the cybernetic model can justify either bureaucratic or libertarian theories of society—the former featuring tight coordination and clear definition of information channels, the latter exposing the loose relations among independent groupings within a social order.

If the primary referent is persons or small groups, then wider social systems must undergo a mutual accommodation with them to accord them liberty. The larger society would be seen merely as an environment with its own, sometimes opposed, interests. One could infer from the cybernetic model that accommodation might mean merely the complete victory of the stronger body. In that case the cybernetic model would be only an analytical device for understanding that conquest. However, inasmuch as the cybernetic symbolic model becomes a bearer of hopes and expectations normally surrounding the perfection and survival of the body, it gains normative power and direction. In that case each body has a right to survive and, in Karl Deutsch's view, to survive with "dignity." This means that bodies should reach positions of mutual accommodation. If they cannot do this, then more expansive systems will arise to adjudicate these disputes and enforce judgments. The possibility of such a pluralistic and libertarian society depends finally on the rationality and essential good will of the disputants, as well as the possibility of rational and mutually enhancing (or at least not mutually destructive) solutions to these conflicts. This seems like an impossible possibility in human affairs. The continual extrapolation of the prime body to ever more expansive levels seems inevitable.

On the other hand, by the very same measure, if the primary referent is the larger society, then the perfection of that social body requires the subordination of its parts for the sake of surviving in the face of the environment. It is important to note, however, that a cybernetic model can be found only with difficulty to legitimate the kind of totalitarian or managed society often associated with body models, beginning with Plato. From a cybernetic perspective there are tendencies for decentralization in the very apparatus of execution and decision making. Without delegation of tasks the central information networks become clogged with trivial messages. In some cases this delegation can lead to the relative autonomy of these subordinate centers. In this case cybernetics suggests that undistorted information transfer is enhanced in societies when the various information centers are related voluntarily. If they do not have appropriate degrees of autonomy, clear messaging tends to be replaced by polemics and propaganda.[5]

However, it is also important to remember that feedback hierarchies have to report about objective conditions, especially those arising from the authorized actions of the system. Demands and desires of groupings outside this purview may go unheeded. In either case, even the most pronounced subordinationism would take care not to damage the functional capacities of the members. The tendency of most symbolic body models, as I have indicated, is to equate forcefully the welfare of selves with fulfillment of the functional needs of the more comprehensive system. It is not yet clear whether the cybernetic model will also be used in that way, despite Deutsch's use of it to endorse decentralization and a wider range of liberties.

We have already seen that a key characteristic of the symbolic body model is the way it expresses the drive for perfection of the self, that is, for ultimate survival. This feature is quite compatible with an essential aspect of cybernetic analysis, namely, the assumption that systems seek survival through accommodation with an environment. Their whole constitution is oriented toward making and executing decisions to achieve objectives within these margins. The prominence of this notion is another way in which the cybernetic model is a compatible modular refinement of the body symbol bearing these survival hopes. However, we then must ask, who or what is going to survive?

5. Karl Deutsch has had long-time interest in political confederation and decentralization. For a recent statement, see Kochen and Deutsch, "Toward a Rational Theory."

In the cybernetic model itself we see several possible answers. In the short run the whole structure of the system is to survive. But this can be altered in accord with the functional needs of the system. Finally, even these functional needs can be redefined in accord with the highest goal of survival itself. "What" survives, then, is sheerly the action of being autonomous and self-directing. To survive means to be in some sense self-controlling. In cybernetics this means to maintain the operation of information processing—in short, of minding. Body, that is, material structure, is taken up into mind, that is, the process of being autonomous.

Thus, we see the extrapolation of ever less tangible goals into the distant future. The idea of survival, which may have started with the simple need for food, has been perfected into a comprehensive and abstract value. This value, lying in the farthest future, can now return into the present as an ultimate value, that of minding and control. When translated into the referent, "society," it means that the central control and communication apparatus has rightful precedence over the simpler, less comprehensive ones. Moreover, just as this minding is the final survival good of the self in temporal extension, so this social minding becomes the good of the self in the immediate present.

In this extrapolation of the referential slippage of the body model under the impact of the self's drive for survival, we see two referential dimensions appear—those of time and of space. The symbolic body model produces ethical implications by being transposed into more expansive realms of time and space, thus creating a condition of subordination of the self to the projected pure self, whose survival is taken to be the precondition for perfection and survival. This is how the symbolic body model has always been a helpful companion to any naturalistic or natural law ethics. It enables us to translate the "is" of the body image into the "ought" of obedience to more comprehensive systems of action. Cybernetics, rooted fundamentally in descriptive science, becomes a model for normative reasoning in public affairs by being incorporated into the body symbol.

Since the cybernetic model accords so well with some of the perennial implications of the body symbol, it is highly likely that it will gain wide use as the contemporary dominant symbolic model. It will receive increasing employment in political discourse, where speakers seek to elicit specific actions from persons. It may also gain use in theories of the church, where it might offer a modern reinterpretation of the exact structural significance of the idea of the church as the Body of Christ.[6] Finally, it will be used extensively within bureaucratic organizations that are seeking to enlist a wider range of strong loyalties from functionaries within the organization.

With this observation we must ask more forcefully, Do all symbolic body models, including the cybernetic one, inevitably fasten upon the most comprehensive social structure, with the implication that they will lead us to make of all mankind a vast cybernetic organization? (Note that "man" is also a body symbol.) Does this tendency always imply a fundamentally anti-libertarian theory of society? Not necessarily. Like Hegelian philosophy, cybernetic body models can receive both right-wing and left-wing interpretations. Once again, this interpretation depends on the specific referential scheme. It is possible to extend the primary referent for the human body beyond mankind or even the cosmos. In terms of symbolic perfection and refinement this is quite a logical step. If this ultimate pure body is transcendent, then it can stand in judgment over any of the purported pure bodies of human societies and organizations. In that case comprehensive organizations inferior to the cosmos could claim only relative and tentative superiority at best. The question would then arise, Are the larger societies irrelevant as intermediaries between the self and this ultimate body? How similar this is to the Reformation critique of the medieval church!

The logic of survival projection would lead us to say that the more comprehensive societies are not irrelevant, for they make possible intermediate periods of survival, whether in securing necessities for us or in guaranteeing the preservation of our accomplishments in the cultural memory. In fact, one could criticize them only if they were not encouraging the survival of selves. Even if particular institutions were murderous, however, others of comparable kind could always claim to be the rightful heirs of their legitimate functions. The jails could always be replaced by "hospitals." In short, the shifting of references could never justify true anarchism, but could always justify at least a tentative legitimation of increasingly comprehensive social structures.

---

6. For examples see Mary Virginia Orna, *Cybernetics*, and Rudge, *Ministry and Management*.

CONCLUSION

In these respects we might say that the symbolic body model is one basis for rational sociality. By attaching human loyalties to larger spheres of action, it binds men in common life. By its modular precision it enables them to form rather precise expectations of one another. By its capacity to change reference points it maintains a certain judgment on any tightly conceived patterns of domination and obedience.

Cybernetics constitutes a formidable modular development of the body symbol because of its rather comprehensive scope of explanation, including in its range selves, machines, and societies. It bridges the old dichotomies between mechanistic and organismic philosophies with the embracing category of communication. At the ethical level it contains some features which militate against its use purely as a justification for bureaucratic totalitarianism, as some might expect of it. Finally, it is a conceptual scheme oriented not only to description but to employment for the sake of acting upon the world. Therefore, it parallels nicely the translation of "is" into "ought" within the body symbol itself.

The symbolic body model appears as a powerful vehicle for translating scientific and other kinds of models into social theories. By being incorporated into the body symbol, with its compelling appeal to desires for self-perfection and salvation, these models become effective means for channeling human activity. The fact that these symbolic models have diverse implications for social policy (in this case both libertarian and bureaucratic) only attests to their broad dominance in orienting men in social affairs.

Finally, the exploration of the symbolic body model is one way to trace the possible impact of development in one area, such as technology or science, on society and public policy. In that sense, this essay has been an effort at prediction and prophecy, as well as at understanding the way men think about society.

2

LITURGY *and* AMERICAN SOCIETY
An Invocation for Ethical Analysis

OVER THE PAST FEW years Americans have paid increasing attention to a wide variety of symbolic actions, demonstrations, liturgical changes and innovations within the churches, and the crystallization of many new ritual practices in such realms as sports, a youth counterculture, and politics.[1] In spite of their heterogeneity these symbolic and ritual actions all appear to have a liturgical quality about them that may be rooted in common causes or conditions. In this essay I want to examine the possible significance these activities might have for Christian social ethics and for our understanding of the relation between the Christian churches and American society.

Exploration of this terrain demands some preliminary definition of the concepts to be employed. Liturgy, ritual, cult, art, and drama indicate significantly different matters. Liturgy, the central concept here, singles out those actions which seek to relate persons and groups to a "higher" or more ultimate pattern of action which orients, determines, or legitimates their public actions.

Liturgy was originally a Greek term denoting "services" performed on behalf of the community or polis.[2] By the time of Aristotle it had come to mean any kind of service or obedience. The Septuagint bound the term

---

1. Cox, *Feast of Fools,* provides an excellent introduction to the concerns of this essay, though our methods of analysis differ markedly.
2. Kittel, "Leitourgen."

closely to the cult and priestly ritual of giving service to God. The term died out in Western Christianity but was revived in the sixteenth century and has received increasing use to this day. In light of its etymology it is appropriate that we employ the term to illuminate relations between persons and their community, especially those involving authority and obedience. "Authority" here pertains to a social relation in which an act ordered by the community or its representatives is obeyed by persons because it accords with their most deeply held beliefs and attitudes.[3] Liturgical action internalizes models of authority, subordination, and participation in people. It achieves some kind of psychological union between the self and the community's deepest beliefs and orientations, especially as these pertain to action.

Liturgy produces and is produced by social institutions. It is more a component of institutional life, even though that life be precarious, than it is a product of personal idiosyncracy. A liturgical action predisposes the participant to be at the service of those authorities or patterns of social relationships symbolized in the liturgy. The habits cultivated in symbolic action flow into the habits and dispositions of public, "real" action. Liturgies articulate the relationship between the ordering powers (whether they be offices, laws, or God) and subjects. Liturgy is therefore deeply involved in the relationship between the societal dimension, with its specific institutions, organizations, roles and offices, and the cultural dimension of values, orientations, overarching loyalties, and concepts.

Liturgy is always an action which is a paradigm for many other actions. It condenses in the form of rituals certain dispositions toward authority, community, and fellow citizens. It is impossible to talk about liturgy without talking about ritual. Liturgy is thoroughly ritualistic, because it must produce stable models for action and response. Ritual is a repeated set of symbolic actions.[4] But not all ritual is liturgy. There can be rituals which do not point to the service, authority, and obedience implied in liturgy.

"Art" and "drama" point to an even further distance from particular social patterns. In their modern form they serve to provide particular meaning patterns by which persons comprehend their peculiar existence. The rich variety of artistic conventions in the modern world makes

---

3. In this definition I am following the ideas of Carl Friedrich, *Philosophy of Law*, and Hannah Arendt, "What Is Authority?"

4. For a general discussion of the various definitions given this term see Leach, "Ritual."

possible meanings in an endless array—an array which is symptomatic of rapid social change but has little necessary relation to the production of large-scale, relatively stable institutions. Street and guerrilla theater as instruments for organizing people around specific goals, though partial exceptions to this claim, still do not present archetypal authority relations which might enable organizations to achieve these goals. It is this latter function which we ascribe to liturgy. Art and drama present the utopian vision. Liturgy articulates the relations within the community which envisions that goal.

The concept of cult accompanies that of ritual. By cult I mean small-scale ritual observance, usually focused around a charismatic leader. The cult's primary emphasis is on the internal life of the group, which usually has all the qualities of a *Gemeinschaft* and therefore has little elaboration in terms of serving the many specialized purposes of a large-scale society. Though the cult has potential to become the liturgy of a large-scale institution, its focus on elaborate esoterica and its exclusiveness greatly limit growth in this direction.

These definitions are familiar landmarks in sociological or anthropological literature. To complete our conceptual map I shall draw on two time-worn sociological concepts. I shall employ the distinction between cultural and societal dimensions of human affairs as it is used in the systems, or structural-functional, theories of society. In those theories the cultural dimension contains the symbols, values, and general orientations of a people as they are revealed in their art, literature, and religion. The societal dimension involves particular institutional forms, such as family, bureaucracy, contract, parliament, and profession. Within these institutional forms appear many specific institutions and organizations. There are not only, especially in American society, relatively clear distinctions between political, economic and familial sectors. There are also distinctions and actual conflict between particular organizations, such as the United Farm Workers and the Teamsters Union. Each institution or institutional sector has a greater or lesser degree of explicitly cultural aspects. Some institutions, such as the church or school, have traditionally had a high degree of cultural importance for the whole society, while others, such as recreation, have had a lower importance. The degree to which any institution exercises a cultural primacy changes over time. At any one time one institution tends to emerge as the central cultural bearer for the dominant institutions, providing them with legitimacy and authority.

The other common concept is that of social differentiation. Differentiation indicates that several organizations can fulfill the same social purpose, as is the case with a multi-firm industry. Differentiation can also mean that various purposes, or functions, have been separated from each other and are fulfilled by separate institutions, such as occurred generally in the differentiation of religion, education, and family.[5] There is no automatic historical process of differentiation or de-differentiation. However, there is in the United States a high degree of institutional differentiation and outright conflict which are reflected in and promoted by various liturgical developments. To understand these liturgical transitions I shall use the idea of "differential institutional association," which means the process by which institutions or institutional sectors relate themselves to each other to fulfill their various aims or tendencies.[6] For example, the ways in which families are related to economic institutions has changed in Western history. At one time families were actual producing units, whether of food, goods, or services. Today they merely produce relatively adaptable labor for large-scale firms. Social history involves continual changes in the distance or intimacy of association among institutions. In a pluralistic society continual complex shifts occur in the kind and degree of relation among basic institutional sectors, such as economy, education, polity, law, entertainment and family. Liturgical developments in America today will be taken as signs and causes of certain shifts in these patterns of inter-institutional association. The cultural sphere provides the bases for legitimating these patterns of associations. Liturgies are essential to the grounding of these relations in basic cultural values and orientations.

Significant liturgical activities are occurring in at least four sectors of American life: politics, the university-sports-entertainment field, social movements, and the churches. There are particular and unique factors present in each of these sectors. I want to specify some of these, but I also want to move the analysis to a level where we can see all of these

5. For an exposition of the process of differentiation in the context of the relation of Christianity to Western societies, see Parsons, "Christianity and Modern Industrial Society." For a statement of the idea that social evolution has entailed increasing differentiation see Bellah, "Religious Evolution."

6. In its original form the concept of differential association described the way in which a person had differing access to one or another means, whether legitimate or illegitimate, for securing a cultural value, such as success or money. See Cloward, "Illegitimate Means." My transposition of the idea to the inter-institutional level merely articulates ideas fairly widespread in sociological literature.

developments in terms of some general changes occurring in American culture and therefore in the relations of churches to that culture.

A rapidly expanding literature has drawn considerable attention to the presence of an American civil religion which provides in part a cultural basis for some central American institutions; namely the political, military, civic, and to some degree economic. Though gaining in its accretions, embellishments, and scholastic elaborations, this civil religion is not a new feature in the American scene. It has been even more garishly pronounced in earlier times than it is now. The civil religion, one might say, has emerged into a stage of formalization and reverent execution unlike its enthusiastic and sectarian beginnings in the Fourth of July, giant patriotic parades, Memorial Day and military pageants.[7] In this respect we might say that the political sphere, whose authority has been the primary beneficiary of the civil religion, has more broadly penetrated every sphere of American life and therefore cannot afford the youthful enthusiasm, eccentric excesses, and unsophisticated emotionalism of the cult or sect. It is and must be more "responsible" to the interests, claims. and needs of other sectors in which it is deeply involved—the large economic and social corporations and the delicately poised military juggernaut in particular. In terms drawn from the sociology of religion, it has become more "churchy" and ecclesial. The increased prominence of the governmental and political institutions has meant an increasingly tight and intricate accommodation between the political institutions and the general society, especially its cultural bases. The political sphere needs cultural legitimation in order to expand its authority. But the cultural traditions of America— whether African, European, Yankee, or frontier—have sought to mold the political to their expectations and values. This has meant a measured and intricate expansion of a fairly homogeneous set of cultural-political assumptions which are embodied in what we call the civil religion. We are, however, far from that nice accommodation of culture and politics known in some European countries, small scale cultures, and totalitarian societies.

Moreover, this very crystallization of the civil religion has meant the appearance of a considerable gap between the common man and the establishment. or as George Wallace puts it, between the working man and

---

7. For a sampling of perspectives see Bellah, "Civil Religion"; Smith, *Religion of the Republic*; Neuhaus, "The War, the Churches"; Mead, "The Nation with the Soul"; Coleman, "Civil Religion"; Cherry, *God's New Israel*; and Moore, "From Profane to Sacred America."

the intellectual slobs. Thus we see the emergence of a priest–laity split in which the roots of individual loyalties are grasped ever more tenuously by the superstructure of the official civil religion. The Vietnam War has shaken even further the ability of the civil religion to elicit people's loyalties. However, in view of the absence of a contender, it seems unlikely that the civil religion will lose its authoritative position with regard to the dominant institutions.

Through the loose interweaving of the civil religion with the motivations of citizens and the authority claims of dominant institutions, we see the important cultural and political conflicts of the American scene. There exist deeply entrenched cultural conflicts which gain expression in institutional conflict—whether it be racial, regional, economic, or political. With regard to these crucially important conflicts the traditional ways of analyzing ritual and liturgy are misleading or inadequate because they pay so little attention to pluralistic societies with a great deal of inter-institutional conflict. Having arisen in the study of simple societies, they offer little analytical help in large scale societies containing a number of cultural traditions or orientations and a number of power centers in an uneasy state of mutual accommodation and hostility. In a simple society liturgy is a way of adjusting social interaction to the meaning pattern shared by all the members of the society. It legitimates social structures and provides meanings which can be internalized into each and every actor. Liturgy is understood in terms of its integrating function within a society. It is the point at which cultural, societal, and personality factors are linked together in a coherent set of models or paradigmatic symbolic actions.

However, in a pluralistic society the very rituals which integrate particular groups sharpen, promote, and define conflict among groups in that society. They contribute to ever-renewed conflicts through which social power is reallocated, rights articulated and defended, and persons are motivated to perform their roles with vigor. The civil religion does not merely integrate American society. It also creates reactions against itself, especially among minorities and the disenfranchised. Liturgies not only order groups internally. They also pit those groups against those who would upset their order.[8] The very ability of a liturgy to condense a group's aims, emotions, hopes, and self-understandings into an easily

---

8. Still a seminal article in this regard is Geertz, "Ritual and Social Change." See also Apter, "Political Religion."

grasped action enables members of antagonistic groups to form and rehearse their identity, thus sharpening the conflict.

Liturgies in American life are means by which movements for social change attempt to legitimate their action not only to their own members but also to the possible future social order in which their claims and needs could be accommodated. Thus there is a tension between the appeal of a liturgy to a movement's members and its appeal to members of the established groups. In the liturgies of the movements for Black liberation we see symbols and rituals which not only define separateness and independence of Black people but which also appeal to some widely accepted symbols of pluralism, participation, or the virtue of the common man. In like manner most peace movements have attempted to retain the American flag for the sake of its legitimation of Constitutional rights and the original vision of a peace-loving and peaceful people. Their utilization on funeral ceremonies including taps for the dead are even more poignant examples of ambiguous and transitional liturgies.

The liturgies legitimating the means for dealing with group conflict in America appear with increasing saliency in sports. Conflict is formalized within rules which promise that the best and the fairest shall win. Racial, ethnic, regional, and supposedly even economic and class distinctions are overcome in the unified action of a team whose clear goal is success. Skill exercised cleverly within the rules emerges as the criterion for advancement and praise. Thus, sports provide a ritual paradigm for all the struggles of life in a land of full participation and of success on the basis of merit and fairness. Not only does the sports performance itself reinforce the legitimations of the political, social, and economic structures. The half-time shows, invocations, and the national anthem all work together to dramatize a complex hierarchy of values, legitimation beliefs, and paradigms of conflict, of reconciliation, and of social order.[9] I assume the varieties of these liturgies and rituals are familiar enough to the reader that I need not go into their description.

The broader sociological and ethical significance of these sportive developments seems to be that in the process of differential institutional association the university-sports-entertainment world is replacing the

---

9. For some provocative, mostly non-sociological insights see Weiss, *Sport*. Arnold R. Beisser uses a Freudian analysis to link sports with the legitimation of family patterns. See Beisser, *Madness in Sports*. For sociological perspectives see Hart, *Sport in the Socio-cultural Process*, especially the articles by Günther Lüschen, David Riesman and Ruel Denny, and David Wolf.

church as the mediator between the economic and political spheres and between both of these spheres and the cultural realm of legitimating values and orientations. The cultural and liturgical task is being transferred from the churches to this tripartite realm made possible by modern forms of communications. What does this sweeping statement mean? How can it be verified?

According to many observers, most famously Max Weber and Alexis de Tocqueville, the church in nineteenth century America functioned to identify and legitimate the economically worthy.[10] The practice of church-advocated virtues and participation in the whole complexus of the Protestant ethic legitimated the economic activity of the small businessman, the entrepreneur, and the independent farmer—in short, the major sectors of the century's economy. In turn, the economy, through these churchmen, showered its benefits on the churches and their educational institutions. The church functioned in a similar but looser fashion with regard to the political order, which itself was less dominant and more loosely structured than it is now. The ways in which Protestant beliefs, congregational polity, scriptural obeisance, enthusiasm, and moral fervor shaped American political life has been researched extensively.[11]

The political and economic spheres related to each other very indirectly, with the notable exception of various government subsidies to important industries all through our history, as well as the converse subsidization of politicians by industries. But this latter dependence had to await the large-scale campaign expenses of the twentieth century before it became acute. Political loyalties were formed through personal contacts, favors, and a laboriously constructed organization more than through mass-media propaganda. This implies that the regulatory relation of the political realm over the economic, to which we have become accustomed, was exercised largely through personal moral suasion on individual managers. This was the task of the churches, which mediated between political goals and values and economic operations by an appeal to the personal morality of office holders in the economy. On the whole the liturgy of the Protestant churches was the liturgy of the dominant powers. The mediation of institutional conflict and the motivating of citizens were grounded in the churches to an extent impossible today.

10. See Weber "Proposal," and Tocqueville, *Democracy in America*.

11. For some well-known materials here see Mead, *Lively Experiment*; Michaelsen, *Piety in the Public Schools*; Miller, *New England Mind*; Wolf, *Religion of Abraham Lincoln*; and Heimert, *Religion and the American Mind*.

With the emergence of the "media triad" of sports-university-entertainment we see a change in these inter-institutional relations. The universities have traditionally supplied certified personnel, both professional and technical, to the economy. Their own legitimation and support were in turn due in great part to their affiliation with the churches. More recently, however, they have become more dependent on their sports programs and government bodies for recognition and financial support than on any church organizations.[12] Moreover, political figures seek identification with the sports world, where they can appear and be marketed to millions of people within a liturgical setting which combines the legitimation of their rational expertise (the university), their production of conflict (sports), and their harmonization of conflicting needs for happiness (entertainment).

Moreover, the conflict of the sports arena, rather than the moral individualism, idealism, and piety of the churches, provides the models for political behavior (even that maverick, Gene McCarthy, is a baseball *aficionado!*). Richard Nixon is not peculiar in his sensitivity to this fact and has detected with his characteristic acumen the primary place occupied by football in this liturgical sphere.[13]

Instead of the church's moral and personal suasion being the means for relating economic, political, and cultural spheres, the media triad becomes the arena in which the demands, claims, values, and orientations of the two worlds are synchronized. Each institution in this triad does this in a different way. Sports establishes the paradigms for conflict and the presentation of claims, as well as their adjudication. Moreover, sports itself is big business and a major vehicle for advertising. It affects the economic sphere by its power for stimulating consumption. It affects the political sphere by legitimating its structure of conflict resolution.

Entertainment presents the harmonious and unifying rituals in which performers and their styles tend toward a *sui generis* categorization that eschews head-on conflict. Each has his or her own justification

---

12. In 1969 revenue from sports programs was slightly greater than the immediate costs of these programs. However, 75 percent of NCAA schools reported a profit from their football operations. See Koch, "Economics of Athletics."

13. According to a Gallup poll released in January, 1972, football has surpassed baseball as America's most popular spectator sport, with 36 percent favoring football and 21 percent favoring baseball. In 1961 the figures were 21 percent for football and 34 percent for baseball. The symbolic reasons for football's supremacy are only touched upon here, but they include corporate hierarchy, teamwork, the pressure of time and weather, violence, and planning.

in terms of a free choice of artistic style or temperament. Each has a clientele whose tastes are not subjected to debate. Finally, entertainment and the mass of advertising accompanying it have their justification in the evocation of pleasant feelings, whose presence and quality go beyond public and rational discernment. At this point entertainment becomes a private or cultic matter and loses the liturgical qualities of seriousness, authority, and publicity.

The universities, which on the whole are obliquely related to the mass-media sector, have played a crucial role in producing sports as a large-scale institution in America. Moreover, they have been deeply involved in producing the media revolutions. Within the triad they function to relate specific information among the sectors of the society. Their methods of inspection, testing, debate, and discussion form the paradigm for most of the organizational life of the society in its day-to-day operation. They are least openly liturgical but are most involved in the actual processes of power in the whole society. The universities have taken over the position of mediating between major cultural values and the economy. Formerly the family and the local school prepared persons and transmitted values from the church to business and industry. Today the universities dominate this bridge between cultural values (now liturgized by sports, politics, and civil religion) and a technological economy.

This transition in the pattern of institutional association could be spelled out much more elaborately. The literature to do so already exists. I can only set forth here the proposal that the development of sports liturgy indicates an important transition in the relation of the economic, political, and cultural spheres—one in which the church is replaced functionally by the media triad. Consequently, the whole nature of Christian liturgy in this situation has been drastically altered.

Social and political movements for wide-spread cultural or political change constitute the third area of significant liturgical developments. With respect to these, liturgies not only provide the means for integrating movements, motivating participation, defining goals, and preparing the way for future legitimation of the group's inclusion among the dominant institutions. They also are means for dealing with failure and frustration in achieving goals. The demonstrations and rituals of the Black, peace, power, and liberation movements are as much symptoms of powerlessness and frustration as they are tools to achieve purposes by gaining mass media coverage and appealing to deeply held ideals of the surrounding society. It is the failure to stop the war in Vietnam which led to increasing escalation of symbolic acts—whether violent or non-violent. These demonstration liturgies are ways of keeping hope and vision alive in spite of the inability to actualize that vision in present human affairs. They are either direct appeals to office holders to conform their official policies to broad cultural values to which they already subscribe or attempts to inculcate into officials new cultural values. In either case, as direct personal appeals they are a sign of actual political weakness.

By way of example, it is interesting to note the salience of victimization themes in both the Black and peace movements. Exaltation or elevation of the victim raises the issue of guilt, especially collective guilt, in a way totally unfamiliar to the sports and entertainment world, in which collective guilt is eliminated because the structure of adjudication is absolutely fair or irrelevant. The victimization of whole peoples or groups calls into question the existing authority that legitimated policies of victimization. The liturgies on this theme seek to form a new authority base rather than recruit persons to participate in the existing one. Obviously, the whole Christian tradition of sacrifice should be related critically to this set of liturgies.

The liturgies of social movements turn our attention to two important matters. First, the symbols, action paradigms, liturgies, and rituals of any society have an ambivalent relation to specific policies or institutional arrangements carried out under their aegis. Not only do we have a plurality of liturgical forms in American life, but each liturgical form is as ambivalent to its mother institution or policy offspring as was the Israelite YHWH to his people.

The inauguration of the President requires that he swear to "uphold and defend the Constitution." On the one hand this act functions as Executive propaganda to persuade people that that is indeed what the President is doing. On the other hand it inculcates in Congressman and citizen alike a standard which ultimately can legitimate actions to limit the President. Similarly, the symbol of "Free Enterprise" can divert attention from the reality of large industrial monopolies, but it can also legitimate attacks on that monopolistic practice. In this way symbols and liturgies can be sources of confirmation as well as of condemnation.

In the case of Vietnam we have seen the flag become a highly ambivalent symbol in which the only meaning on which all Americans can agree is that the flag stands for the blood of dead men. Even the primordial feeling of revenge for the death of one's buddy is transformed into a general state of guilt or pity, since the identification of the dead men or their cause is avoided. We have passed from the symbol of the Unknown Soldier to that of the Unknown Dead—a purer and perhaps more universal symbol. The conflict over the policies that caused the military limbo of the POW and the MIA, rather than being reconciled, are simply obliterated by a purely human pity. At the highest level liturgies and symbols which once had very specific meaning and impact have become mute groping for meaning and norms. Social movements have seen their liturgies evolve as a sign of frustration. The established liturgies, instead of looking forward to an anticipated eventual victory, look back on a failure which has robbed them not only of their efficacy but of their very meaning.

From a Christian theological point of view we might inquire into the tension between the linear eschatology of social movements and the cyclical liturgies of sports. The seasonal sports cycle goes nowhere historically except in terms of setting new records. Not only does it tend to reinforce the present social structure, it also drains off energies that might be channeled into movements for social change. Though sports helps fuel the economy, its symbols and liturgics also tend to contribute to cyclical orientations not in keeping with linear growth views of the economy. Moreover, the ecology movement threatens to undermine the linearity espoused in the expectation of infinite economic growth by pulling the economy back into the cycling and recycling of nature.[14] The conflict between these cyclical liturgies and the linear social movement betrays and defines a profound power conflict over the allocation of resources as opposed to the cleanliness of their production. The upward social mobility of individuals within the present system, as liturgized in sports, conflicts with the effort to change the social structure by group action in social movements.

From the standpoint of the Christian churches, this movement away from an economy with a linear eschatology means that the capitalistic growth economy has lost its centrality in American life. This development in turn rebounds on those Protestant liturgies which have so assiduously cultivated the linearly oriented economic man. The legitimacy of the productive worker is being challenged by the leisured playboy.

This concern brings us to the final area of liturgical change—the churches. Liturgical change in Catholicism differs markedly from that occurring in Protestant, especially mainline, denominations. Much of

---

14. For an ethical analysis of the ecology movement from this perspective, see Neuhaus, *In Defense of People*.

this is due to the differing way they have participated in the transitions of institutional association in this country.

In official Catholic circles the liturgical movement, which dates back to the late nineteenth century, first began as a return to the sources. In the rediscovery of ancient and traditional liturgies, a pluralism of forms in the present was justified. Appeal to tradition made possible innovations in the tradition. This resuscitation eventuated in the reforms of Vatican II. Sociologically, especially on the American scene since 1950, liturgical reform has become symptomatic of a wide-ranging alteration in the role church membership plays for American Catholics. The reason that official reforms are not enough for many Catholics is that they presuppose a fairly traditional understanding of the way the church is to relate to its cultural and social environment. The determined way in which The Liturgical Conference in the USA, with its annual North American Liturgical Week, moved from antiquarianism to its eventual self-immolation on the 1969 theme of liturgy and revolution can be understood as an attempt to redefine the relation of the church to the dominant structures of power.[15]

The very notion of an official liturgy is closely tied to the understanding that the church is to dominate the whole society by working from the higher institutional levels down. The hierarchical subordination whose paradigm permeated the traditional Catholic liturgy and priestly officiation promoted in turn a rather tenuous relation with most Protestant-influenced institutions in America, with their models of individual initiative, conflict, and associational pluralism. Catholic liturgy was a legitimation of kinship structure and its related ethnic associations, and the hierarchy of bureaucracy and factory. This hierarchical liturgy was meaningful as long as these two spheres encompassed the lives of most American Catholics.

However, as many Catholics emerge from this epoch they seek liturgies which will relate them to decision-making, to political conflict, to democratic and voluntary forms of association, and to affirmation of individual conscience in matters of public policy. Thus we see an even more

---

15. The Benedictines of Solesmes initiated much of this reform in the late nineteenth century. The Liturgical Conference began publishing the papers from its annual conferences in 1940, with such subjects as parish worship and the realization of "the ideal" in the city and country parish. The last three conferences dealt with "worship in the city of man," "experiments in community," and "liturgy and revolution (1969)," the last of which did not receive publication because of the dissolution of the conference.

radical, even utopian, democratization of the liturgy (which is partially obscured by the appeal to the semi-organic kinship notion of brotherhood) along with salient appeals to peace, justice, and a love unconnected to kinship obligations or patterns—whether sexual or parental. These liturgies reorient persons to a broader participation in a highly mobile, individualized, and rapidly changing society. This participation is legitimated by appeal to the perfection of peace, justice, and love. Much of this development reminds Protestants of the social gospel movement—its earnest search for a tangible and imminent peace and justice, its enthusiasm, personalism, and subjectivity. That is, unofficial Catholic liturgical developments, as well as in part official ones, serve to thrust believers into action to control or affect public policy.

Protestants, and at this point I deal only with white mainline denominations, have always assumed a dominant status with regard to all spheres of public life. Their worship services and congregational business meetings (of equal importance in congregational polities) have served to reinforce those habits of self-control, methodical rationality, goal striving, and efficiency which prepared men (and to some extent women) to participate in decision-making in a mobile society. The recent liberal or radical movements for social change in America are fully in line with this Protestant tradition. Church-spawned movements for peace and justice have lineage that goes back to the Puritans. These movements have experienced tremendous frustration in achieving their goals. Moreover, the tenuous line of argumentation which reforming Church leaders used to lead the man in the pew from a simple devotion to an idealized Jesus to an engagement in structural reform and revolution has broken down for a number of intellectual and social reasons. The denominational officers and leaders seem to have exhausted their devotional capital in expenditures on far-flung efforts at social reform. Both this break and the perceived frustration of many reform efforts are the seed bed for present Protestant concerns with liturgical action.

Thus, I expect to see increasing numbers of activist white mainline Protestants spending more time appreciating liturgical actions in all their many dimensions as impetus for action, laboratory of action models and paradigms of authority, and as a means for absorbing defeat and the delay of achieving goals. In doing this they will give increasing attention to the relation of the church to the cultural dimensions of American society—especially its popular liturgies.[16] It is in these liturgies, rather than the "high" culture of art, drama, and literature that the crucial loyalty and authority symbols are being forged and internalized.

The theology of play and its associated liturgical innovations reflect an increasing affinity between the churches and the media triad, whether it is in the discussion format of the sermon, the entertainment by balloons and singers, or the playing of games. From the standpoint of authority paradigms, play liturgy reinforces the media triad as a cultural source of legitimation for American society.

In particular, this new association of church with the media triad involves a shift from pedagogy to entertainment, especially in Protestant worship. The preacher no longer teaches in worship. His model is increasingly formed by entertainers and the recrudescence of the revivalist which they often promote. The way in which entertainment motifs dominate pedagogical ones in the liturgy of Billy Graham exemplifies this situation. In Catholicism entertainment motifs replace those of the moralist. The importance of a priest's "appeal" to the congregation is heightened by the increasing democratization of Catholic polity and finance.

Association with the media triad also evinces the churches' shift from legitimating economic production (the classic Protestant man) to legitimating economic consumption, which occurs most saliently in the area of recreation, entertainment, and leisure. That is to say, church worship no longer functions to inculcate those dispositions, action paradigms, and values that prepared persons for factory, management, and bureaucracy. Moreover, identity formation is not seen in terms of working—an assumption shared by capitalist and Marxist alike—but in play.[17] Identity (or the salvation of one's soul, if you prefer) arises in the distinctive patterns of consumption, use of leisure, recreation and play. The theology and liturgy of play tends to reinforce this shift in the churches' pattern of inter-institutional association.

However, the societal implications of play liturgy are not without their ambiguities, for some of its exponents see it as a rejuvenation of the

16. The argument I am offering here is very similar to that made by Donald Meyer, *Protestant Search*, in which he shows the increased relevance of Tillich's work in the theology of culture as a response to the shortcomings of political action as analyzed by Reinhold Niebuhr. The problem with Tillich's work is that its terminology and initial residence in the high culture of Europe made it difficult to apply his ideas to the popular liturgies of the American scene, where, in my opinion, they begin bearing their fruit.

17. See Neale, *In Praise of Play*, and Moltmann, *Theology of Play*.

political sector. We could, in fact, draw a clear line of opposition between the political players (Cox, Moltmann, and Hugh Duncan) and the media players (Neale, Keen, Miller).[18] The former employ concepts from anthropology (Cox), sociology (Moltmann), and socio-literary analysis (Duncan). The latter turn to poetry (Keen) and psychology (Miller, Neale). The former oppose play to coercion, a political problem, while the latter oppose play to death, a psycho-medical problem. However, among the political theologians we also see strains that betoken inter-institutional changes. Generally, when play is perceived in relation to politics, it is seen to function as criticism, a criticism which calls operating patterns of political legitimation into question. But these are usually critiques without alternatives (to paraphrase the German title of one of Leszek Kolakowski's works). Or they reinforce existing structures by mystifying them with high drama.[19] Cox's *Feast of Fools* attempts to escape this oscillation by presenting play as social fantasy—a bearer of utopian visions. The social context for this view seems to be the social movements so familiar at the end of the sixties. This represents a shift in the church's association from one with law and constitutional polity to one with social movements. The church functions to mediate or even create a cultural vision for minority groups excluded from established structures. But without a renewed association between the church and government, law, and management, social movements of minorities will founder on the rocks of conventional ties between church and economy or the new individualism nurtured by the tie between churches and the media triad.[20]

Play liturgy and theology is symptomatic of the overall associational shift I have been describing. Jürgen Moltmann criticizes exponents like Keen and Neale for not appreciating the "serious" nature of play.[21] The

---

18. Cox, *Feast of Fools*; Moltmann, *Theology of Play*; and Neale, *In Praise of Play*. For the late Hugh Duncan's elaboration of Kenneth Burke's ideas in a dramatic and sociological direction see Duncan, *Symbols in Society,* and Duncan, *Communication and Social Order*. For the autobiographical and poetic see Keen, *To a Dancing God*, and for the psycho-anthropological see Miller, *Gods and Games*.

19. For Duncan's concept of "mystification," by which social conflicts are "resolved" at higher levels through liturgical drama, see Duncan, *Communication and Social Order*, ch. 23, and Duncan, *Symbols and Social Theory*, ch. 22.

20. Cox's neglect of these structural issues arises in good part because of his reliance on Ernst Bloch's defense of utopian thinking in the Marxist tradition. However, Bloch has only tempered, not refuted, the traditional Marxist criticisms of utopian thinking.

21. Moltmann, *Theology of Play*, 33, 112. For David Miller's concept of seriousness

sociological issue behind this criticism is whether play shall be defined in terms of entertainment or in terms of reconciling major inter-institutional conflicts. Shall the church embrace the media triad or renew its ties to the political sector by means of other institutional areas? At this point in time, play liturgies have not been related to the patterns of authority organizing the sphere of economic production and political structure—two traditional affinities of the church. In developing its implications in these sectors we could gain a more precise and ethically responsible view which takes seriously the struggle for and the preservation of a more just social order in which play could be more than a mockery of human pretension and a balm for failure. In liturgy as the paradigm of authority, we can have the depth of judgment, guilt, and forgiveness which constitute the classic Christian movement of faith. In this connection it would seem that the point of richest explosion would lie in the recovery of the liturgical traditions of the militant Black church, with its expansive ritualization of despair and longing, struggle and anticipation, personal guilt and collective acceptance. Here we see the essentially "serious" qualities of liturgy.[22]

This sociological analysis, along with its limited attempts at prophecy, sees in liturgical developments evidence of long-run shifts in associations among the institutional sectors in America, especially with regard to establishing which sector is to be the prime mediator between the cultural sector and the societal spheres of polity and economy. In this regard the media triad of university–sports–entertainment is replacing or has replaced the Christian churches. The Protestant churches, having cashed in as much of their motivational capital as possible for the sake of specific social reforms, will be turning their attention to the way they relate to cultural tasks of defining basic models for personal and organizational relations, for initiation and sustenance of action, and for the ends of human life. By affecting these authority and action paradigms the Church can make a long range impact on the whole society. In relinquishing immediate power it gains long range authority.

see Miller, *Gods and Games*, 103–8, 123–26.

22. The critical recovery of Black liturgical traditions is a complex matter involving tensions between the formal services of the middle class denominational churches and the many sectarian and cultic movements among Black people. I have in mind the direction taken by pioneering works like Miles Mark Fisher, *Negro Slave Songs,* and the more recent work of Henry Mitchell, *Black Preaching.* See also Adams, "Some Aspects of Black Worship," and Soulen, "Black Worship and Hermeneutic."

This shift appears to be a matter of historical and sociological necessity. But it can also be seen as a profound recognition of the most appropriate way in which the Church relates to any society. A message about ultimate loyalties, goals and visions must perforce relate primarily to the alternative ultimacy symbols and liturgies in the society. I think that in this relationship the fundamental tension between Christianity and "the world" is to be formulated in the coming decade. If we follow this direction to its ecclesiological outcome, we see a severe streamlining of the Church down to its liturgical and cultic center. Then, moving out from this center stem a vast array of voluntary associations to act in accord with the values, visions, paradigms, and loyalties inculcated by the liturgy. The question will not be whether or not the Church's "essence" is its liturgy, but whether that liturgy is articulated with respect to the dominant patterns of authority in the society. In particular, what response will the churches make to the emergence of the media triad as the cultural base legitimating patterns of authority in the polity and the economy?

If, then, the primary interchange between the Church and its social environment occurs at the point of liturgical articulation of authority relations, and if this fact is seen in terms of associational shifts among institutions in American life, then we can say that the primary subject of practical analysis for Christian social ethics would be the way in which various liturgical patterns shape persons' actions and perceptions of their place in history and human affairs. On what basis ethically do we criticize liturgies? What kind of social or ecclesiological patterns do the various liturgies legitimate? In what pattern of inter-institutional association are these Christian liturgies participating? How is a particular liturgy translating motivational and loyalty appeals into rational and publicly trustworthy models of action? How do we move from the intimately inward decision to live thankfully to the very public assertion of the worth of one pattern of action over others? Again, liturgy is the area where personal sense of worth can be translated into paradigms for public action. This analysis of the significance of various societal transitions in terms of their liturgical aspects is a call for ethicists to deal with liturgy with the same depth and precision they have devoted to finding Oughts, defining appropriate decisions within official jurisdictions, and evaluating specific strategies for social change. It is also a call to experts in liturgy to conceive of their study in terms of these implications for action and for the Church's mission in an altered social landscape. It is a call to them, whether Catholic, Protestant, Evangelical, or Orthodox, to conceive of liturgy within the broad pattern of human authority and institutional relations.

3

ECCLESIOLOGY *and* POLITICAL AUTHORITY
A Dialogue with Hannah Arendt

THE APPROACH OF AMERICA's bicentennial has prompted many American Christians to re-examine the religious significance of the American revolution and the subsequent framing of a constitutional government. Central to this significance has been the long interplay between constitutionalism and American church structures, especially those congregational types of churches that we often associate with peculiarly American forms of religion. Just as we begin our memorial celebration we are haunted however by an omnipresent suspicion that the very foundations of authority on which church and republic both stand have been undermined. We are faced with the task either of recovering constitutional and religious authority or of creating a whole new foundation for politics and faith. A reinvestigation of the significance of the American Revolution can be the first step in discerning and carrying out the formidable task before us.

In *On Revolution* Hannah Arendt, the noted contemporary political theorist, distills from the ferment of modern revolutions the concept of a ward-based federal republic. Every modern revolution, from the American to the present day, has been characterized by the spontaneous emergence of councils (soviets, *Räte*, or communes) through which the people sought to take charge of the public realm after the collapse of the old order.[1] In every European case some kind of party or other hierarchical organization exterminated these councils because they threatened the interests of the revolutionary party. These aborted structures of freedom tended to vanish from memory because no theory emerged to legitimate their presence. In the American case these local publics were safeguarded more self-consciously, but in the absence of a general theory were unable to direct other revolutions or modern political thought in any decisive way. Arendt has tried to provide a theoretical framework through which the peculiar political contribution of the revolutionary era can be remembered and preserved.

Revolutions cleared spaces of freedom which enabled people to manifest the unique human capacity for action. These spaces, these councils, are the heart of the revolution and of Arendt's theoretical effort.[2] The essential components of these councils have been:

Their non-partisan quality. Participants did not "represent" parties or interest groups but only themselves. There was general debate about opinions regarding the whole shape of the public realm, rather than simple horse-trading for votes or immediate power. Moreover, the exchange of opinions was concerned more with mutual instruction than with awe-inspiring persuasion.

Their public spirit. Persons participated in the councils because they offered the peculiar enjoyments Arendt calls public happiness. Their action was governed by a public spirit which sought the general good rather than the success of a narrow self-interest.

Their political character. Participants assembled to debate matters of policy rather than to handle problems of a basically administrative or bureaucratic character. They did not appear as managers who had to handle technical matters, but as citizens debating matters of opinion and judgment.

Their small size. The councils had a limited size, certainly not exceeding the 5000 allotted by Plato to a genuine polis. But they were not interpersonal groups in the modern sense. The participants interacted in terms of the public and worldly matters that brought them together.

Their self-selected "elitism." The councils were composed of those who desired to participate. There was no pre-arranged list of eligible participants based on status. The only criterion for continued participation

---

1. Arendt, *On Revolution*, 260.

2. This summary is taken from various parts of *On Revolution*, principally Part 5, ch. 3, 4.

was a concern for public matters and a rudimentary ability to engage in the debate, judgment, and decision of the assembly. They were an elite not in the sense that they ruled a mass, but that they gathered according to a distinct criterion for participation.

Their conciliar and federal interrelationships. Councils tended to associate with each other in ever-expanding federal arrangements. Each council, when this tendency reached expression, sent deputies to councils concerned with the issues impinging on them all. And these in turn sent deputies to more encompassing assemblies. As deputies, or delegates, they acted according to their own judgment rather than according to the self-interests of a constituency which they would only "represent."

Authority came from below. The "higher" assemblies had their authorization from lower councils. Matters for adjudication were referred to the constituting councils, not to a higher person, office, or council.

In the emergence of these councils, and not in party dictatorships or economic programs, Arendt finds the "principle"[3] of revolution. In the American case this principle, and by implication the basic principle for any public founded in the modern age, is that of mutual promise and common deliberation.[4]

Theologians should be struck by the parallels between Arendt's concept of conciliar federalism and the theory legitimating the "free churches" in America. If indeed the general pattern of authority of these American churches is deeply congruent with the principle of the Revolutionary republic, then exploration of these similarities should illuminate the implication of Arendt's thought for the church. The many parallels between the organization of American denominations and the American polity offer a web across which a peculiarly American articulation of Christianity can engage Arendt's own thought. Her own concepts might provide an illuminating way of grasping the significance of the American religious experience. Conversely the theology and ecclesiology of these

---

3. "Principle" has a specific meaning for Arendt. "The absolute from which the beginning is to derive its own validity and which must save it, as it were, from its inherent arbitrariness is the principle which, together with it, makes its appearance in the world. The way the beginner starts whatever he intends to do lays down the law of action for those who have joined him in order to partake in the enterprise and to bring about its accomplishment. As such, the principle inspires the deeds that are to follow and remains a parent as long as the action lasts." Arendt, *On Revolution*, 214. See also Arendt, *On Revolution*, 202, 249, 271; and Arendt, *Between Past and Future*, 152.

4. Arendt, *On Revolution*, 215.

free churches might help fill out Arendt's own intimations about the possible political meaning of Christianity in the modern world.[5]

Both the Revolutionary and the free church tradition contain problematic aspects. The problems raised in this comparison will be traced back to the central problem of authority. In this case also we will see striking affinities between Arendt and the free church tradition. But first let me lay out briefly the basic characteristics of congregational ecclesiology, whose similarity to Arendt's conciliarism should be immediately apparent.[6]

In this "free church" tradition organization is created by an elect who form a network of mutual promises and covenants. Policies are voluntarily decided by mutual agreement rather than coercively imposed by a superior power. Each congregation is autonomous, never exceeding the maximum of a face-to-face group, but can form alliances with other congregations on a pattern of increasing federalism. Moreover, the deputies to general conventions or assemblies do not merely represent the interests of the constituting assemblies. They are chosen for their general wisdom and good judgment, just as the elders of each congregation are selected on the basis of their spirituality and overall vision. Authority moves up from the bottom in the sense that all conflicts are referred back to the constituting assemblies, where each member consults his conscience within the context of the debate of the community. Such a decision, after prayer and discussion, is authoritative, because it is closest to the will of God as mediated through the elect. Bureaucracies and other organizational machinery are only instruments for carrying out the decisions of these

---

5. Max Stackhouse's discussion of the relation between ecclesiology and social structures provides a suggestive introduction to this approach. See Stackhouse, *Ethics and the Urban Ethos*, especially ch. 7. See also the articles by F. S. Carney, M. Novak, R. T. Handy, and D. E. Smucker in Robertson, *Voluntary Associations*, and follow the footnotes to works of Sidney Mead, Franklin Littell, and Winthrop Hudson. I do not think we should be hindered in this task by her claim that the principles of conciliar federalism are solely political and have no application in art, culture, and economics. See Arendt, *On Revolution*, 283. Her discussion of this point does not take up or explicitly exclude religious organization and its specific kind of authority. Since she introduces a typology of kinds of authority to distinguish these spheres of activity, I shall also use concepts of authority to explore the relation between congregational ecclesiology and conciliar federalism.

6. Nomenclature among denominations varies. The "free church" tradition includes Congregationalists, Baptists, Adventists, Friends, and innumerable sects. The Methodists and Presbyterians form an intermediary blend between them and the Episcopalians, Orthodox and Catholic.

various publics and have no authority themselves to emerge in the public as agents. As a concluding analog we see that in both theories the actors find an intrinsic worth to sheer membership in the council or church, but the meaning of their participation lies in the serious struggle to affect the world in which they live—to preserve, perfect, and to beautify it.

This kind of congregationalism has played as extensive a role in sustaining authority in America as Catholic Christianity did in the history of the European continent. These churches and associations have developed rituals, patterns of nurture, and educational processes reinforcing this particular concept of authority. Many students of American religion would lend support to the claim that the churches have been the primary school for politics in American culture. The public schools themselves have generally been oriented toward the economy and have educated persons to job-holding and civic obedience. It is in the church business meetings, the discussion groups, the factional feuds, the preaching and the revivals that many Americans have traditionally learned the skills, reflexes, and habits for public and political discourse. Indeed, even to this day it is church politics that constitutes the real politics of the small town and rural country. These churches, because they were "free" (not merely free from the state, but free in the sense that authority flowed "upward"), could be the schools for freedom, that is for genuine political and public association on a voluntary basis.

This school for politics was ritually reinforced through heavy emphasis on adult baptism, on comprehension of the written word, and on persuasive preaching as the high point of the worship service: And finally, in the free churches the skilled professional, that is the minister, has been subordinated to the council composed of the "amateurs," that is the ordinary church public.

These are merely some indications of the close interaction between this peculiar American form of Christianity and the principle of action abstracted by Arendt from the revolutionary experience. In highlighting them in this fashion I do not mean that power has always imitated the urgings of authority or principle.[7] I am speaking here about a theory, a principle, a pattern of authority, that has arisen in and dominated broad sectors of American Christianity and which is greatly affecting papal

---

7. I cannot go into empirical studies of the interaction between various polities and other sectors or patterns in American society. Obviously, the affinities between institutions bearing similar forms of authority have been interrupted by racial, economic, and geographical divisions.

and episcopal forms of Christianity in the United States. It is this pattern of life and thought that has preserved in practice the revolutionary principle, while American political theory as such has tended to become either a theory about government or has become sociology and social reformism.

This ecclesiastical pattern is not without its critics. Paul Harrison, in his study of the American Baptist Convention, shows how bureaucratic powers have grown up to override their legitimate authority.[8] Other critics trace American Christianity's weakness in challenging economic, governmental, and military structures to this voluntarism, or as the Marxist would say to this *volunteerism*. A similar dilemma appears in Arendt's thought. Either bureaucracies dealing with "social problems" and economic necessities have obliterated genuine publics, or the publics themselves have been too powerless to execute their policies.[9] In both cases what seems to be criticized is their utopian quality—their adherence to a principle which drives them beyond the historical conditions which gave them birth.

But this detachment of theory from practice implies that these organizational patterns are significant primarily for the concepts of authority which they bear. They are primarily symbol-bearers for legitimating action. In both cases they have been deeply committed to the rhetoric and sometimes the reality of freedom—whether through Christ or through revolution. For Arendt true freedom demands a framework of authority. Christianity has made similar claims (Gal. 5:1, 13; Rom. 6:22). Let us then turn to the question of authority in Arendt's thought and in the free church tradition.

In general, the revolutions of the modern age have failed to achieve their expressed aim of founding new orders of freedom. This failure arose from their inability to grasp an authoritative absolute which would enable men to engage in the persuasion of political action rather than in the coercion and warfare of dictatorship, purge, and terrorism.

Western Christianity bears an ambivalent relation to this failure and attempted recovery of authority. On the one hand, according to Arendt, it conserves notions and symbols that legitimate violence in politics, as if human affairs were a thing to be made. God is conceived of as a Creator

---

8. Harrison, *Power and Authority*. See also Gustafson, "Voluntary Church."
9. Arendt, *Human Condition*, 292–97; Arendt, *On Revolution*, 54–6, 247.

who fabricates a world.[10] God is a Legislator who imposes laws.[11] In this tradition contemplation, rather than action, is the highest good,[12] and Platonic idealism and organicism are the dominant structures of thought.[13]

While Christianity was relevant to the traditional pattern of authority it appears irrelevant to the revolutionary era. But, we must ask, what does Arendt mean by "religion" and "Christianity?" She seems to have in mind that "ecclesial" (or in Troeltsch's terms "church") type of Christianity characteristic of Medieval Catholicism, Lutheranism, and in diminishing power early Calvinism. Its primary contribution to the structure of authority was an elaborate myth of heaven and hell, which, as Arendt rightly points out, is tangential to original Christianity.[14] She implies, however, that Christianity as such, having lost this social position and this theological myth, can make no further contribution to the question of authority. This accords, no doubt, with her earlier claims in *The Human Condition* that Christianity's concern with contemplation and Jesus' concern with love and goodness are inherently other-worldly or even anti-political.[15]

On Revolution is thus an attempt to find a new ground for authority in a world which has lost the old Roman trinity of authority, religion, and tradition. Authority for Arendt has a quality of distance from the present—a distance gained through foundation in the past or through appeal to a "higher" dimension of value in the present. This distance is further reinforced by the preservation of a cultural world whose permanence provides stable points of reference for the actors who emerge into it. But transcendence as an appeal to the past, to higher values, and to a durable cultural world, for many complicated reasons which Arendt explores in *The Human Condition* and in "What is Authority," are either inadequate to or unavailable for an articulation of authority today. In religious terms, the Jesus of Logos, Lordship, and Retribution is not enough to handle the problem of providing for human action a transcendent framework

---

10. Arendt, *On Revolution*, 207.
11. Arendt, *On Revolution*, 190, 199.
12. Arendt, *Human Condition*, 15.
13. Arendt, *Between Past and Future*, 127, 138.
14. Arendt, *Between Past and Future*, 129, 133.

15. In her response to this paper, Arendt, though not quarrelling with this presentation of her thought, was very pessimistic about the possibility that free church ecclesiology could help us find a form of authority that could undergird a large-scale political order.

ECCLESIOLOGY AND POLITICAL AUTHORITY

by which to judge, legitimate, and preserve actions. In historical terms, the tremendous human artifact of "Christian Culture" has for all political purposes disappeared.

The revolutions of the modern world could emerge because the structure of authority legitimated by Christian tradition since Gregory I and Charlemagne had collapsed. The American revolution succeeded in founding a new order because it reached back to antiquity to grasp anew the essence of political authority known to the Romans. That was, that authority is the augmentation and development of the "principle" contained within the acts of foundation. It is on the quality of these foundational acts that the future of that new order depends.[16]

Authority in the revolutionary experience finds its ground in the very condition of action itself.[17] This ground is not merely the particular principle which emerges in the action founding a new body politic. But each act carries its own principle which acts as the authority for subsequent action. The qualities peculiar to genuine action are also those peculiar to authority. To act is to begin. Beginning itself manifests the quality of freedom unique to being human. It is that miracle which intervenes in the moribund inevitability by which the past oppresses the present. In this sense authority is always a loyalty to the founding action—but not necessarily to the principles contained in the Roman foundation. In the modern world, she seems to say, we have the possibility of making action itself the founding principle. The conditions of the modem world have made it possible for the reality behind all previous forms of authority to emerge in a kind of quintessential revelation for which none of our previous experience prepared us.

Arendt recognizes that the Founding Fathers were greatly influenced by the Christian heritage of promise making and common agreement as formulated by the Pilgrim Fathers.[18] But she questions whether Christianity and its institutions can make a positive contribution to new forms of authority and organization, and especially to a form of authority consonant with the revolutionary principle. She does indicate that Biblical religion contains some powerful insights into action and authority—that genuine creation is *ex nihilo* (not violently wrested from pre-existing

---

16. Arendt, *On Revolution*, 201–5.
17. Arendt, *On Revolution*, 205.
18. Arendt, *On Revolution*, 166, 171.

materials),[19] that God is made manifest in the birth of the Bethlehem babe (indeed that God *is* that new beginning),[20] and that covenants are the fundamental form for living together.[21]

On several occasions Arendt interprets both the birth and the teachings of Jesus in terms of this miracle of beginning again—whether through natality or through forgiveness.[22] The glad tidings of the Gospels, that "a child has been born unto us," reinforces belief in the possibility of new beginnings. It is forgiveness which enables us to bring this principle of natality into the web of unintended consequences and wrongly conceived promises, enabling us to start over with each other.[23] In such observations she opens up the possibility that Jesus himself, as the guarantee of this basic freedom, can become in a new way an authorizer of present action. Her shifting of attention from Christ the Founder to Jesus the miraculous Actor repeats a shift which characterized the development of the American free church tradition. While distortions abound in practice, the figure of Jesus, much as she portrays him, has stood at the center of American religious traditions. It is this figure, more than Christ, the imperial messiah, who has legitimated action for American Christians.[24]

Arendt herself understandably does not develop this opening. But her own thought helps us to illumine the ecclesiological consequence of this centrality of the active, historically concrete, Jesus. The decay of the monarchical Jesus of political sovereignty has led many Christians to a renewed focus on the active career of Jesus—the immanent side of incarnation—to legitimate public or quasi-public action. But this immanence is not merely his presence as a Founder in the Roman sense.[25] That particular implication of his historicity, though not I think irrecoverable, has languished in the rigid arms of patriarchy and traditionalism. The mode

19. Arendt, *Between Past and Future*, 167–71.

20. Arendt, *On Revolution*, 212; Arendt, *Human Condition*, 222.

21. Arendt, *On Revolution*, 171–76.

22. Arendt, *Human Condition*, 222; Arendt, *On Revolution*, 212; Arendt, *Between Past and Future*, 167.

23. Arendt, *Human Condition*, 214.

24. This would be typified by the social gospel, whether at its most pedestrian (Charles Sheldon, *In His Steps*) or its most brilliant (Walter Rauschenbusch, *A Theology for the Social Gospel*). For other helpful perspectives on this issue see H. Richard Niebuhr's classic exposition, *The Kingdom of God in America*, especially ch. 3.

25. Arendt, *Between Past and Future*, 125.

of legitimation introduced by the free churches and explored extensively in American Christianity is the presence of Jesus in the Spirit. It is this form of legitimation which appears to be very close to Arendt's concept of the revolutionary form of authority.

Sketched broadly, Jesus's presence in the Spirit has meant that church decisions and policies are reached through a debate including all inspired participants. While the claims of each are weighed conscientiously by the other members, initial admission to the assembly rests on the individuals' experience and voluntary affirmation, a pattern very similar to Arendt's self-selected elite. The fundamental character of this experience is a compelling and vivid encounter with Jesus. This experience arises in a number of ways, sometimes through private meditation, prayer, self-inspection, and sometimes through the public encounter of preaching, worship or teaching. The authority of Jesus is mediated to the group primarily through individual consciences and then refined by the insights of the others, insights including the wisdom of the earlier generations and stretching back to the Biblical writings themselves. In short, the "principle" of this kind of Christianity *is* Jesus. Moreover, this active initiator is present "in the Spirit" in the ongoing movement of people inspired by him.

Jesus' presence in the Spirit inevitably implies pluralism of organizations and authorities. It supports a conception of the church in which the principle of voluntary association becomes paramount. But to claim that the principle of organizational voluntarism is the specific genius of American Christianity is not to exhaust the case. Of equal importance is the principle of confederal or conciliar cooperation in order to achieve specific goals which flow from the claims of Jesus as preserved in scripture and in the confirming experience of successive generations of Christians. On this basis we have seen the growth of innumerable associations, councils, and federations within and among denominations.

One reason that Arendt does not explore this connection more assiduously is the fact that the Founding Fathers were almost exclusively classicist in their orientation. They turned to ancient Rome to legitimate their action and to articulate their experience. But, as Arendt demonstrates, it is precisely these traditions which blinded them to the unique discovery of the Revolution. In seeking a stable order they neglected those structures which might preserve the revolutionary spirit.[26] Thus,

---

26. Arendt, *On Revolution*, 242.

the officially recognized authority of the Constitution has always been in great tension with the actual principle of the Revolution, which at heart was much more populist. Arendt implies that the Hamiltonians have been nourished by classical theory while the Jeffersonians have languished for want of a theory. But it is my claim that the free churches have preserved this revolutionary principle. Therefore, an exploration of their ecclesiology provides us with an entree onto the operative form of authority "discovered" in the Revolution. The Constitution gave no formal, governmentally relevant place to the spontaneous assemblies lauded by Jefferson. So the impulses and principles of the revolution found an alternative home, not among the cultured strata or the leisured class of "public servants," but among the common people and their vast proliferation of religious associations.

This juxtaposition of Arendt's thoughts with analogous strains in American Christianity implies many possible changes in the way church persons might think about the church and its relation to American society. Foremost among them is a reformulation of the basic lines of division among Christian groupings and a concomitant reformulation of how the churches are related to the emerging new cleavages in American society.

Arendt's description of the conflict between the revolutionary principle and the Constitutional form of authority helps us to see more vividly the tension between Jeffersonian and Hamiltonian traditions in American life. These tensions are mirrored in the distinction between the free churches and the more hierarchical polities. It would seem that once we adopt Arendt's view of politics in the revolutionary era we should side with the free church ecclesiologies. As a consequence we could go on to say that the primary contribution of the churches to American culture is not their views on individual morality but their peculiar ecclesiologies—that is, their nascent political theories. The task of the free churches therefore emerges as one of founding, augmenting, and preserving conciliar publics in American society. This they can do through preserving their own polities and articulating the principles behind them, by preserving rituals and procedures which inculcate these principles in new generations, and by direct action against those parties and powers who would extinguish these conciliar publics. In this way the basic distinctions among the churches, seen from a political standpoint, is their distinctive ecclesiologies, which for some time have been of peripheral concern for most Christians as well as many sociologists of religion.

In this light the primary social significance of religion in America is not its positive relation to the economy but to the polity. Whether we follow Marx or Weber we are still under the spell of their inquiry into the roots of capitalism. Consequently, we are also persuaded, sometimes almost subliminally, that the social significance of the churches is their impact on the economy through their legitimation of certain kinds of personal morality. Conversely, the most important distinctions among them are their class character. But perhaps that era is over and in the emergence of serious inquiry into the civil religion and into the relation of ecclesiology to political theory we will perceive the primary importance of the relationships between the churches and the polity. This would of course reflect the overall shift in American power from the corporate to the governmental sector.

But Arendt's further point is that within the political sector, of which the governmental is just one aspect, there exists a conflict between two kinds of authority—the Constitutional and the truly revolutionary. This raises the possibility that what we have in the "silent majority" and the alienated middle class is a new kind of proletariat—a political proletariat—and that the mission of the church is to further that kind of authority which will legitimate their own participation in political life.[27] In a time of increasing governmental attempts to constrict the public sphere, the decision over this task may prove to be the major line of division among the Christian groups in the next decades.

---

27. The organizing methods of Saul Alinsky and similar community organizers have been adopted by many churches as a major method for relating to the communities around them. While there are significant differences between Arendt and Alinsky, the direction they take is strikingly similar. At his death Alinsky urged that organizing now be taken to the silent majority of the middle class. Not organizing them would leave the field clear for a genuinely fascist reaction by the powerless.

4

Vocation *and* Location
An Exploration in the Ethics of Ethics

ALMOST EVERYBODY ENGAGES IN ethical self-criticism from time to time. Am I a good parent? Was my decision to disobey the draft law right? Are the ends I am pursuing genuine goods? Ethicists, who study the right and the good systematically, raise similar ethical questions about their own vocation. Is my activity as an ethicist contributing to justice? Does my work do anyone any good? Is the institution in which I operate just in itself and conducive to justice in the society? What is the ethical impact of my professional associations?

The argument heard recently in the American Society of Christian Ethics between "actionists" and "reflectionists" is an important example of ethical self-criticism among ethicists.[1] The actionists declare that ethicists must engage in activity to transform the social order to fit the social ideals and goals which they have formulated. Sincere ethicists must not only discern the good and the right, they must seek to do it. In turn, social action is seen as a necessary stimulus for perceiving the world ethically. The actionists' call has its counterpart in other professional associations as well. For example, the American Sociological Association has had for many years a radical caucus with a journal and a section in the *Sociological Abstracts* of its own.[2]

---

1. Gustafson, "Burden of the Ethical"; Clark, "Calling"; and Long, "Christian Ethics and Credibility."

2. *Social Problems* emerged from an earlier reformist uprising. The section of

The reflectionists argue that ethics is a scholarly discipline—a process of thought, not of action. They do not discourage action, but they do assign that task to other groups or institutions. Ethics, being a rational activity, must be detached from the passions and interests of worldly affairs. It must establish a standard against which to measure all the achievements of power and politics. Ethicists must be, in some sense, ideal observers to whom all parties in a dispute can come for fair and credible judgment.

This debate is an important one. It affects the vocational understanding of individual ethicists as well as the policies of their professional associations and teaching institutions.

In the strict sense, the reflectionists would seem to have the better part of the argument, for ethics must be at least a process of rational inquiry into questions about just, good, or proper action. But much more than a problem of definition lies behind the action-reflection debate. There are at least three areas of controversy:

A. Disagreements about social theories and the function of ethics as a social activity;

B. Differences in the social location from which different ethicists speak and think;

C. Disagreements about the fundamental value or set of values by which the activity of ethicists is to be judged.

This essay is an attempt to clarify the issues involved in each of these three areas and to present some proposals for resolving the debate. Thus, it will begin with description and clarification and move to prescription and advocacy. Furthermore, much of the debate is determined by assumptions about social theory and social location which remain implicit and unexamined. Someone needs to lift them up for explicit critical attention. Therefore, the essay will explore some of the sociological factors that condition ethical reflection and ethically motivated action. My method of analyzing the social theory and social location implicit in a

---

"radical sociology" in the *Abstracts* reflects movements of the late 1960s. The American Political Science Association also has its radical caucus and journal, *Politics and Society*, as well as its *Caucus Cable: Newsletter of the Caucus for a New Political Science*. For the phenomenon of conflict between practitioners and theoreticians in other professions, see Barber, "Some Problems." For a stimulating engagement with this tension in the context of profession and occupation, see Krause, *Sociology of Occupations*.

theoretical viewpoint may appear unusual to some ethicists, but it has ample precedent in Ernst Troeltsch and the Niebuhrs.

THE PROBLEM OF SOCIAL THEORY

The disagreement between actionists and reflectionists may stem from differences in the social theories they espouse. In the first place, this may affect their understanding of the relation between ideas and action. At one extreme are those who view ideas as merely the product of existing social relationships. At the other extreme are those who view all actions and social structures as the immediate outcome of certain ideas held by agents.

Secondly, the choice of social theory will affect one's assessment of strategies for achieving ethical ends. Some strategies may be highly appropriate if society is viewed as rigidly polarized but completely irrelevant if it is seen as highly homogeneous or integrated. Consequently, ethicists with different social theories will evaluate the relative merits of their occupations, associations, or personal acts quite differently.

Three Descriptive Social Theories

Broadly speaking, ethicists employ three different perspectives from which to analyze their society: the systemic, the dualistic, and the pluralistic. Each of these perspectives implies a certain attitude concerning the relation of ideas to action, and each tends to endorse a particular range of strategies for changing or maintaining social relationships.

Systems Theories

"Systems" or "structural-functional" social theories describe societies as fairly well-integrated structures working to fulfill the "functional prerequisites" for societal survival.[3] The maintenance of any one group, institution, or structure serves (in a particular way) the maintenance of the whole society. Action is therefore not a matter of unique initiative but of patterned behaviors. From this point of view, ethicists who favor

---

3. For an examination of this concept, especially in the context of the debate between systems theorists and various conflict theorists see Demerath and Peterson, *System, Change and Conflict.*

reflection can say that they are not inactive although their action tends to be confined to established roles and institutions. In chastising them for not "acting," the critics actually betray their own espousal of an alternative social theory.

Dualistic Theories

The usual alternative to a systems view is some kind of sociological dualism, Marxist theories being the pre-eminent representatives. The distinction between multiple social systems, such as "economy," "polity," "culture," and "society" is obliterated in favor of a single cleavage separating powerful and powerless. The function of any group is determined by which side of this cleavage it supports. Whereas ethicists working within the systems viewpoint might claim that ethicists function to expand the range of decision-making in economics and politics, the dualists would claim that ethicists provide credibility for the ruling class. The "educative" function extolled by systems devotees is excoriated by the dualists as a parlor game for the affluent.

Dualists have contended that ethical ideas are manifestations of the interests of various groups engaged in conflict. As such, they are not universal norms to be brought to conduct in order to judge or guide action. Their abstract form does not produce a comprehensive approach to social injustice but only disguises the actual world of economic, military, and political structures. Ideas are not plans for acting but acts themselves—symbols used to mobilize and arouse people, or disperse them. Like Presidential TV addresses, ideas are more concerned with effectiveness than truth.

Abstract ideas tend to function either as ideologies or as utopias, to use Karl Mannheim's classic terms.[4] As ideologies, ideas tend to maintain

---

4. Mannheim's distinction between ideology, a set of relatively detached ideas that tend to reinforce the existing power structure, and utopia, a set of relatively detached ideas that tend to undermine the existing structure of power, is important to this essay. See Mannheim, *Ideology and Utopia*, Parts 2, 4; and Mannheim, *Essays*, especially the essays "On the Interpretation of Weltanschauung," "Historicism," and "The Problem of a Sociology of Knowledge." His legacy has been brought up to date in Berger and Luckmann, *Social Construction of Reality*. For representative efforts, see Hodges, "Class Significance of Ethical Traditions," and Hodges, "Class Analysis"; and Ossowska, *Social Determinants of Ideas*. For the problem of ideology, a bibliography by van der Bent "Ideology and Ideologies," is available. For some more general treatments of the sociology of knowledge, see Curtis and Petras, *Sociology of Knowledge* (especially Dibble,

the existing structures of domination. As utopias, they might serve to undermine the existing structures if they contained resources to guide practical action, but most are nothing more than an abstract vision of the ultimate goal. Thus, they are likely to peter out in isolated, small-scale communes rather than leading to societal transformation. And those utopias that do make significant contributions to the emergence of a new order can, in turn, become ideological.

Ethicists tend to speak in terms of abstract principles, whether they see these as a universal standpoint for judging actions or as a set of enduring symbols to legitimate an ongoing structure of authority which can be the basis for adjudicating disputes peacefully. Marxists attack this tendency to abstraction because it obfuscates social reality ideologically, thus preserving an unjust balance of power. Moreover, in order to gain the leisure necessary for "distance" and reflection intellectuals become dependent on the affluent class, and this leads to a class identification which distorts their understanding of social reality.

The corrective that is usually prescribed for this intellectual bias is to encourage teachers and students to read widely in the tradition of ethics, participate in voluntary associations concerned with public policy, commune with an active religious body whose scriptures or beliefs emphasize judgment by a transcendent Giver of moral law, or travel to foreign cultures. The goal of these activities is impartiality.

However, even this adjustment for bias can be shown to have in it a bias against those who seek to change institutions or society. Impartiality is the cool reason which supports the subtle equilibrium of responsibilities, interests, and duties that make up large-scale institutions. Innovators, reformers, and revolutionaries must speak simply and one-sidedly, no matter how just their grievance and cause, if they hope to evoke moral commitments strong enough to lead people to sacrifice themselves in order to inaugurate new orders of justice.

Needless to say, the systemic approach views ideas and symbols in a fashion opposite to the Marxists. Systemics believe that social relationships depend on the strength and consistency of the social symbol system. Ideas have their own logic and power independent of the desires of interest groups or the brute certainty of economic forces.

"Occupations and Ideologies,") and Reynolds and Reynolds, *Sociology of Sociology* (especially the articles by McKee, Mills, and Gouldner).

Pluralistic Theories

Forms of the systemic and dualistic orientations have received strong support among students of American society. But both viewpoints and the strategies associated with each have also been severely criticized.[5] In contrast to systemic theories, which emphasize the importance of cultural factors, and dualistic theories, which emphasize economic ones, the third orientation focuses on political factors. Certainly this view is also an "interested" one and has received considerable attack from its friends as well as its enemies.[6] However, it seems to me that theories of institutional pluralism come closer than any other to an adequate descriptive theory for America. Pluralists are willing to acknowledge that social power is highly integrated, especially through finance and management; but they are also aware of considerable cleavages among institutions, groupings, and associations within these sectors. These cleavages make it possible for outside groups to amass various forms of power. A choice of this descriptive model helps us to go beyond the debate between "reflectionists," with their systems affinities, and "actionists," with their frequently dualistic assumptions. Moreover, it begins the response to the dualist critique of ideology at the point of the adequacy of the descriptive sociological assumptions underlying it. The criticism of ideology must go further than assuming a simple division of Western or American society into two classes.

Social Theories as Strategies

Each descriptive social theory implies a set of strategies for action. A systems theorist, for example, would define social problems in terms of a lack of integration among institutional sectors or between individuals and the socio-cultural system. Universities, government, business, the family, or religion are not performing their proper functions in the society. This lack of integration, in turn, is produced primarily by a lack of authority—either because the system of authority is fragmented and incoherent or else because it is out of touch with the social organization whose activity it is to regulate. Further, this lack of authority is fundamentally a cultural problem. What is needed is a stronger value and symbol system in the

---

5. Domhoff and Ballard, *C. Wright Mills*; Gouldner, *Coming Crisis*.
6. Connolly, *Bias of Pluralism*, is a good collection of critical essays.

society, more cogent ritualization of its loyalties, concepts, and beliefs, and their more effective inculcation through a system of education.[7]

A systemic orientation, thus, tends to emphasize strategies of changing public opinion, improving education, or strengthening the authority of elites who have expertise and information. In other words, the belief that the society is a fairly integrated system leads to an individualistic focus on the opinions of its members as well as to a comprehensive focus on legislation or planning by a fairly centralized elite. However, in all cases strategy is concerned with various forms of persuasion—whether of the man in the street or of the planning elite. Systemic strategies of this sort tend to be favored by those who see ethics as an enterprise in articulating a structure of legitimate authority.

However, persuasion is possible only when disputants share common social and cultural ties and thus do not have to rely on power confrontations Therefore, systems theorists open themselves to the criticisms of dualists, who seek to define and bring into play some fundamental societal contradiction—whether of class or race—in which the fundamental antagonism between the two groups overrides anything they have in common. The more radical dualists resort to armed force in this situation. The more sophisticated dualists, including most Marxists, find various ways in which the fundamental dualism inevitably determines the downfall of the oppressing group. In either case, the dualist relies on the pitting of power against power.

The pluralists also appreciate power struggles, but they speak of countervailing rather than contradictory powers. Sometimes these countervailing powers are supplied by elites whose various intentions are in a continuously unstable state of only relative consensus. At other times pluralists overcome stark dualism by seeking a third party to intervene in a particular conflict. This intervening party is mobilized either by persuasion and appeal to common cultural values, or else by bringing into play its own self-interest. Pluralists may also appeal to the legislative processes to define and resolve conflicts.

Pluralists can appear similar to systemics because of their search for some means of reconciling a multitude of seemingly opposed interests. Like most systemics, they tend to believe that there is a solution which will enhance the power of all groups concerned. Conflicts are not merely

---

7. See Horton, "Order and Conflict" and "Dehumanization of Anomie"; Merton, "Social Problems"; Rubington and Weinberg, *Study of Social Problems*; and Spector and Kitsuse, "Social Problems."

a matter of opposed power, but also of conflicting appeals to some common authority on the basis of which the conflicts can be adjudicated. But pluralists are also dualists in perceiving that there is some inherent struggle over scarce resources and that the search for justice is more a matter of distributing power equally (or at least equitably) than of adjusting individuals to a proper social order.

Power and Authority in Three Perspectives

Each of these three views approaches the reality of power and authority in a different way, and this difference brings the debate over the ethics of ethics to a sharp focus. The actionists tend to be concerned with power. For them, successful strategies for social change result from an effective use of power. An evaluation of ethicists' actions from this standpoint involves an assessment of their place in a set of power relations.

The reflectionists are concerned with issues of authority.[8] Some, for example, advocate an ideal observer conception of ethical reasoning in an attempt to create a structure of authority and legitimation. Authority arises, they say, when persons have internalized certain commitments held in common with others. It is this structure of assent that makes society possible. In fact, people regularly forego their own immediate interests because they are committed to the survival of the social order. They yield to authority because to do otherwise is to deny the basis for their own existence. One of the chief means by which rational shape is given to this structure of authority is the practice of ethical reasoning.

The distinction between culture and society parallels that between power and authority. *Culture* constitutes the basic symbolic patterns of a society. It undergirds and creates the particular relations of authority in the society. *Society,* on the other hand, requires power—the more stringent use of sanctions and rewards to create and maintain social relations.

---

8. The definition of authority I am employing here has been greatly influenced by Carl Friedrich's claim that authority means "capacity for reasoned elaboration." It is the "augmentation, the reinforcement by reason, of the communications which bind men into a community; as such reason expresses itself in the judgment of the old, the learned, and the wise . . . Only by reasoned elaboration in terms of the community's ideas, values and beliefs, is authority added to the exercise of power, whether autocratic or democratic." In Friedrich, *Philosophy of Law,* 203–4. See also Arendt, "What Is Authority?"

Power arises in the immediate actual relationships of social institutions rather than in the ground of their ultimate appeals.

The systemic view places emphasis on the presence of authority and the effects of culture. For reflectionist ethicists, who share this view, "action" is cultural action. It focuses on the slow and long-range transformation of the ultimate grounds of the society's authority structure. The actionist ethicists, on the other hand, tend to hold pluralist or sometimes dualist views. For them, power and society are central to any concept of action.

If they are pluralists they do not conceive of either power or authority as a single fabric within society. Power exists only in discrete groups and institutions, whether they be community organizations, ethnic movements, labor unions, business firms, or political parties. Correlatively, authority exists only as the particular pattern of voluntary compliance within each of these power groups. Neither power nor authority exists for the society in general but only for the groups which make it up. Therefore, the ethicist can deal with society as a whole only through one of these groups by affecting the structure of authority within it to the extent that they act to enhance the power of the group. Nothing is to be gained by standing aside to adjudicate disputes among groups. Consequently, ethicists affect society only as they take sides.

Taking sides reaches its logical conclusion in the dualist perspective. Here authority is seen purely in terms of ideology or utopia. Power is conceived in terms of force. Finally, the society polarizes into two armed camps, and the ethicist's speech and thought is turned to the immediate tasks of mobilization and warfare. Reasons become slogans which are used as weapons in the struggle.

The action-reflection debate is greatly shaped by the adoption of positions from across this sociological spectrum. It is a debate about the nature of society, the most effective and appropriate means for changing it, and the specific kinds of strategies appropriate to ethicists. Any evaluation of the actual social effects of ethicists' activity involves evaluation of these sociological choices in particular situations.

Ethics as Cultural Action

If a strict definition of ethics as a process of thinking and speaking about normative matters is adhered to, the reflectionists must be given the advantage in the argument. However, once the ethicist tries to take into account the actual effects of this cultural activity, he may move beyond a purely systemic view of society. Though ethics is a cultural action concerned foremost with authority, it need not necessarily work through a systemic sociology. In the light of actual conditions in the society, ethicists might adopt a pluralist view. This would give them a different view of how culture and authority are to be mediated or created, which might lead them to devote their energies to establishing a structure of culture and authority within particular ethnic, community, or economic groups. Alternatively, they might reject a systemic approach if the society is so polarized that systemic approaches to culture and authority are seen to be ideological window dressing for an oppressive elite.

The dualist choice is hardest on ethicists because it is hardest on the independent function of speech and reason, but it is not barred to them. In this case ethics may be transmuted back into the prayer, longing, and visionary prophecy characteristic of ancient Israel or the sectarian church. At this point the separate activity of ethics has disappeared and rejoined the broader concerns of "religion."

LOCATION AND VEHICLE

The identification of ethics as a cultural activity in no way concludes the discussion. One might adopt a dualist understanding of society on empirical grounds and, on that basis, challenge the pretensions to "ideal observation" taken up by systemic reflectionists. This challenge would have two points. First, it can be pointed out that the social position of the reflectionists greatly affects the issues, methods, language, and conclusions they espouse. For instance, they accept the social system largely as it is and turn their attention to personal matters of medicine or sexuality. Second, the systemic strategies pursued by reflective ethicists might be shown to have sociological and empirical consequences which contradict the claims of justice.

Both of these challenges are directed to the social position chosen by the ethicist. No one can escape the inevitable choice of standing, thinking, and acting at some particular place in society. Whether or not we choose it freely and self-consciously, each of us occupies a place that conditions us culturally and socially. Inasmuch as it affects the ethicist as a cultural creator, we shall call it a *location*. Inasmuch as it becomes a channel for power and effectiveness, we shall call it a *vehicle*.

An examination of the ethics of ethics forces us to evaluate the locations and vehicles in which ethicists live, think, and work; and this leads to two kinds of questions. First, what institutional locations serve as *primary* vehicles through which ethicists exercise their vocation? What are the institutions through which they earn a living? Second, what social locations constitute *secondary* vehicles for the expression of ethicists' vocations? With what organizations, movements, and institutions do they affiliate to express their sense of vocation as ethicists independent of earning a living?

The first question seeks to discern the ethicist's economic base. While there can be variation in the strength of the connection between an ethicist's personal acts and the interests of the primary institution, we cannot ignore the restraints imposed by this host's interests. Not only are there boundaries to the ethicist's discretion, but more importantly there are boundaries defining whom she encounters, controls, and affects. Finally, as a guest, an ethicist does much less to change the institution than to support it if only by her presence and prestige. Inasmuch as ethicists have some choice about hosts, they have a degree of responsibility for their hosts' impact as part of their own vocational impact.

The second question asks what associations, societies, and other voluntary endeavors ethicists support in order to pursue their wider vocation. These range from labor unions to community organizations, from churches to movements. Some of these groups expire at the completion of their specific tasks. Others are as enduring as the problems they seek to confront. All are voluntary in the sense that the ethicist's livelihood does not depend on them directly.

Primary Vehicle

According to a survey of the membership of the American Society of Christian ethics, about 90 percent of its members belong to faculties of higher education.[9] Within this institutional sector there are significant differences among specific locations. Only about 8 percent occupy

---

9. These figures are drawn from an analysis of the 1973 membership list of the American Society of Christian Ethics. The proportions are only approximate, since it was impossible to ascertain with certainty the locations of 26 percent of 335 total names.

positions in state universities. Private colleges account for about 22 percent while church-related colleges (an admittedly elastic designation in our time) account for another 22 percent. The largest number, 38 percent, belong to seminary faculties. About 9 percent are denominational officials, ministers, or staff members of church agencies. Only 1 percent occupy positions in associations, policy-research organizations or public bodies.

State universities differ significantly from private and church-related universities and colleges in their relation to politics and government, but they differ much less in their relation to the economy. Universities and colleges, for instance, generally accredit persons and channel them toward jobs, alter the technological base of the society through pure and applied research, coordinate various elites through the social life and intermarriage of their students, and form the opinions of people who have favorable attitudes toward education. This historical affinity between universities and the business world has been buffered in recent times by intervening charitable foundations, and it may have been obscured by the recent displays of political action during the Vietnam period. On the whole, universities have been rather distant from political structures and much closer both to the economy, whose structure they may affect only marginally but whose personnel they greatly influence, and to culture, through which the university-cultivated values and beliefs have exerted long-range societal impact. Their economic role leads them to develop an increasingly specialized departmental organization of knowledge to match the pattern of a technological economy. When this specialization is taken into the cultural arena it produces a series of esoteric and rarefied cultures far removed from the popular culture of any existing or incipient mass movement. Universities reward faculty who can establish their own prestige through specialized publication and recognition by the technological elites rather than through popular, practical, or religious action or influence on social change.

The economic impact of educational vehicles may be somewhat more pronounced in the large universities than in the liberal arts colleges. The "capital-intensive" university is not only better equipped to serve industry, it is also more dependent on external funds to support its research enterprise. Colleges too, of course, are seriously concerned about good relations with funding sources, but their greater dependence on local and regional communities undergirds their historic concern for liberal education and the broad values envisioned by their public-spirited founders. Ethicists in these positions would have more concern for culture and its implicit structure of authority. Both of these emphases—the economic one of the university and the cultural one of the college—are more congenial to systemic theories and strategies than to any others.

Seminaries generally serve as vehicles for maintaining and reforming church organizations by training and influencing their staff and leadership. Secondarily, they affect the orientation of many service and action organizations in the public voluntary sector, both through their field education programs and through graduates who enter these institutions as professionals or charismatic innovators. Given this pattern of relationship to churches and to social agencies, seminary ethicists would find attractive those strategies combining the appeal to elites and general opinion that are typical of systemics, and the nurture of countervailing powers typical of pluralists. On the whole, their relative distance from the economy leads them away from a systemic emphasis and toward a pluralist emphasis congruent not only with their political and constitutional position, but also with the practices of denominational ecumenism. Cultivation of pluralistic social theories and strategies is fostered not only by their social location but also by their self-interest.

In sum, the preponderant influence of the social, economic, and vocational situation of faculties leads them toward symbolic, cultural, systemic understandings of society and of strategies for changing it. The long years of expensive training lead to a dependence on the professional class and the resources it can command. Their social and economic status leads them to identify with this class. Their primary conversation and peer influence come from fellow faculty members rather than from the poor, the politicians, the angry, the agitators, or the economic power brokers.

Much the same can be said of the churches and denominational organizations. They differ from seminaries, however, in focusing greater attention on local power structures than on the national and cosmopolitan issues which gain more immediate hearing in the seminaries. Regional variations influence the degree to which location in a church fosters attention to pluralistic strategies rather than to competing or planning elites. Independent agencies and voluntary associations vary greatly. While they are all relatively independent of governmental and business institutions, their theories and strategies are greatly affected by the nature of their funding, the societal issues and conflicts they deal with, and their own size and structure. Their common location in the "voluntary sector" cultivates a pluralistic outlook, but this can be taken either in a systemic direction, as when they serve to advise planning elites, or in a dualistic one, where they polarize opinion over racial or economic justice.[10] Moreover, the degree to which they espouse systemic views depends on whether their primary role is to affect general legislation (by lobbying as well as by trying to affect public opinion) or to affect special policy-makers in business, industry, the professions, or governmental bureaucracies.

Secondary Vehicles: Sociological and Vocational Critiques

To complete the description of ethicists' locations, we must pay attention to the secondary organizations through which they express their vocation. Here we encounter much more difficulty, for there is no adequate information to locate ethicists in the vast geography of voluntary associations, movements, cultural experiments, and isolated charismatic groups, not to mention their memberships in families, sexes, races, classes, and nations. Nevertheless, it is probably safe to assume that leadership in voluntary associations concerned with public policy constitutes a secondary vehicle for the vast majority of ethicists. However, it is difficult to assess the degree to which they exercise their expertise as ethicists in these associations. They may be more important as prestigious figureheads, skilled office workers, or persuasive spokesmen. Thus, rather than flowing directly from their vocation as ethicists, their selection of their particular kind of vehicle may flow from a particular moral judgment or from espousal of a pluralistic social theory.

Participation in voluntary associations is most congenial to pluralistic views. Therefore, its effectiveness is questioned by dualists and systemics alike, who stress in different ways the importance of involuntary dimensions of social life. Many voluntary associations can be criticized for equating "ethical" activity with altruistic endeavor. They divert attention from the ethical significance of the central driving forces in human

---

10. Sometimes, as in industrial missions, both theories may be present in an antagonistic way. On the one hand these missions tend to advise elites and to bridge the gap between management and labor, as in the Detroit Industrial Mission; but they also cater heavily to dualistic views that emerge either from analysis of racial conflict or from traditional Marxist analyses of the opposed interests of labor and management. Accurate analysis of the transformationist impact of these missions presents observers with exceedingly torturous problems. See Clark, *Ministries of Dialogue*, and Jack, "SANE as a Voluntary Organization."

affairs and lead people to affiliate with less effective groups in the social arena.[11]

Our analysis is not, of course, the only way to assess the influence of location and vehicle on the thinking, action, interests, and values of ethicists. Others may want to supplement or replace this analysis with their own understanding of the ways in which societal conditions affect the selection of problems, methods, theories, and strategies by ethicists. Such a concern stood at the center of the work of the modern founders of social ethics—people like Emile Durkheim, Max Weber, Ernst Troeltsch, and H. Richard Niebuhr. The importance of these factors, and of paying critical attention to biases arising from location and vehicle, can be ignored only at the price of un-self-critical bias.

It is clear from this cursory examination that ethicists' locations and vehicles do not promote a dualistic approach to society. Regardless of its empirical validity, dualism finds very few homes among American ethicists since it is not confirmed by the day-to-day experiences of those in churches and universities. Instead, ethicists' locations foster pluralist and systemic approaches to society. Sociologically, then, the action-reflection debate is one between pluralists and systemics.

If the debate were simply an opposition between contrasting primary locations, with their accompanying theories and strategies, we might see no rational way of resolving it. However, ethicists occupy an additional location with its own place and power in American society—the professional association—and this may offer a way of transcending this opposition. The association of professionals generally exhibits a peculiar mix of power and authority which mediates a basic tension. It also occupies a third place somewhere between primary and secondary locations. Therefore, exploration of the professional association provides a fruitful basis for resolving some aspects of the debate.

11. While voluntary association as a practice and actual participation in such associations as a value enjoy wide support among theologians, as evidenced by Robertson, *Voluntary Associations* (the Festschrift for James Luther Adams), criticisms can arise not only on theological grounds (Dillistone, *Structure of the Divine Society*, ch. 9) but also on ethical grounds (Gustafson, "Voluntary Church") and on sociological or political theory grounds (see articles by McConnell, McBride and Chapman in Pennock and Chapman, *Voluntary Associations*). Much of this critical debate is further controverted by unclarity about the definition and use of the term, or by unclear specification of the type of association under consideration (i.e. party, union, public interest group, etc.). For attempts to clarify these matters, see Smith, "Formal Voluntary Organizations" and Rose, *Power Structure*, ch. 7. I am indebted to Max Stackhouse for criticisms of earlier drafts of this section.

ETHICS AS A PROFESSION

The dominant way in which the tensions that influence the debate have been institutionalized in the American scene is the profession.[12] The power of a professional resides primarily in a complex relationship of authority between the professional with his skills and the client with his needs. A client–professional relationship demands great trust from the client and great personal integrity from the professional.

To be a professional is to have a clearly established, publicly authorized organization defining the legitimate exercise of professional powers. Jurists and lawyers have the court system. Doctors have hospitals. Until the last century ethicists (and more precisely moralists) had churches and their related schools. As the Protestant churches built colleges, the ethicists slipped out of the pulpit and into the classroom. The nineteenth century saw the rise of professors of moral philosophy, who were usually college presidents and clergymen. This shift coincided with the churches' gradual loss of their responsibility for local community control. In the Catholic Church, the role of moralist was always more clearly defined; but similar trends have developed there as well. Thus, the church has been increasingly lost as the institutional vehicle through which ethicists make a difference for human affairs. Moreover, the attempts by ethicists to return to this institutional vehicle have been directed not at the older task of disciplining members (and thus indirectly affecting the whole society) but at the social action task of directing the institutional resources of the churches at their environing institutions.

To the extent that ethics has secured the university as its primary vehicle it has become a *discipline* rather than a *profession*. To say that ethics is a discipline is to understand it primarily as a subgroup within the university, which involves issues of faculty size and composition, departmental obligations, standing in curricular programs, and other aspects of university organization and politics. Ralph Potter's discussion of the discipline of ethics occurs within such a context. Joseph Hough's paper in this issue[13] is an attempt to reconstruct an appropriate relationship

---

12. See Jackson, *Professions and Professionalization*; Turner and Hodge, "Occupations and Professions." Discussions of the "credibility" of ethics should be seen as questions about the selection of location and vehicle. See, for example, Long, "Problem of Credibility"; Krause, *Sociology of Occupations*, 174–95, 296–315; and Goode, "Limits of Professionalization."

13. *Journal of Religious Ethics* 5:1 (1977).

between ethics as discipline in the university and ethics as profession in the church.

The very existence of the American Society of Christian Ethics inevitably fosters among ethicists a more professional self-understanding, which pulls against a purely disciplinary approach to the vocational question. But this development also cultivates increasing conflict over selection of a proper primary vehicle. Is ethics a profession or a discipline? If it is a discipline, is it a sub-unit of a larger profession of educators? Or is it a sub-unit of the clergy? Or of social service workers or city planners? If it is a profession, what is to be its publicly legitimated jurisdiction and code of ethics? While ethicists are suspicious of any proposal to try to return to their traditional primary vehicle, they are well-instructed concerning the significance of the professional vehicle as a peculiarly appropriate way for working out the sociological implications of their vocation.

Societal Requirements of an Ethics Profession

Historically, ethics has been pursued for many purposes. Some purposes concern individuals and their virtues rather than social order and societal consequences of action; but most include a secondary concern about societal consequences, since, for example, the virtue of being a just person implies a concept of justice and an attempt to pursue it in specific actions.

Other purposes for ethics are directly concerned about societal consequences of action. One example is transformationist ethics, pursued for the sake of transforming a whole social order. Whether we find it in the social gospel of Rauschenbusch or in Marx's claim that "the philosophers have only interpreted the world in various ways; the point is to change it," the thrust is to end institutional injustice or to attain an ideal vision of the good society. Far from being sheer enthusiasm, it usually involves methodical, long-range, organized effort to transform societies.

The transformationist purpose is so strong in American circles that many ethicists seem to think it is the only valid reason for pursuing ethics. From this myopia arise some misunderstandings that aggravate the debates between theoreticians and practitioners. If an ethicist happens to be pursuing a more individualistic purpose, he may choose a "practice" which does not look at all ethically justifiable to someone impelled by the transformationist cause.

To this point, we have presupposed that transformation is the proper goal of ethics. Thus, the evaluation of social consequences has been most important for our analysis. Our examination has been concerned with the contemplative purpose of ethics only secondarily. If assumptions about the purpose of ethics are not further clarified, however, the debate wallows in confusion. To determine success in transformation of society is the task of sociologists, and indeed a final evaluation is possible only from the standpoint of the end of history.

Thus, it would be useful to consider an alternative that might help knit together the claims already made concerning cultural authority and professional association. This alternative would be to evaluate ethics from the standpoint of an "ecclesial" purpose, i.e., one which sees ethics as inquiry into the minimal conditions for the existence of the community of ultimate commitment. In the history of the Christian church, this kind of ethical reflection has attempted to stake out the boundaries of the church's minimum social requirements, the space it must have in order to proclaim the gospel and conduct its sacramental rituals. This "ecclesial" purpose is seen as early as Augustine and more recently in John Courtney Murray. In the present analysis the focus will be not on the church but on the professional association of ethicists as "ecclesia."

Institutional Deontology and the Ethics Profession

The ecclesial approach emphasizes an institutional focus in which the shape and impact of the institution carries greater ethical weight than the particular acts and intentions of the individual. Thus, ethicists assess the minimum as well as optimum societal conditions for the professional association's own life. This kind of ethical reasoning can be characterized as "institutional deontology." It seems very close in method to what Max Stackhouse calls an analytical ethic.[14] The deontological requirements of the profession could be summarized as leisure, rationality, information, plurality, freedom, and publicity. The first four of these seem to be essential for any kind of ethical inquiry, whether by lone individuals or groups. The last two requirements seem to be necessary only when ethical inquiry is a matter of professional discourse. Perhaps other requirements could

---

14. Stackhouse, *Ethics and the Urban Ethos*, 61–65, 176. Though this step looks very close to Durkheim's (Durkheim, *Professional Ethics*, 1–41) approach to the delineation of "positive morality," it is quite far from Durkheim's organicism and concentration on the integration of self with social structure.

be named. Moreover, these requirements are not unique to the ethics profession; the press, the university, or even a truly competitive market may also seek and depend on them.

First, ethical inquiry demands an adequate supply of economic resources to produce the leisure for extended analysis. The professional society thus has an interest in promoting jobs for its members in which they have the time and opportunity to think through the ethical problems with which they are concerned. This interest would lead some ethicists to defend or condone a highly unequal distribution of wealth, while others might seek to further a much more equalitarian or socialist economic order to secure these benefits. In either case, from the standpoint of the professional association, these directions would have to be adjusted to the other ethical requirements.

Second, ethical inquiry presupposes rational discourse. As an ongoing structure for discourse, the profession, like most institutions, demands predictability (and in that sense rationality) in its social environment. The meaning of "rational" is somewhat diffuse, and may be taken either intellectually, where one must differentiate between technological, speculative, consensual, and logical meanings of reason, or sociologically, where the cultural value, "reason," can legitimate legal or commercial procedures to constrain the arbitrary exercise of power. The professional society requires enough agreement to prevent a fixation on static positions that have become irrational in their obscurantism.

Third, ethical inquiry requires information. Action, which is the subject matter of ethics, is always particular and historical. It comes to the thinker in the form of discrete information which is constantly renewed in the course of events. The members of the professional society need to know and to transmit to others accurate information about particular, changing cases of ethical action. They would therefore be particularly concerned with freedom of the press and information.

Fourth, ethical reflection requires the existence (not merely theoretically, but actually) of alternative courses of action. Ethics arises when people have to consider alternatives and decide among them. Thus, a plurality of courses of action is an intrinsic requirement for the ethics profession. The richer the plurality, the more complex and sophisticated ethical discourse will be. Plurality here affects not only the logical complexity but also the depth of feelings, sensitivities, and perceptions that figure in ethical reflection. The professional society functions best when its members are neither so committed to one course of action that they cannot consider alternatives, nor so detached that they are mystified by the depth of feeling that motivates advocates of one or another ethical alternative. It is thus in the association's self-interest to cultivate and defend publicly recognized proponents of a number of alternative policies for the society.

Freedom, the fifth condition, goes beyond plurality. It means here the capacity to associate with others in order to act for common purposes. Without freedom in this sense, there can be no "profession." The capacity to form voluntary associations is also in the self-interest of the professional association, regardless of the "good" it may do society as a whole. The association therefore has an interest in cultivating the disposition and courage in people to act through the formation of associations seeking the public good.

The sixth condition, publicity, carries the condition of freedom to its fuller implication. Those who seek to discourse professionally, that is in an association based on trans-personal criteria for excellence, demand a public in which they can find each other and make an appearance.[15] The construction, preservation, and enhancement of publics is thus an intrinsic necessity for the ethics profession. Those publics include the professional society itself, public associations, including governmental ones, and a culture that values and understands moral discourse.

These survival conditions constitute the self-interest of the professional association. They can also be supported as values in themselves. The defense of societal structures congruent with these values will be, in fact, a large part of the function of the profession; and this will influence the ethics of its members by predisposing them to a cultural concern for authority in the context of pluralistic assumptions about human affairs. They will seek to strengthen the institutional means for promoting the beliefs, values, and commitments according to which disputes might be publicly aired and discussed, and adjudicated on grounds other than the relative power of the protagonists. The profession is itself one institution among a number of competing groups engaged in a struggle over jobs, influence, status, and publishing, but these self-interests also serve distinctive structures of justice. It encourages its members to be informed and involved in a plurality of movements and cases that are the subject matter of ethics. Thus, ethical reflection in the professional association

---

15. For the concept of "public," see Arendt, *Human Condition*, 45–52, and Arendt, *On Revolution*, ch. 6.

embraces the authority concerns of the systemics and also takes into account the realities of power acknowledged by pluralists.

Even though they have entered a more pluralistic framework, ethicists need not limit their impact to the authority dimension of the dominant powers. Ethicists can strengthen the dimension of authority in movements struggling for social power—whether they be farm workers or Black businessmen. By helping to articulate the structure of authority and its cultural base, they can enhance the organizational power of the insurgent group as well as clarifying the ways it might appeal to authority in the dominant society.[16] In this sense, ethicists are always committed to non-violence, not because they are absolute pacifists in the Christian sense, but because they function practically in terms of authority and culture and because long-run organizational effectiveness depends on the non-violent impact of authority.

SOME CONSTRUCTIVE PROPOSALS

A number of claims have been acknowledged in this exploration:

1. The ideologization of ethical speech and thought demands a careful attention to the location in which ethicists work;
2. A pluralistic theory of American society is both more accurate and more appropriate for the associations which ethicists employ as vehicles for their vocation;
3. Ethics has a reflective core, but this basically cultural activity can and perhaps should be exercised through practices encouraged by pluralist theory;
4. The professional association is a fulcrum for understanding the ethical significance of ethics as it is practiced in a variety of locations;
5. The "ecclesial" purpose for ethics suggests a perspective from which to assess the ethics of ethics.

What then do these claims imply for the organization of the lives of ethicists? What kinds of practices flow from these premises and theories? Can we find a balance among the demands of actionists and reflectionists, on the one hand, and, on the other, the demands of varieties of

---

16. I have discussed this function of liturgy in Everett, "Liturgy in American Society," in this volume.

locations? Perhaps we can. Our analysis of the impact of institutional bias on ethicists' thought and action, combined with a normative preference for correcting bias, leads to a proposal for a *triadic* location for ethical practice.

A triadic arrangement of ethicists' institutional commitments works both to correct the bias in their thinking and to relate their activities as ethicists to a realistic social practice. This triad of social locations is constituted by the primary vehicle, the professional association, and two kinds of secondary vehicles—one "cultural" and the other "societal."

It is assumed here that the primary vehicle will usually be a large institution with a systemic orientation to society. The host institution will by itself exercise considerable impact on the thought and practice of ethicists. It will tend especially to insulate them from dualist as well as many pluralist claims.

The professional association offers a point of stability and independence from the pressure of primary vehicles. Professional associations develop varying degrees of power relative to primary vehicles, with doctors and lawyers having the greatest powers of jurisdiction, and teachers (especially ethicists) having less. The professional association gives ethicists at least some independence and power in seeking alternative primary locations, in formulating patterns of ethics instruction congenial to the understandings and conversations among ethicists, and in finding a point of entry into other sectors of society. Finally, in the professional gathering, ethicists can correct one another's biases—an important move toward correcting the ideological distortions of ethical language. However, lest they merely correct systemic and dualistic theories in favor of the institutional pluralism congenial to the professional association, ethicists need yet other locations in cultural and societal associations.

This need for mutual correction of biases also affects membership policies in the professional association of ethicists—in this case, the American Society of Christian Ethics. Professional criteria have usually been established with an eye to performance in a particular primary location, namely, the university and seminary. But a greater attention to the intrinsic quality of ethical reflection as worked out here implies the need for alterations in these criteria. Specifically, membership should be drawn from a much broader variety of primary locations in order to produce a discourse defined by a broader range of social theories, stratagems, and intellectual biases. Ethicists located in communication organizations, newspapers, the arts, action organizations, and even labor unions and businesses could be invited into the professional association. Their common bond with current members would be their concepts and methods for examining ethical questions, which are mutually understood even if not accepted by all.

The need for correction of biases, and the fact that strong biases tend to be rooted in one's social location, suggest that a third sort of location is also important for ethicists. The third corner of the triad is composed of cultural and societal associations.

Cultural associations attempt to generate and sustain the long-range orientations, symbolic and ritual actions, and fundamental belief systems that underlie authority. Through these associations, ethicists can grapple with the issues of authority that are intrinsic to ethical reflection. For some, the appropriate group would be the church or the university; for others it might be any of various associations concerned with religion, education, film, or literature.

Societal associations are concerned with the redistribution of power. These might be political parties, unionization movements, or efforts to organize elderly pensioners. Here the pragmatic and strategic efforts to achieve power and change serve to balance the long-run concern for authority, order, and continuity.

Ethicists whose primary location leans more toward cultural authority are probably most in need of a secondary location in a societal association. Those whose primary location is societal may need a secondary cultural location. In any case, our analysis suggests that ethicists need to be aware of the influence of their own primary locations upon their thought and action; and they should become self-critical toward the resultant bias. It is very difficult to be aware of one's own implicit locational ideology without a strong involvement in a contrasting location.

This triadic proposal and the analysis underlying it have revealed a complex of issues that is bound to be obscured by any simple debate between "action" and "reflection." I have presented the vocation of ethics as a task, not only of reflection, but also of participation in a variety of social locations. An ethical evaluation of the lives of ethicists demands not only an ethical theory concerning purposes of ethics, but also a social theory through which one can evaluate the actual impact of those lives and the institutions they support. Those with dualistic tendencies will be unhappy with my acceptance of some reflectionist claims and of a pluralistic view of America. Those more attuned to a systemic sociology and a reflectionist stance will probably feel that I have given short shrift to the transcendent power of rationality. In adopting this mixture of pluralism and triadic location, I have tried to lay out an arena of assumptions which might establish the truths in each claim and reorient the ongoing argument.

5

Land Ethics
Toward a Covenantal Model

CONFLICTS OVER LAND USE and ownership pervade public policy in the United States today.[1] They are an aspect of any major problem, whether that of food, urban planning, or ecology, for nothing is more basic to our life than the land on which we live. Attempts to resolve these conflicts inevitably lead us back to fundamental ethical questions. In this paper I want to set forth a basic framework for understanding the ethical issues and for moving toward an appropriate ethic of land use and ownership within the biblical context common to Jews and Christians.

I cannot, therefore, take up the many issues concerning the present setting of these conflicts. Grasping these issues is crucially important, however. We cannot understand the nature or urgency of the question apart from a grasp of the dynamics by which land has moved from being the major factor in economic production to being simply one factor among others in the commodity markets. We have to be able to see how speculation in land is related to the uncertainties of the stock market as well as to the press of population. We should be aware of the intense cultural norms which drive people to preserve the family farm or erect its expensive symbol in suburban ranch haciendas. While all these

---

1. For a sample see Barnes, *People's Land*; Clawson, *America's Land*; Harriss, *Good Earth*; McClellan, *Land Use;* and Timmons and Murray *Land Problems*. For bibliography see Institute of the Church and Urban Industrial Society, *Land Use Issues*; and Charles Smith *Bibliography on Land Reform*.

explorations lead us to the ethical theme, they cannot be the focus of this brief paper.

Debates over land return continually to definitions of parties, claims, and models shaping land ethics. Who or what are they and how are they to be related to each other? Within this web of relationships we can identify the fundamental issues at stake in any land ethics.

THE PARTIES

Who are the parties to the land? In general terms they are God, Nature, Society, and Persons. Much ethical debate has suffered from a truncated vision of the range of parties making claims to the land. The history of human struggle over land perennially marshals forth representatives of these four parties.

God

Walter Brueggeman and W. D. Davies have shown us how the Hebrews viewed God as the real owner of the land—His to give and His to take away.[2] It was God who took the land from the Canaanites and gave it to Israel in the conquest. God gave it in trust and remained the owner. This is why the first fruits of the land were to be offered to Him. This tithe symbolized Israel's recognition of God's sovereignty and His laws for Israel's life.

The land was the means by which God bonded Himself to Israel in promise and in commandment. The promise which preserved Israel in hope was realized through a particular land. The commandments by which Israel was to honor its Lord were largely conditions for living in the land. Central to these commandments was the observance of Sabbaths by which the land and the people rested, just as God, the Creator and Lord of the land, had rested.[3] The land, like the people, shared in a common holiness arising from its consecration to God. The land, like the people, was to be the expression of God's graciousness. It was to blossom forth with good fruits. It was to stay within the right order of God's creation, wisdom, and law.

---

2. Brueggeman, *The Land;* and Davies, *Gospel and the Land.*
3. Blenkinsopp, "Structure of P"; von Rad, "Promised Land," 85.

The land, therefore, was a medium through which God revealed himself, not only to Israel but to the whole world. In this sense it could be called a central sacrament of God's relation to the world. As the medium of a relationship the land was the means for expressing faith in God. Care for the land and its right ordering were the signs of Israel's faith in God as Creator, Redeemer, and Judge. Faith in God was inseparable from faithfulness toward God's land.

There is, then, One who stands apart from the land and whose claims are to determine its use and meaning. While this perspective contained numerous features assembled over many centuries, its dominant thrust is fairly clear. In Judaism of the Roman period and in Christianity[4] "the land" came to symbolize a transcendent arena for God's revelation and our response. However, the particularist meaning has never disappeared and has continually arisen in Judaism, as with Zionism, and in the numerous Christian communities who "return to the land" in order to experience God's presence and exercise a life of greater holiness and righteousness. Moreover, through the land's transcendentalization and through the work of these groups Biblical believers have increasingly come to see all land as having this peculiar relation to God.

Nature

The second party to the land is "Nature." Nature embraces not only the physical condition of the world but the processes and laws which seem to govern this environment. While technology has robbed nature of its aura of immutability, these underlying processes still take their toll on human intention and pretension. There is some objective presence whose claims must be admitted.[5]

In our own time this party is dressed in the costume of ecology, a term pointing more closely to the total interlocking system within which we must live if we are to survive. Nature is a moving equilibrium of forces and processes striving for harmony. The land, the water, the air, and living beings of all kinds must seek a balance in accord with nature's laws. And

---

4. Davies, *Gospel and the Land*.

5. For this emphasis see Austin, "Three Axioms"; Dunlap and van Liere, "Land Ethic or Golden Rule"; Commoner, *Closing Circle*; Leopold, *Sand County Almanac*; and Stone, *Should Trees Have Standing?*

indeed, these laws grind out their relentless ecological retribution on all who violate them.

Within this natural system there is neither "The Land," as with Israel, nor "land," but only lands of various types. Rocky soil, clay, loam, and glacial moraine all have unique characteristics with their appropriate uses within the ecological scheme. Each set of physical characteristics sets boundaries for its appropriate use. It makes a claim on its user. Each kind of land has its own hierarchy of best uses which must be honored if the total ecological scheme is to be maintained.

Society

The third party to the land is Society. Here we enter the murky complexities of human affairs. Some might want to use the term Community, while others might want the broader one of Culture. However, Society points more directly to the total complex of human relations laying claim to the land. Society indicates the organized character of these human relationships. Through societal organizations the claims of the relevant human group are made articulate and effective.

Israel claimed the land as a society before it claimed it as separate tribes (Josh 13–14; Num 26:34; Lev 25:23). In our own time the claims of the society as a whole are recognized as being distinct from those of individuals or lesser groups and even communities. Society embraces the whole complex of activities by which people sustain themselves. Even in our increasingly interdependent world, in which a single world society may be developing, we still see a plurality of societies, each of which is relatively self-sustaining and lays claims on the land for that sustenance.

But land is not merely a means of physical sustenance through the crops and products it yields. It is even more importantly the arena in which human action occurs. The very contours of the land set up the type of society which will be preserved there. The very places are hooks on which to hang the hats of social memory. Land once bloodied in battle becomes a point of sacred recollection by which societies preserve their identity and the values by which they live. As Walter Brueggeman points out, land is not merely a neutral space, it is a place. It is a theater of memory. It yields visions of heaven as well as hell. Land sustains society culturally as well as economically.[6]

---

6. Brueggeman, *The Land*, 5; Hyams, *Soil and Civilization*; Berry, *Unsettling of*

Person

We think of the fourth party, that of "person," as less comprehensive than society." Here, even more obviously than with society, we deal with a whole category of claimants, though the kinds of claims they register tend to be similar. There are two kinds of persons—natural and corporate. Natural persons include individuals and families. Corporate persons include "private" and "public" bodies. Anglo-Americans are most familiar with individuals as the prime persons making claims upon the land. They have perceived the matter from this standpoint, and, as with the Jeffersonian tradition, have sought to maximize this party's claims.

However, there are other persons who are party to the land. Israel was not alone in apportioning the land according to families and clans. Individuals only held the land in trust for the family. They represented the family, as Fustel de Coulanges points out, as owners of the land.[8] In the effort to preserve family farms the real party is seen to be the family, not the lone individual.[9] Even beyond the family, and often seeking to replace it, are various corporations that are persons before the law. Businesses, non-profit corporations, trusts, and cooperatives are corporate persons laying claim to the land.

Finally, there are the corporate personalities of the public realm, principally the state, but also including quasi-public utilities, authorities, and public corporations. They frequently lay claim to being the primary if not the sole corporate expression of the society. This assertion, however, is a controverted one, since the whole individualist tradition sees society as best served through the efforts of the other persons we have presented.[10]

These persons are as much at conflict with each other as with the other parties to the land. In our effort to lay out a comprehensive framework we should not overlook this fact. Each of these categories of person sees its relationship with the other parties somewhat differently. However, they all lay claims on particular parcels of land and seek to encompass as

---

America; Leopold, *Sand County Almanac*, 196–97, 242; and Jackson and Dubos, *Only One Earth*, 218–20.

7. Kelso, "Resolving Land Use Conflicts."
8. Fustel de Coulanges, *Ancient City*, 149.
9. Ackerman and Harris, *Family Farm Policy*.
10. See McClaughry, "Future of Private Property"; Lippman, *Method of Freedom*; Friedman, *Capitalism and Freedom*; and Heyne, *Private Keepers*.

many as possible of the rights and goods I shall set forth below. In our discussion we shall have to remember both their similarities and their differences.

Having set forth the four parties to the land, noting that the last two are really categories of specific parties, we see one possible category that has been omitted—that of past and future parties to the land.[11] Past generations of persons find their claims registered through the honor and reverence we pay toward the land on which they lived or in which they are buried. Promises, deeds, covenants, and wills bear their claims into the present, often in bewilderingly complex ways, as many a title search can reveal. Future persons or generations are sometimes represented in the claims of individuals and families. Other times these claims are taken up as part of the claims of Society, God, or Nature.[12] Sometimes people think it is enough simply to preserve the natural world or the society in order to satisfy these potential parties. Others think we must represent their claims more concretely, empathically, and specifically. In short, there are several ways of conceiving the presence of these future or potential parties.[13] However, they are present to us through one or another of the four parties already specified. While consideration of these potential parties affects our conception of the claims of the parties and the relationships among them, it does not seem appropriate to consider them as a separate category.

These four parties to the land function in an ethical theory in two ways. First, they set up a theory of obligation. People differ over questions of land use and ownership because they sense an obligation to one party more than to another. Some feel obliged to yield to the claims of God,[14] others to the claims of Nature.[15] Still others believe that the society is the central party to whom we are obliged. Finally, some root their sense of obligation in the dignity or ultimacy of persons. Most theories of land ethics try to combine these sources of obligation in some way, assigning to each a primary or subordinate role. While theories of obligation play a critical role in land ethics and efforts for change, we cannot concentrate on them here. We must proceed to discerning the specific claims that

11. Feinberg, "Rights of Animals."
12. Black, *Dominion of Man*, 109–24; Derr, *Ecology and Human Need*, 89–95.
13. Rawls, *Theory of Justice*, 284–92.
14. Derr, *Ecology and Human Need*, 95.
15. Leopold, *Sand County Almanac*, 223–30.

each party makes on the land. These claims can be conceived in terms of the traditional distinction between rights and goods. Registering these claims is the second function of the parties in a land ethic.

Rights are fundamental claims arising from the very being, nature, and needs of a party.[16] In contemporary discussion they tend to refer to the minimum entitlements—the right to housing, education, etc. A portrayal of rights in land ethics discussion, however, tends to revolve around matters of power, authority, and law. Rights articulate the just relation among persons and institutions as they seek various goods.

Parties seek certain goods through the exercise of their rights. They seek conditions, relations, or things which will benefit them. Goods are the objects of interest, want, and desire. They take the form of ideals as well as needs. With this rough distinction we can explore the dynamic among the four parties to the land.

RIGHTS IN LAND

Most of the questions over ownership finally devolve around the matter of party rights to the land. Who finally ought to have control over the land? In the Bible God is seen to be the final owner of the land. It is God's rights that must be honored. Contemporary ecological views tend to see Nature as the bearer of these rights. Most conventional views find in society or persons the repository of rights to the land.

What, then, are these rights that relate to the land? They tend to coincide with the rights of "ownership." The rights of ownership are those of use, income, transfer, and alteration. Ownership is not a single right, but a bundle of separable rights. It is important to distinguish them so that public policy and ethics can take a more nuanced approach to the question of ownership.[17]

Use

Use entails occupancy or exploitation. The occupant uses the land as a space in which to act or exist. Land becomes a home as well as a theater in which the party exercises itself. Normally we think of persons "occupying" the land, but conservationists may assign the right of occupancy to

---

16. Feinberg, "Rights of Animals," 43.
17. Darin-Drabkin, *Land Policy*, ch. 18.

"Nature" while those in search of spiritual sustenance see God or sacred powers as the occupants of the land. All four parties may have this right, just as all four may exercise the others.

Exploitation does not mean the ruthless rape of the land, but simply that utilization of the land which does not alter its continuing use. We usually think of agricultural use, in which we consume the products of the land without consuming the land itself. Exploitation may also mean its use for hunting, transportation, recreation, or construction. At a certain point, as with construction, we enter into uses which are in fact the relatively permanent alteration of the land, a separate right altogether.

Income

The second land right is income. We normally think of income from the products of the land or from its use in general. The former are income proper, the latter are rents. However, as Henry George and many others in his tradition have pointed out, taxes are social rents upon the land. They are rent paid to society rather than to persons. George, along with earlier proponents like Thomas Paine, thought that all rent should go to society, with the income from products being held tax free by the land developers or users.[18]

We must not forget other kinds of income as well as ways other parties receive income from the land. I have already referred to Israel's sacrificial tithe as a recognition of God's ownership. The earlier American practice of setting land aside for church, school or other nonprofit purposes expresses similar sentiments. Income from land may arise not merely from its use but from its transfer or development in the form of commissions. To overlook this fact would be to ignore the foundation of real estate brokerage—an additional factor in the escalation of land prices.

---

18. In George, *Progress and Poverty*, 405, George sets forth ideas espoused by many people to this day. See Geiger, *Land Question*; Neilson, *Cultural Tradition*, 206–24; Netzer, *Property Tax*; and Woodruff, "A Comparison." For the history of a George-inspired community see Alyea and Alyea, *Fairhope*.

Transfer

The right of transfer is the third right adhering to the land. Land is normally transferred through sale, gift, and inheritance. It is also transferred through outright confiscation, as many native peoples will attest. In transferring the land, parties actually are transferring one or more rights to the land. When YHWH "gave" land to Israel (the Canaanites probably viewed the matter somewhat differently), He did not transfer the right as well. This inheres permanently in YHWH. YHWH only transferred limited use, income, and alteration rights. The concept of Jubilee honored YHWH's rights by seeking to return land to the original clan-holders.

Ecologists could similarly argue that Nature has only yielded to societies and persons limited sets of rights appropriate to the ecological role played by these parties. In conventional practice persons often try to assign transfer rights in perpetuity to some trust, public body, or even to society, in order to limit the discretion a party can exercise over the land. Transfer is therefore a kind of master right over the other rights. The party holding this right is in a primary position to control the land. The choice of where to locate this right flows directly from our theory of obligation, for it reflects our belief about the "real owner" of the land.

Even when we do assign this transfer right to a person, for instance, we normally hedge about the exercise of this right by assigning other rights to other parties. The exercise of the right to transfer is deeply affected by taxes, commission structures, legislation and zoning regulations.

Alteration

Alteration is the fourth right of parties to the land. Alteration means changing the use possibilities of the land. Surface mineral extraction, processing and industrial activity, construction, depositing of wastes, and removal of various land components all involve the relatively permanent destruction of certain land uses. Alteration may also mean introducing land components, such as topsoil, that make possible new land uses. The more neutral term "alteration" indicates that "development" always entails destruction of certain use possibilities in favor of others. Because of the economic model dominating our present approach to land, development has come to be synonymous with greater income yield to persons. In fact, of course, "development" for the industrialist is destruction for the farmer. Development for Nature means destruction for the industrialist.

Because alteration of land is so crucial in a land ethic, many efforts to safeguard the rights of nature, society, or God involve removing development rights from persons and assigning them to a public body. The owner is then reimbursed for the monetary value imputed to the development right.[19]

The right to alter land inevitably means the right to alter it with regard to the goods or interests of the particular party exercising that right. Therefore, we have to be clear not only about how the four rights to land affect each other, we have also to be clear about how the exercise of these rights affects the goods of each party.

GOODS IN LAND

No land ethic can avoid the question, What is the land good for? What goods are we to seek in owning and using the land? Assessing and ranking these goods are crucial to any evaluation of land. Land enables us to pursue the goods of nurture, shelter, recreation, culture, and industry—in short, most of the goods of life. The question of the land's goodness is therefore exceedingly complex. In order to guide our thinking we must return to the distinction among the four parties to the land—God, Nature, Society, and persons. Each of these parties, in laying claim to the land, seeks its own good. Therefore, we have to ask first of all, Which parties are pursuing which goods through the land? The goods that any of these parties seek are security, expression, enjoyment, and perfection.

Security

Security entails minimal well-being as well as sheer survival. The strain toward security is most obvious with natural persons and families. Here we find the nub of the family farm ideal. While the family farm may be a cultural ideal, industrialization has practically wiped out the economic basis for assuming that survival of persons is directly linked to the land. More recently, ethicists focus on the survival of the society or even of nature, with personal survival devolving from the welfare of these more comprehensive parties.

Nature also evidences the drive toward survival, whether we see this in the constant re-equilibration that responds to outside changes or

---

19. Costonis, "Development Rights Transfer."

in the struggle among the fittest of the species.[20] We do not often think of God as striving for security or survival. In spite of efforts of process theologians,[21] we tend to ascribe security to God by definition and see in God only the effort to make Divine Will and Law effective.[22] The most intense relation of God to the land in Biblical writings is found in YHWH's marriage to Israel. The honoring of the marriage covenant fructifies the land, its violation pollutes it (Hos 2–3; Jer 3:1).

Expression

It is much easier to think of God's good in terms of expression. Biblical theology stresses the ways in which Israel's land is the basis for YHWH's expression of his call, faithfulness, mercy, and judgment.

Nature also seeks the good of expression through the land. We generally think of the sheer beauty of nature's ordered intricacy. We marvel at the processes by which natural systems preserve themselves in the face of catastrophic changes. We are awed by the sheer superfluity of sensation that can crowd upon us. Nature is at least a wondrous machine. Some see it as their very god. Natural expression can become the highest good of a land ethic.[23]

Societies also seek to express themselves upon and through the land. Their demands for splendor result in monuments. Their drive for more intense communication creates roads and cable corridors. Our land use reflects the values we place on privacy and sociality, on the nature of work, and the nature of family life. Societies shape the land to their own desire to express their fundamental values.[24]

Finally, we can look at land as the means for expressing what is proper to persons. Indeed, our own concept of property derives from the sense of persons' "proprium," that which is intrinsic to them. Property is indispensable for self-expression.[25] Without it, there is no sphere of

---

20. Geiger, *Theory of the Land Question*; Leopold, *Sand County Almanac*, 240.
21. Cobb, *Is It Too Late?*
22. Derr, *Ecology and Human Need*, 40–57.
23. Leopold, *A Sand County Almanac*.
24. Wagner, *Environments and Peoples*.
25. John Black surveys the development of Christianity's twin notions of personal development and responsibility to God (Black, *Dominion of Man*, 44–72). See Wilber, "Role of Property," for a contemporary treatment. For the history leading to Jefferson see Harris, "Private Interest"; Clawson, *Land System*; and Macpherson, *Possessive*

freedom in which persons can articulate their calling or their uniqueness. In this way, as C. B. Macpherson and many others point out, property, especially land, became identified with political and economic systems characterized by freedom. This is the source of the Jeffersonian ideal in American culture. Political freedom will be as widespread as is the ownership of land by individual persons.[26]

Enjoyment

Inclusion of enjoyment as a good may strike some as odd. In many respects it is the converse of expression. One party's expression is or can be another's enjoyment. However, it is important to stress enjoyment as a good in itself.[27] Goods arise in relationship. The ethic of land arises in relationships among the parties we have delineated. The land is not merely a means for a party to express itself. It is also a means for a party to enjoy that other party through the land.

We often think of natural persons and families enjoying the land. They may enjoy the land "in itself," or they may enjoy another party, whether God or nature, through the land. When we emphasize the way land is a theater we find it as the medium by which we enjoy each other. We dispose our buildings, streets, parks, and even farms as means for structuring the way we are to enjoy each other. Land, especially "natural land," provides a distance from work and activity by which we can sense the presence of the Divine Other. This is the good underlying parks and retreats as well as wildernesses.

By lifting up the good of enjoyment we also emphasize that one party, for its own good, may enhance the capacity of other parties to express themselves through the land. This drive toward enjoyment can emerge as the root of altruism in land ethics.[28] It grounds our obligation in the mu-

---

Individualism. Sakolski, *Land Tenure,* provides a good history of the history of tenure patterns in America.

26. The Jeffersonian ideal seems omnipresent. For an excellent history see Brewster, "Jeffersonian Dream." For present arguments see Barnes and Casalino, *Who Owns the Land?*, McClaughrey, "Future of Private Property," Austin, "Three Axioms," and Griswold, *Farming and Democracy.*

27. This is contained in Austin's second axiom in Austin, "Three Axioms," and in Santmire, *Brother Earth,* 190.

28. Dunlap and van Liere, "Land Ethic or Golden Rule."

tual goods of the parties. By combining self-interest and self-sacrifice this good is often involved in the struggle for land conservation and reform.

Perfection

Finally, we find in land the medium for perfection of the parties. Just as land, especially the promised land, is the medium for God's expression, so it is for God's perfection. We do not need process theologians to guide us to this understanding, though the concept of God's perfection is dear to them. The Bible sees in Israel's faithfulness and in the land's renewal the perfection of God's will for the world. Deuteronomy as well as the later prophets see in this flowering of the land the completion of God's purpose in creating the world. The perfection of the land and of Israel's care for it thus entails God's perfection.

Similarly, people may ascribe to Nature the drive toward self-perfection.[29] This perfection is usually conceived of as the achievement of perfect equilibrium among all the factors of nature. The land can become the focal point for all efforts to advance the perfection of nature. All land use and ownership concepts must revolve around the achievement of nature's perfection as a perfectly balanced system.

There is a variety of societal ideals dependent on the land. We have already noted how the Jeffersonian ideal of republican democracy required widespread holdings by individuals and families. Under industrial conditions this ideal has to be translated into some notion of minimum social security, but Jefferson's ideal still lies behind many efforts to sustain the family farm. Most ideals of socialism or a social welfare state lead to the demand that most rights to the land be vested in society or the state in order to guard against the land speculation, destruction, or misuse that serves the interests and goods of only a few people.

Proponents of agricultural visions of the good society struggle constantly with exponents of industrial visions of perfection.[30] While those devoted to technology may find this to be a struggle among dinosaurs, it certainly appears in much of the literature on land ethics. Those pursuing industrial efficiency as society's central good find themselves pitted against those who find in labor-intensive agriculture the proper condition for humane socialization.

---

29. Passmore, *Man's Responsibility*, 32–38.
30. Austin, "Three Axioms."

Some visions of societal perfection turn away from the totalitarian tug of state and society in order to pursue the perfection of individuals, families, communes, and corporations.[31] The independent farming family is often extolled as a model derived from religious teaching.[32] The personalities raised in such a context evidence the virtues of initiative, industry, familial care, and thankful openness to God's blessing. In this sense, a particular pattern of land use and ownership becomes indispensable to the evangelization tasks of the Church.

We find a similar argument among those pursuing a cooperative model of human perfection. They believe land should be owned and operated by cooperatives and communes. Control over the land ensures the relative independence of the group from the social pressures of urban industrial society. Moreover, since the cooperative corporation never dies, it can achieve the longevity of land tenure necessary to cultivate cooperative ideals in the course of successive generations.

The transnational corporations also find that perfection of their purposes requires control over the land. This is notably true of oil companies, which combine oil, fertilizer, land, food production, and marketing. We find it among other conglomerates based in mineral extraction, wood products, and other land intensive industries. Though their purposes may not seem as lofty as those of religious communes, their drive for perfection through land control is very intense.

MODELS OF LAND ETHICS

We find, then, the four parties trying to realize various rights and goods through the land. We can realize the complexity of land ethics when we see that the four parties, the four rights to land, and the four goods of the land can combine in many ways to produce various visions of the land. We obviously cannot explore all these possible models. In keeping with my numerology, I can set forth four basic models competing for our attention today. Some are dominant and very powerful, others are marginal. From the standpoint of Biblical religion they are not equally worthy. My presentation of them will not be merely descriptive, but will

31. Berry, *Unsettling of America,* and Borsodi, *Flight from the City.*

32. See Harris and Ackerman, *Agrarian Reform,* 1–35. Catholic social philosophy has persistently echoed this theme. See Cronin, *Social Principles,* and various publications of the National Catholic Rural Life Conference, such as "Toward a Greater Rural America" (1968).

argue toward the virtue of a particular model that Christians and Jews should advance in our society.

The Market Model

The land ethic presently governing land decisions in this country is dominated by a market model of relationships.[33] The primary claimants to the land are persons, whether natural or corporate. Land is distributed among persons and used by them according to their own perceived goods, especially those of security and expression. I could call this an exchange model, but that would deflect us from the economic nexus through which land value is established. Land has inexorably been drawn into the commodity and capital markets. Even though land's position as the dominant if not sole basis for economic welfare has eroded, it is still an important factor in capital. More importantly, because of its increasing scarcity it becomes a speculative investment to buffer the shocks registered by stock market stagnation and other long-range weaknesses in the American economy.

The ethical defects of this model are manifold. As land is drawn into the capitalist nexus, control over it follows the concentration of capital in fewer hands. These controlling persons are increasingly corporate and transnational. Increasing numbers of natural persons cannot even enter the land market to exercise rights over the land—even those of use. The market model is inevitably limited and self-destructive.

In reaction to this development persons turn to governmental bodies to restrain the market. The dominant instruments for regulating the land market in this country are zoning and taxation.[34] These seek to channel rather than eliminate the dynamics of the market. Like all regulatory devices of course, they tend to become dominated by the interests they seek to control. In addition to these instruments, other laws, deed restrictions, governmental ownership, control over development rights ("alteration"), and planning commissions also affect the unrestricted operation of the market.[35]

---

33. For discussion of various economic models in ethical perspective see Wogaman, *Great Economic Debate*. Some still see the market as the best means for allocating and managing land, as with Gramm and Ekelund, "Land Use Planning," and Siegan, "No Zoning."

34. Delafons, *Land Use Controls*.

35. Bosselman and Callies, *Quiet Revolution*; Darin-Drabkin, *Land Policy*; Healy,

Those committed to the market model often support these restrictions inasmuch as they maintain wide participation in the market. Less powerful persons can gain some leverage over against corporations and foreign persons. Others endorse this regulatory apparatus because they actually support another model and would like to replace the market model altogether. While joined in a common struggle today they must part ways down the road.

The Societal Model

Both the increasing involvement of government and the increasing concentration of land control in corporate hands point to the societal model for a land ethic. In this case Society is the primary party claiming the land. These claims can be administered through the state or other bodies representing the society. Socialists are only one of the groups using this model. English and American reformers like Spence, Paine, and George, who were not socialists in the usual sense, have espoused it as well.[36] The important thing to remember is that society is the party exercising rights and achieving its goods. The market is replaced either by a political process or by communal tradition—the former being a "progressive" stance and the latter a reactionary one. In this debate the progressives clearly have the advantage, since technology devastates the conditions for traditional society.

Society, of course, delegates certain use and even income rights to persons in order to serve the common good.[37] These can be withdrawn when such arrangements no longer serve that societal good.

The practical problem of the societal model is that corporate concentration can simply be replaced with governmental concentration. The good of the governmental bodies, rather than the good of the society, is served.[38] The other problem, which runs far deeper and which opens us up to the next model, is that political decisions, like the economic ones they replace, often don't respect the claims of nature. Societalists criticize marketeers for not respecting societal needs, not to mention the needs of

---

Land Use; Tobin, "Some Observations"; and Berchin, "Regulation of Land Use."
36. Davidson, *Concerning Four Precursors*.
37. Cronin, *Social Principles*; Speltz, "Restrict Ownership."
38. Black, *Dominion of Man*, 73–89.

the majority of natural persons, but ecologists criticize both for ignoring Nature's claims.

The Ecological Model

Within the ecological model, land is a part of nature and nature is the primary party. Decisions regarding the land are controlled by the needs and limits of the ecosystem. At the practical level this tends to mean that land ethics are shaped by those who know nature best, in this case the geologists, agronomists, and ecologists. Planning replaces politics and economics.

The ecological model focuses on the right of land use and on the good of nature's perfection.[39] While it seeks to contain and limit the market and society, it does not provide much help with the disposition of the other rights and goods. Nature mystics provide us with a personality ideal—oneness with nature—or the harmonistic model usually associated with traditional society, but these are usually associated with bucolic utopias which tend simply to reject the conditions of modern life.

In fact, of course, the claims of nature are constrained by various personal and societal claims. These parties seek security as well as expression and perfection.[40] They cannot be content with the enjoyment of nature in itself. Moreover, nature's retribution for transferring and altering the land is usually too slow and obscure for these other parties to discern or abide by.[41] Because of these limitations we are drawn to our final model, which stands at the center of Biblical tradition—the covenantal model.

The Covenantal Model

In the covenantal model God is the primary party and the land is a trust God establishes with His people. It is not a gift, which implies the transfer of all rights.[42] A trust transfers certain rights according to certain terms.

---

39. Merrill, *Radical Agriculture*.
40. Neuhaus, *In Defense of People*, 161, 256.
41. Passmore, *Man's Responsibility*.
42. The "gift" notion of land focuses on God's transfer of land from the Canaanites to Israel in the Exodus. This tends to be Brueggeman's focus in *The Land*. The trust concept, in which the focus is on land and "man" in general, (*adamah* and *adam* in Hebrew), comes from Genesis.

God entrusts us with use of the land and with the income from it. God establishes this trust first of all with humanity as such ("Adam") and then by subsidiary claim to societies residing in specific lands. The God-Society trust exemplified in Genesis must be qualified with the affirmation that God is the creator and savior of every person. Therefore, use and income rights must be arranged so that all persons can achieve at least the minimal human goods. In this model corporate persons have a subsidiary position as means to promote personal and societal goods within the trust.[43]

The covenantal view understands Nature as God's creation rather than as a machine separate from human goods and rights. God's claims to the land are mediated through nature as well as through persons and societies. Nature's claims are derivative and not absolute. As a medium of God's covenant, nature provides gracious boundaries for our dominating lust as well as projects and wonders to overcome our slothful isolation.

Having assigned a relative weight to the various parties to the covenant we can see that certain rights pertain primarily to persons, who may also constitute corporations for that purpose, while alteration rights would be assigned to nature and right of transfer to God. Simply the roughest sketch of the covenantal model displays a sharp contrast with our present conceptions of land ownership and use.

The covenantal model has equally important implications for our ranking and disposition of the goods of the land. The good of perfection would tend to fall to natural persons. This personal perfection is linked directly to God's expression of His creativity and goodness through the land. Therefore, the ideal of personal perfection here revolves around the concepts of steward and trustee.[44] The trustee pays attention to the honoring of rights attached to the land and the steward helps participants in the trust achieve the goods provided by and through the land. The motifs

---

43. Santmire, *Brother Earth;* Schumacher, *Small is Beautiful,* 241–53.

44. Sider, "A Biblical Perspective," Byron, Ethics of Stewardship," Swartley, "Biblical Sources," and Derr, *Ecology and Human Need,* 101–25, provide contemporary religious views. Black, *Dominion of Man,* 44–57, provides their history. Dubos, "Franciscan Conservation," finds Benedict's views of stewardship more appropriate than Francis' model of conservation. Santmire, *Brother Earth,* 187, finds that "caretaker" better expresses his ethic of responsibility, and Cobb, *Is It Too Late?,* 124, prefers "participant" to express his process commitments, while Passmore, *Man's Responsibility,* 29–31, argues against the Christian origins of the concept as the basis for a land ethic. He holds that Christian notions of stewardship applied to obligations to the community, and only indirectly to land. Stewardship toward nature is a Classical notion, revived in Fichte and Teilhard. Passmore rejects the use of pre-lapsarian models (Gen 1–2), claiming they are incongruent with orthodox Christian emphases on sin.

of trustee and steward therefore affect public policy as well as personal character. This is the context in which we can talk of responsibility and care. It is in this activity that we are drawn out of ourselves toward the giver of life and toward those with whom we share life.

The stewardship of the land is not dumb awe before an absolute. The enjoyment of the land involves its transformation as well as its preservation. Transformation, in turn, demands deep respect for the finite conditions within which we must act. Land is a trust. It is more than a commodity, but less than sacred substance. Therefore, the claims of nature, societies, and persons are all relative to those of God. In the exercise of our stewardship and trusteeship we can come closer to that traditional concept of human perfection, that we glorify God and enjoy him forever.

The covenantal model provides the fundamental ground for a Biblical land ethic in our time. It has particular prophetic sharpness in the North American context. Moreover, the definition of parties, rights and goods delineated here can help us clarify the ethical issues regarding land use. However, it does not resolve in specificity the many issues of public policy which lie before us if we are to approximate a covenantal model more closely. Lest our pragmatic temper not be served, however, let me finish with some practical observations concerning land policy.

TOWARD COVENANTAL INSTITUTIONS FOR THE LAND

The model most inimical to covenant and trust is the market. While markets may be appropriate for the exchange of marginally important commodities, they are not appropriate for the essentials of life. The peculiar characteristics of land and its special role in our life demand that it be removed from the dynamics of the "free" market.

In a sense, this is happening willy-nilly, as corporations consolidate land holdings for long-term interests rather than short-term speculation.[45] However, these corporations have relatively narrow interests and goods in view. Their interests are rarely societal in scope.

Public ownership through various governmental bodies secures the land even more permanently from the market and tends to broaden the interests served. Societal goods can be envisaged more explicitly.

---

45. Pignone, "Concentrated Ownership"; Barnes and Casalino, *Who Owns the Land?*

However, even this public ownership must be corrected by the limits and goals of covenant. The societal bodies are only one party to the trust God gives to all parties. While rights of alteration and even limited transfer might reside in public bodies or even corporations, use and income rights must devolve more fully to natural persons in keeping with the type of economy dominant in an area.

For this reason the land trust movement exerts a peculiar attraction.[46] Its very name is symbolic of the deeper theological foundation for an adequate land ethic. In this respect it alters the cultural views of our society, disposing them away from a market model to some combination of the other three. Land trusts remove land from the market and also engage persons in the stewardship of the land.

The trust acts to produce as well as sustain community, for it demands some degree of community organization for its establishment and maintenance. The trust movement is therefore closely bound up with the cooperative movement. The means for exercising rights to the land are cooperative and the means for using the land can be cooperative as well. Moreover, a trust must contain some terms under which it is executed. It must set limits and goals of use. These can set forth some ideals for persons and communities as well as act to ensure security, expression, and enjoyment of the land. The trust agreement can set forth some relatively permanent expressions of larger covenant within which it stands.

Lest this portrait verge too much on the romantic, let us turn also to the trust's weaknesses. At this point trusts rely more on altruism than necessity. Public spirited citizens are seeking ways to put their lands in trusts rather than subject them to the caprice of the market and its "development."[47] However, the trust movement must depend on economic power to purchase the trusts. Land is not tied in this country as closely to political forces, since it is not as directly tied to the immediate economic interests of the majority of the people. Therefore, the path

---

46. This movement utilizes various structures. The Trust for Public Lands (San Francisco) receives or purchases land which it then places in trust with public agencies or community corporations. The Center for Community Economics (Cambridge, MA) advises and helps establish local trusts for holding land. See International Independence Institute, *Community Land Trust*; Blackmore, "Community Trusts"; and McClaughry, "Model State Land Trust Act." As a variant, communities can take direct control of the land (Grant, "Community Land Act"; Duerksen, "England's Community Land Act"; Bosselman, Callies, and Banta, *The Taking Issue*).

47. Barnes, "Buying Back the Land"; Open Space Action Committee, *Stewardship*; The Conservation Foundation, *National Parks*.

through government confiscation is not as likely in the North American context as it would be in agricultural, non-urban countries. This means that trusts must depend on the economic power of churches, foundations, pension funds, and wealthy individuals. The tax-deductibility of land gifts also assists the trust movement. The question, then, is whether the trust movement can gain enough momentum, not only to entrust a majority of land, but to affect public policy so that trust concepts are extended to public and private land as well. This is the major task before us and the one to which the covenant model calls us.

6

Stewardship Through Trust *and* Cooperation

CHRISTIAN RESPONSE TO THE call to stewardship has always been deeply shaped by the economic, political, and social conditions of the time. The concept of stewardship always needs other ideas in order to form this appropriate response. In this article I want to point out "trust" and "cooperation" as helper ideas to implement the call to stewardship. While stewardship focuses on our use of things, trusteeship brings in the patterns of ownership. While stewardship has often received a very individualistic interpretation through our reading of the parable of the talents, cooperation helps fill out the communal theme actually implied by that parable.

To unlock the elements of trust and cooperation implied by stewardship, I shall investigate the famous parable of the talents and trace its changing interpretations. I shall show how trust and cooperation are vital ingredients to a faithful view of stewardship. Finally, I shall describe how the legal concept of trust can enable us to relate a stewardship spirituality to the ownership and use of land.

STEWARDSHIP AND THE PARABLE OF THE TALENTS

One of the most popular Biblical sources for a concept of stewardship is the parable of the talents in the Gospel according to Matthew (Matt

25:14–30). In this parable Jesus tells his disciples that a master entrusted various amounts of money to his stewards to administer while he was gone. On his return he found that some had invested it to good success. One, however, had done nothing, returning only the amount he had received. This timid fellow received condemnation, while the faithful, entrepreneurial servants received their master's blessing.

This parable has exercised such an impact on our culture that our very word "talent" results from it. Talent originally was just a monetary unit but has come to mean all the abilities and capacities individuals discover or develop within themselves. Our whole understanding of personality has been shaped by the belief that these abilities are entrusted to us for the purpose of glorifying God or serving our neighbor. At the very least they should be exercised simply because they are there. God, not myself, owns them.

This individualistic interpretation of talents gave some religious basis for the rising businessman of the eighteenth and nineteenth century. One's talents were expressed in some kind of market where individuals allocated goods and services according to purchasing power. It is this very marketplace model of life which turns everything into a commodity, including the very land which sustains us. A religious effort to go beyond this kind of marketplace individualism must return to the parable of the talents itself.

Close examination of the parable will show that this individualistic interpretation is mistaken. First, we must set aside the automatic equation of talents with personal abilities and see that the talents referred to in the parable are the Gospel itself. Second, therefore, the judgment which is upon us and for which we must take great risks concerns how we handle the Good News entrusted to us. Third, we must see the connection between the preaching of this Good News and the building up of the company of the saints.

In this respect the parable of the talents is like that of the sower and the seed (Matt 13:1–9). Here we find a similar concern for preaching the gospel to all the world in order to create a community of faith. This central theme in Matthew deeply shapes our understanding of the parable of the talents. (See Matt 18:12–14, 24:14; 28:19–20.)

Moreover, all the other uses of trust or stewardship in the New Testament refer to the mysteries or grace of God (1 Cor 4:1; 1 Pet 4:10) or Gospel (1 Cor 9: 17; 1 Tim 1:11; Titus 1:7) entrusted to us for the salvation of the world. The church is the trustee of this Gospel of which we are agents. Our "talent" therefore is primarily the Gospel entrusted to the whole church. This understanding is very different from the modern conception of talents, by which we distinguish ourselves from the community and enter the marketplace as producers and consumers.

STEWARDSHIP AND COMMUNITY OF FAITH

By interpreting talent in terms of the Gospel we can clarify the relation of faith to stewardship and trust. First of all, stewardship has to reflect the Good News of God's coming perfect community. Moreover, this Good News nourishes the community which awaits this new creation.

Not only is trust related very closely to faith, but also to communal faith—that is, "faithing together," the original meaning of "con-fidence." This communal focus of faith, trust, talent, and stewardship takes us to the final helping idea: cooperation. With cooperation we move away from the pattern of master and slave to that of friends and partners. Through Christ we enter into a life of equality before God and of cooperation with God. We pass from the obedience under God's wrath to the cooperation in God's fellowship. The acts of stewardship, therefore, because of their close relation to faith and trust, are necessarily acts of cooperation. Stewards are no longer slaves but trustees acting in conscience and equality before God. Proclamation of the Good News of God's presence and coming always implies an ever-widening extension of this cooperative network.

Stewardship, trust, and cooperation form a constellation of terms which illuminate each other. To remove one is to darken the face of the others.

STEWARDSHIP: THE LAND AS TRUST

A nagging question then confronts us. If stewardship is a matter of "accounting for the faith entrusted to us," what are we to think about all these "material" things in our care? At least the individualistic interpretation of talents talked about our part in the transformation of creation. Is this stance a new kind of "mere spirituality" with little of the earth in it? Not at all. This original meaning of stewardship implies a fundamental change in the way we own and use land.

Let us use the old legal concept of a trust in order to see those implications. We have to remember that faith and the sharing of faith stand at the center of stewardship. "Trust" expresses the meaning of faith for us as members of a community of faith. We are trustees of the promises of God guaranteed to us through the life, death, and resurrection of Jesus.

TRUST, THE FRAMEWORK OF STEWARDSHIP

How then does the concept of trust help illuminate our relationship to the land, to God, and to each other? What broader context might it give for our concept of stewardship? To pursue the trust analogy we must see its five basic elements: the settler, the trustees, the instrument, the property, and the beneficiaries.

The Settler

The trust is established by a settler who grants his or her property into the hands of others to benefit some third person or group. From a faith standpoint God is the settler of a trust with us. The God who has settled with Abraham, Isaac, Jacob, and Jesus is a covenanting God calling us to "con-fidence" in Him. God also covenants with humanity ("Adam"). Not only does God covenant with us about a land in which we already dwell but He also promises us yet a better land into which He will lead us. The very term "settler" adds to covenant and trust the connotation of settlement, dwelling, and of being secure in the land.

The Trustees

What does it mean to say that we are the trustees? "We" might be interpreted as simply the sum total of individuals, each with a direct relation to God. But that would be inadequate, for a trust also implies a number of trustees bound to each other. The "we" is simultaneously many individuals who receive God's promises in faith and also a corporate person bound together by "confidence." Communities, corporations and even whole societies might be trustees if they are willing to take on the obligations of the trust.

Just as people of faith are no longer slaves but friends, so trustees are not slaves of the settler but free persons entering into a confidence, or covenant, with the settler. They are persons of conscience, whose selfish interests are to play no part in administering the trust property and executing the trust.

Trusteeship therefore requires a particular ethos or spirituality, in which the needs and interests of beneficiaries come before self-interest, in which the will of the settler is to be respected, and in which prudential management supplants reckless speculation. Finally, trusteeship implies a commitment to consensus and cooperation among the trustees in the execution of the trust.

The Property

Some kind of property is intrinsic to any trust. Land has always been one of the most prominent. Our word for property comes from the Latin "proprium," meaning pertaining to the very presence of a person in the public order. Our "proprium" is an extension of our personality, as we would say today. When we say that the land is God's property, we are not only saying that land ultimately "belongs" to God but is also in a special way that which is "proper" to God. It is not merely a thing over which God exercises some rights. It is a means for God's full manifestation, whether in its sheer awesome power and beauty or in its nurture of the people to whom God is faithful as parent. When we deal with the land as God's property we are confronting God's very "persona," a way God is present to us.

The Beneficiaries

The beneficiaries of the trust are all the future generations. Since the end is in God's hands we cannot limit this obligation to the future. For our descendants' sake the principal, that is the land, is to be kept intact. If at all possible it is to be enhanced to the extent that we can anticipate. Savings from the present can be set aside when they do not infringe on the terms of the trust or the legitimate needs of present users.

The Instrument

The intent of the settler is expressed in the trust instrument, which states the values and nature of the stewardship the settler expects. Trustees supervise the stewardship of the property. The instrument is a kind of covenant to which they must conform.

The instrument also specifies the kinds of claims that can be registered against the trust property. With regard to the land I see four claims that need to be apportioned and satisfied—use, income, alteration, and transfer. This bundle of rights, or claims, is usually seen as a single right of "ownership," but in fact they are separable, both in theory and in practice.

Use entails direct occupancy and resource exploitation. Land is used for a dwelling and workplace as well as a theater for action, recreation, and contemplation. It is exploited to provide food, shelter, clothing, arid mineral-based products.

Income is similar to use but occurs at one remove. It is really indirect use of the land by which parties benefit—in money or "in kind"—from the work of others as well as of themselves. Thus, a portion of the proceeds from the use of the land can go to parties quite distant from it—other persons, corporations, and the state. In ancient Israel the "tenth" went to the Lord as a way of acknowledging His ownership rights on the land.

Alteration carries use of the land a step further than those actions which do not deplete it. In alteration, pursuit of one use (say mining or building construction) radically alters for some time to come the possibility of other uses. We use the term "alteration" here in order to escape the prejudicial connotations of "development," which assumes that greater income for some parties is better than a use more appropriate to the land's character.

Transfer is a kind of master right over the others. It is the right to assign the rights of use, income, and alteration to specific parties. While normally it is combined at least with the right of income, it can also be independent, as when the state transfers land into the public domain for public purposes.

An ethical inquiry into the nature of trusteeship demands that we sort out these rights among the various parties to the trust. The use right, it would seem, should be quite broadly apportioned among individuals, families, and smaller groups, though societies have some rights here for land use that supports the common good—whether for preserving social memory in sacred places or enabling the public business to be conducted. A trust property can be stewarded by tenants or other agents. However, it is the nature of trust that the dwellers on the land would not "own" the land. They would use it for the purpose of their livelihood.

The *income* from the land can be distributed among the beneficiaries, the settler, as well as the trustees themselves. The trustees deserve some reimbursement for faithful administration of the trust. Future beneficiaries may receive some income in the form of savings from the present. God may also receive a tithe in the sense that some values of the land are set aside to enhance our worship of God, our conduct of wider social life that expresses the love and fellowship God desires for us, or for giving due accord to the manifestation of the land as God's "*proprium*," as with the preservation of natural wonders.

Among the trustees themselves income might also be variously distributed, according to the contribution they have made. For instance, individuals contribute their ingenuity, labor, and imagination to make the land fruitful. Communities and societies may make land more desirable by the kind of public realm they maintain or simply by sheer increase in population, which forces up land values. The increased value they create should redound to them rather than to specific individuals.

The right to *alter* the property is spread between the settler and the trustees but is also deeply conditioned by the nature of the property and the purposes of the trust. The simple use of the land does not necessarily involve its alteration. But alteration of the land within a trust framework should be hedged about by many constraints—many more than presently exist in American politics.

First, we must remember the purposes of land as given by God. These include our very survival, our wider enjoyment of the land, and the perfection of our bodies and spirits in recreation and contemplation. Moreover, land provides places for sustaining our collective memories, arenas for public action, and spaces in which to mark out the patterns of our usual relationships. Alteration of land must first respect these purposes.

Second, it must respect the very characteristics of the land. The "best use" of the land must not be subordinated to the market but the market to the wider values of trusteeship. Finally, any alteration of the land must raise prudential considerations of long run effects over short run speculation. In this respect the general concept of trust overrides the mis-use of the parable of the talents that justified high-risk speculation in land as a commodity.

The right of transfer ultimately belongs to God. The exercise of this right by individuals, corporations, or communities should reflect the wider values of trust, stewardship, and cooperation. We transfer land by a delegated and therefore limited authority. The entrustedness of all land calls into question the blithe way we transfer land control. Suspicion of easy land transfer has always characterized covenantal traditions, whether in Israel (Lev 25), medieval Europe (where the monarch held the land in trust to God), or in American agrarian reform movements.

In short, exercise of the four ownership rights should not be exalted as a supreme value apart from the context of trust. Though fee simple ownership of land, in which an individual holds all the rights, is dear to the American dream, it is becoming inimical to proper trusteeship. It is also rapidly becoming an impossibility. While invocation of the family farm ideal or the suburban ranch home has great emotional appeal, we simply can't afford it as individuals or as a society. Ownership will become more and more collective. The only question involves the nature of that collective ownership. The concept of trust helps point the way to a possible answer in line with Biblical faith.

The trust provides a context for our understanding of stewardship. It helps us see the nature of stewardly obligation more precisely. It shows how stewardship, which primarily concerns the use of things, must be understood in the context of ownership, with its complex of rights to use, income, alteration, and transfer. Because trust is a form of social relationship familiar to American ears it can mediate the religious values of faith to the practical realities of land.

Trusteeship itself helps us move from a master-slave view of stewardship to a partnership in which cooperation among trustees prevails. The trustee shares faith with the settler as well as the beneficiaries. Trusteeship is a role that stimulates a broader sense of obligation than narrow self-interest. Not only is it a role demanding con-fidence but it also demands con-science—a common knowing of life as well as a common faithing of God. It presses us to a faithful way of holding and using the land entrusted to us. Stewardship, trust, and cooperation are pillars rooted in the Gospel and supporting a spirituality for action in the world.

7

SHARED PARENTHOOD *in* DIVORCE
The Parental Covenant and Custody Law

INTRODUCTION

Ethical discussions of divorce have usually taken up the question of whether or when two spouses may divorce and remarry. They have focused on the dissolution of the marital bond. Almost no attention has been given to the ethical issues involved in negotiating a divorce or maintaining the parental bond in divorce.[1]

Emerging efforts to maintain parental bonds in divorce rest on many diverse considerations. Let us begin with the social changes demanding this concern and then move to some religious, ethical, and legal responses to this challenge.

---

1. Charles E. Curran presents a progressive Catholic view in Curran, "Gospel and Culture." See also Wrenn, *Divorce and Remarriage*. Protestants have tended to look on divorce as a tragic failure rather than a sin. Karl Barth presents a Protestant version of annulment theory in Barth, *Church Dogmatics*, 3/4, 211. Elizabeth Achtemeier puts the matter in a more contemporary Protestant form in Achtemeier. *Committed Marriage*, 109–31. Ethically sensitive treatments of the divorce process can be found in Young, *Growing Through Divorce* and Whitehead and Whitehead, *Marrying Well*, 347–70.

Recent ethical attention has been mostly philosophical. See especially Blustein, *Parents and Children* for an historical and constructive argument. Relevant articles are assembled in Aiken and Lafollette, eds. *Whose Child?* and O'Neill and Ruddick, eds. *Having Children*. Stanley Hauerwas has stimulated theological discussion of the procreative fidelity of the family in *Community of Character*, 155–95.

PARENTHOOD IN DIVORCE: THE SOCIAL CONTEXT

For the past decade about 1.1 million children have been involved in divorce each year in the USA. According to some estimates about one-third of all children will spend some time in one-parent households before they are eighteen.[2] Clearly, the decisions involved in providing parental structures for these children are emotional and important. In order to clarify the values and principles that should guide us we need to understand three social changes.

First, over the past century we have seen the crystallization of a general cultural pattern in which women and men do not have to be married in order to have social position and status. Moreover, couples do not necessarily have to be parents, nor do parents necessarily have to be married to each other. Some were never married, many are divorced, and some are single parents by adoption or death of the spouse. Finally, parents and children need not constitute a single household. There has been a pronounced differentiation of parenthood from marriage and household. One need not entail the other. Clearly, the exercise of parenthood takes on a number of forms under these circumstances, and these forms demand ethical and legal attention.

Second, we experience considerable lag between our legal institutions and cultural myths on the one hand and social reality on the other. Culturally, we suffer from an inadequate language[3] to describe these new arrangements, including religious and legal language. We speak of "one-parent families" when we really mean "one-parent households." Or we speak of "single parents" without clarifying whether they are their children's only parent or whether they simply are not married.

Legally, most states have moved to a no-fault divorce pattern in which a divorce is granted, with provision for property and children, on petition by one of the partners that the marriage is irretrievably broken.[4] However, in fact, in most states the divorce is still treated as a crime for which the parents yield their parental rights to the court, which

---

2. For recent statistics see Spanier and Glick, "Marital Instability." Levinger and Moles, eds. *Divorce and Separation* offers the best coverage of the social-psychological aspects.

3. Cherlin, "Remarriage," claims that lack of adequate role language undermines second marriages.

4. For overviews see Eisler, *Dissolution*, and Freed and Foster, "Divorce in the Fifty States."

then restores them as it sees fit. Legally, felons still retain their parental rights, such as decisions about their children's marriage, education, or medical treatment, while noncustodial parents lose all of these rights. Increasing numbers of noncustodial parents are seeing this as an inequity and an injustice.

According to the tradition of the last 100 years, the rights and responsibilities of parenthood are usually "awarded" to the mother. The father is given responsibilities of financial support and the right to "visit" with his children. This follows the general "two-spheres" doctrine of the nineteenth century, in which the mother ran the family and household while the father provided financial support.[5]

A third social change is that a model of shared parenting is taking hold in which both parents are expected to share the tasks of raising the child. Both parents are to have a public and occupational life as well as a parental life. This model entails, in cases of divorce, that they will go on sharing their parental rights and responsibilities. The model is reflected in the recent rise of "joint custody" statutes in California, Florida, Louisiana, and elsewhere.[6] In general they provide for custodial arrangements in which both parents have legal and physical custody of their children. Sometimes both parents must agree to this arrangement, other times the court can assign it in "the best interests of the child." Joint custody need not require that the child live half the time with each parent. A wide variety of arrangements are possible in carrying out a joint custody order.

In an increasing number of cases, then, both parents are trying to exercise their rights and responsibilities apart from their marriage to each other or residence in the same household. Yet the distinction between the marital and parental bond increasingly characteristic of our society is not yet reflected in our cultural and legal forms. Shared parenthood

---

5. Carl N. Degler sets forth the two-spheres history in Degler, *At Odds*, 249–78. Mel Morgenbesser and Nadine Nehls provide some history of custody determinations in Morgenbesser and Nehls *Joint Custody*, 5–26.

6. *California Civil Code* §§ 4600, 4600.5; *Florida Statues Annotated* § 61.13(2)(b); *Louisiana Civil Code Annotated*, art. 146. 147. One of the earliest appeals for a joint custody law appeared in 1964 from a psychologist, Lawrence Kubie, "Provisions for the Care of Children." For comprehensive updates see Folberg and Graham, "Joint Custody"; Gouge, "Joint Custody"; and Miller, "Joint Custody." By 1983 The National Conference of State Legislatures, "50 State Overview," could report that twenty-nine states had provisions for joint custody, thirteen with preference, six if one parent requests it, and eight if both agree. For the most recent statutory overview see Folberg, *Joint Custody*.

in divorce is not yet a widespread phenomenon, but the high incidence of divorce coupled with a new parental image for fathers means that it deserves our careful attention.

The major question in connection with this development is whether joint custody should be granted only when both parents agree to it or whether the court can order it in its historic role of *"parens patriae"* ("parent of the country"). Should the law establish a framework of *presumption* for joint custody or for sole custody? While we have had only a brief experience with shared custody arrangements, initial studies have already shown that they reduce post-divorce litigation and contribute to the child's healthy adjustment to divorce.[7] The more that society and the law presume joint custody arrangements, the less parents are tempted to use the "custody chip" as part of the retributive arsenal of the marital divorce.

Some critics of joint custody claim that fathers have used it to decrease their support payments or to continue abusive relations with their children or former spouse. These, however, would not be defects arising intrinsically from such an arrangement, and could be dealt with by providing safeguards against abusive parents and guidelines for meeting financial responsibilities equitably.

Other resistance rests on demurrers that shared custody should not be a presumed starting point. It should only be used when both parents agree.[8] However, resting the custody decision on the emotional state of divorcing people subjects the matter to their desires for revenge or vindication. It is precisely at this moment that the law should provide an objective framework for balancing the legitimate claims of children, parents, and society.

Joanne Schulman, a leading critic of joint custody, leads us from these utilitarian calculations to a more fundamental ethical dispute when she claims that:

> The current joint custody trend is, in effect, an attack on women who have been, and wish to continue to be, the primary

---

7. The central psychological study is by Wallerstein and Kelly, *Surviving the Breakup*. See also Hess and Camara, "Post-Divorce Family Relationships." See also Ahrons, "Joint Custody Arrangements." For specific focus on fathers and children see Hamilton, *Father's Influence on Children*; Winkler and Winkler, *Fathers and Custody*; and Jacobs, "Effect of Divorce on Fathers." John W. Miller provides a theological focus in Miller, "Contemporary Fathering Crisis."

8. See Morgenbesser and Nehls, *Joint Custody*, 68.

caretakers of their children. Their past assumption of the daily care and responsibility for children is denied any value or credit.

The current joint custody trend will not lead to equality between the sexes. Sexism does not end when women lose rights or lose custody of their children. Forced joint custody, like forced sterilization and forced pregnancy, is a denial of women's right to control their lives.[9]

This viewpoint asserts clearly the dominance of the mother's rights over all other claims, even though still drawing on elements of the argument from best interests. It is time, however, to move the discussion away from the primacy of parental claims and ask what rights the children have in divorce—specifically their right to both of their parents. The question of joint custody forces us back to a re-examination of the emotional bonds underlying our cultural, ethical, and legal forms. We are dealing here with a reconstruction of some of our most basic relationships.

To reconstruct these cultural and legal forms we need to turn first to the religious and ethical concepts underlying them. We must examine the implications of this social change for our religious conception of marriage and parenting. How does this change affect our root metaphors and convictions? Then, what implications does this transformation have for our ethical perspective on parenting in divorce? Moreover, how can our historic convictions shape these social developments and the legal changes they require?

THE RELIGIOUS FRAMEWORK

Our ethical reasoning is rooted in our fundamental worldview. Moreover, our capacity to live out an ethical life derives its energies from our fundamental commitments. This world view and these ultimate commitments find formulation in our theology. Therefore, let us first examine the relevant theological considerations underlying our approach to parenting in divorce.

Our theological formulations assume a fusion between marriage and parenthood, family and household. The theological values of fidelity, permanence, and service have all resided in the composite reality of "marriage." The differentiation of marriage from parenthood and household

---

9. Schulman, "Second Thoughts." For a more measured research-oriented review of problems in joint custody see Steinman, "Joint Custody." For a rejoinder see Kelly, "Further Observations."

forces us to revise the way we have tried to advance these religious values. The primary religious values churches have tried to advance through "marriage" are actually parental values.[10] This fact was obscured when parenthood was an automatic concomitant of marriage. Religious values were attached to the act of marriage, that is to the wedding. The fact that the marital relations between husband and wife usually maintained the father-daughter pattern of parental subordination further obscured this distinction. However, the covenantal relationship between God and Israel and between Christ and the Church are parental relations. To the extent that we seek a worldly exemplification of covenant fidelity we must seek it in the parental bond. The parental covenant is not only distinct from the marital covenant but also more resonant with historic symbols of faith. Likewise, for churches that treat marriage as a sacrament, the real sacrament lies in parenthood—raising up new members for participation in the redemptive life of the Church. Of course, the marital bond is not devoid of theological significance! But it is a bond of communion resting on the natures of the two people whose union rests on more than their will to construct an ethical achievement. It is a peculiar kind of friendship rather than a worldly institution. It is a garden of experience throwing up metaphors for faith, but it is not primarily to be an instrument of the faith community.

Though marriage itself should not be seen necessarily as a vehicle for the Church's proclamation of the Gospel, we can indeed see marriage as an expression of vocation. The couple themselves can have a vocation, such as missionary couples have had in many Protestant churches. But historically the basic vocation of the couple was to have children. The public vocations were restricted to men. Women's vocation was to care for the household and its children. However, even the use of this vocational symbol already lifts up the way parenthood is a project needing a public form and public support, even apart from the marital bond.[11]

Failure to see the theological centrality of the parental covenant has hindered our efforts to support parenting. Theologically we have given more attention to weddings than to births. Moreover, when we did attend to the parental bond apart from marriage we could only allow one parent to fill this divine role. Before 1900 it was the father. From 1900 to the

10. The full version of this argument can be found in Everett, *Blessed Be the Bond*.
11. The responsibility of public institutions for parenting, with a Puritan model in the background, is argued extensively in Grubb and Lazerson, *Broken Promises*. For a detailed British proposal see Rapoport and Rapoport, *Fathers, Mothers, and Society*.

present it was the mother. We are now in a position to see that parenthood, whether as vocation, covenant, or sacrament, demands the active and equal participation of both parents.

The implication of this reassignment of religious values is that churches ought to give as much attention to preserving and enhancing parental bonds as they gave to marital bonds in the past. This demands changes in church priorities, direct assistance to parents and children, and efforts to change legal and economic forms in order to advance parenthood.

Given this reordering of our basic symbols and values as well as their social implications, what shape should our ethical concerns take and how should they affect reformulations in the law?

THE ETHICAL PERSPECTIVE

The parental bond consists not only of a mutual emotional orientation between parent and child. It is also a set of public rights and responsibilities. However, because a small child is little able to comprehend and exercise rights and responsibilities, most ethical reflection begins with the goods which children are presumed to need. In the time-honored legal phrase, it begins with "the best interests of the child."[12] Our ethical reasoning usually revolves around the interests of the child as an individual rather than the parent-child relationship. It does not reflect the relational pattern of Biblical or ecclesiastical metaphors but the individualism of economic society and its disposition to interests.

We must remember, of course, that even the step to seeing that the child is a kind of person with interests was a positive advance. Prior to that the child was a kind of property of the marriage-family-household. As such the child belonged to the final owner—the father. Discovery of the child as an interested being sprung him or her loose from the household into the bourgeois world—an emancipation first elucidated by Locke

---

12. Aside from Grubb and Lazerson, *Broken Promises*, see Coyne, "Who Will Speak for the Child?" 193. For the tension between natural rights and best interest, see Levine, "Child Custody," 232. Children's right to the parental bond combines both the "protective" and the "autonomy" rights described by Wald in "Children's Rights," 255, because it seeks protection for the bonds which can nurture eventual autonomy. "Best interest" is the controlling criterion for custody determinations in the *Uniform Marriage and Divorce Act* § 402 (1970) and in the *United Nations Declaration on the Rights of the Child*, Sec. 2 and 7, as discussed in O'Neill and Ruddick, *Having Children*, 111.

and Rousseau.[13] It soon became clear that these presumed "interests" went beyond the child's need for food, shelter, clothing, and education. Individuality produced a psychological focus on temperament and feeling. The child's interests came to revolve around his or her need for his or her parents. Thus, the child needs not only their love but their image. It is through the parental bond that the child welds his or her emotions to a workable pattern of life manifested by these parents.

Lacking a relational viewpoint, however, bourgeois society had to derive the child's relationship with the parent from some estimation of its individual interests. In the nineteenth century its interest lay in the financial means for his or her education. This conception still bore the marks of the child as the property of the father. In divorce, the child was usually given over to his custody.

In the late nineteenth century and then even more strongly with the Freudian emphasis on the "tender years," the child's best interests—that is, its emotional needs—were seen to lie with the mother. The child's psychological interests demanded continuity with one parent and only one parent. As Goldstein, Freud, and Solnit stated only ten years ago, children "will freely love more than one adult only if the individuals in question feel positively to one another. Failing this, children become prey to severe and crippling loyalty conflicts." Moreover, "the non-custodial parent should have no legally enforceable right to visit the child . . . to protect the security of an ongoing relationship—that between the child and the custodial parent."[14]

On the basis of this kind of viewpoint, courts would have to award custody to the mother, with the father providing ancillary essential financial support to both. Further, any sharing of custody would have to be a very special case in which the mother was totally unfit to be this "psychological parent." Goldstein, Freud, and Solnit, as appalling as their demands may appear to be, did advance beyond a mercantile individualism of "interest," but did so on the basis of a psychoanalytic model which only reified the isolated maternalism of the Victorian family.

---

13. See Blustein, *Parents and Children*, 80.

14. Goldstein, Freud, and Solnit, *Beyond the Best Interests*, 12, 38. See Wallerstein and Kelly, *Surviving the Breakup*, 311, for a scientific rebuttal of this position. For a review of the diverse legal and limited judicial responses to Goldstein see Crouch, "An Essay."

In fact, judicial, cultural, and ethical thinking shifted away from this model even as it was being enunciated.[15] The shift to our current understanding involves a number of factors. Interest is still central to the popular argument, simply because children cannot enunciate and stand up for their rights. Moreover, governments, which always want to exercise their power as "*parens patriae*" to socialize children to the civil order, step in to defend children's interests. However, the definition of "best interest" is shifting under the impact of different psychologies which stress the child's resourcefulness and flexibility, his or her need for both a male and female model, and his or her cognitive need for the differing environments that might be provided by the two parents.[16] In short, children do better through ongoing contact with both parents than with only one. In fact, what emerges is that children have rights to be defended. Specifically, they have rights to the maintenance of their bonds with their parents. The language of interests begins to shift to the language of rights.

A second factor is the recovery of a theory of parental rights that does not treat the child as household property which one adult can hold in custody, but as a person with whom both parents have a right to maintain a particular kind of relationship intrinsic to their status as citizen parents. Recently jurists have begun to argue that the right to exercise these parental rights is a constitutional one that cannot be taken away simply because of the dissolution of the marriage that gave rise to the child.[17] Parental rights are among those inalienable rights to life, liberty, and the pursuit of happiness. Conversely, children have equal rights to the preservation of this relationship. These rights can be taken away only for substantial cause, such as abuse of the child.

Thirdly, to augment these appeals to interest and constitutional rights, we need to draw on central metaphors of our religious tradition to lift up the parental bond itself as something worthy of special care, protection, and enhancement. Theological awareness of the centrality of

---

15. For the collapse of the "tender years" doctrine see Roth, "Tender Years"; Jones, "Tender Years Doctrine"; *Schall v. Schall*, 251 Pennsylvania Superior Court 262.380 A.2d (1977), 478; and *Johnson v. Johnson*, 564 P.2d (Alaska 1977), 71.

16. See Wallerstein and Kelly, *Surviving the Breakup*, as well as M. Hamilton, *Father's Influence on Children*, and various articles in "Children of Divorce."

17. For an argument in the liberal tradition see Canacakos, "Joint Custody." For some rulings in this direction see *Santosky v. Kramer*, and *Beck v. Beck*. The argument for termination only on the basis of abuse is argued in Kiser, "Termination of Parental Rights." For a general argument for constitutional protection of joint custody see Bratt, "Joint Custody."

the parental bond provides the foundation for any effort to reformulate the language of rights as distinguished from interests. In ethical terms we need to readjust the relationship between deontological claims (matters of right order and rights) and teleological claims (considerations of interests, goods, and utility) by giving priority to the right of preserving the bond over any putative goods to be served in severing it.

Here we see the emergence of the parental bond itself as a kind of holder of rights. The language of rights moves from an individualistic claim to certain goods, to denoting the right to participate in a certain kind of relationship, namely the parent-child bond. We see here the inevitable re-emergence of the need to defend the bond itself. In Roman Canon Law the Church appointed a "defender of the [marital] bond." Now we need to appoint defenders of the parental bond—a relationship with its own special dignity and claim to rights. Perhaps the use of the guardian ad litem exercised in some states to protect the child's interest should be extended to care for the bond itself.

By focusing on the relationship we see that parents also have obligations which the state can enforce—to feed, shelter, clothe, and educate the child, at least to the extent he or she would be cared for in an "intact" family. Thus, we see more and more efforts to enforce child support orders, payment of which has limped along at a dismal forty percent.[18] Increasingly, however, people are asserting that the real means for enforcing responsibilities is to enable parents to invest themselves emotionally in their children through continuing contact.

At the same time, by focusing on the rights of children as well as parents to preserving a relationship, we provide the basis for concepts of shared parenting. Not only must we break through the myths of *the psychological parent*, "single-parent families" (even though there are two parents), and "absent fathers" (even though he lives around the corner), we also need to lift up models for sharing parental rights and responsibilities so that we can move beyond maternal martyrs and paternal absentees. To do this we don't have to indulge in myths of the "friendly divorce;" we must simply attend to the importance of parental bonds.[19]

What ethical perspective should we follow in establishing the way this parental covenant should be maintained and reordered in divorce?

---

18. For an in-depth study of the problem see Cassetty. *Child Support and Public Policy*. In Irving, Benjamin, and Trocme, "Shared Parenting," the authors report 100 percent support compliance by joint custodial parents.

19. Practical advice is available from Ware, *Sharing Parenthood*.

According to what hierarchy of values is this to be done? There are two kinds of considerations here. First, the ranking of rights, duties, and goods. Second, the alteration of these in light of the development of the child.

The basic principle is that the parental structure must first reflect the rights of the members and then be adjusted in light of the goods and interests of the parties—first of the child and then of the adults.[20]

Rights are first of all rooted in congruence with our general grasp of reality. They arise from the basic order of things—in this case the parental bond. We first try to preserve this basic order. This is the importance of grounding the parental covenant in the general worldview of faith. We then turn to calculations of goods and interests, because these rights also arise from our need for certain primary goods, as John Rawls would say. In addition to the usually cited goods of food, shelter, clothing, and education, we need to add the cultivation of a character adequate to life's normal tasks. Here is where we come upon the varying psychological theories of development. They are ways of clarifying our understanding of the interests at work.

Theories of development have changed radically in only a few decades. and consequently so has our perception of what is in the best interests of the child. Since these have fluctuated quite markedly they should not have the primacy many have given them. Nevertheless, it is clear from a great deal of emerging literature cited earlier that the maximal articulation of the parental bond conduces best to child development. Both parents are indispensable for normal development at any time.

In addition, with regard to psychological grounds for establishing the rights and goods sought by the family members, we must acknowledge more forcefully the destructive impact of separation from children on mothers and fathers. Aggrieved parents, cut off from such an essential good, cause enormous destruction every year. The "best interests of the child" need to be augmented by an attention to the best interests of parents and society as a whole.

We have then a series of considerations for qualifying the rights of the parental bond. The major bases for attenuating parental contact lie in the frustration of a child's developmental interests. Thus, an abusive

---

20. Blustein, *Parents and Children*, tries to find an appropriate balance for these considerations. John Rawls' work shapes a good deal of his argument, in which he seeks to give priority to parent duties over parental rights. The duties, in turn, revolve around pursuit of the child's interest in autonomy.

parent, for instance, should have curtailed or supervised contact with the child, but only to the extent necessary to protect the child. Moreover, our public monies would be better spent on support and rehabilitative programs for parents than on their exclusion from relationships with their children, with all the social costs that such exclusion entails in the long run.

These religious and ethical changes mean nothing if they fail to find legal expression. What then, in conclusion. does the legal framework for such an ethical perspective look like?

TOWARD A NEW LEGAL FRAMEWORK

An appropriate legal framework for shared parenting in divorce rests on changed assumptions, definitions, and provisions. Just as the law is not to treat the child as a kind of property, so it should not treat the parents as a species of criminal whose divorce has shattered their parental bond and delivered it into the hands of the court. This means that the child is not some indivisible "thing" over which a person is to have "custody" or not. If "custody" is a problematic term for criminal suspects, how much more is it unsuited with reference to the care and nurture of children? Similarly, children are not guests to be "visited" by their parents. Visitation is for funeral homes and hospitals.

The law's focus should be on preservation of a relationship—the parental covenant. This parental bond is a bundle of rights and responsibilities shared by parents and their children. They involve care, contact, and nurture. It is society's right and responsibility to sustain this parental bond through the law, not only so that children will be socialized adequately, but so that adults will have an anchor of concern binding them in common cause with others.

It should be the presumption of the law, therefore, that this bundle remains intact in time of divorce, unless there is good reason that it be altered. That alteration should be as slight as possible.[21] For instance, courts have frequently severed the rights of "legal custody" from those of "physical custody" in a shared custody order. However, this focuses entirely on the parents' rights without realizing that the child's right to

21. *Shelton v. Tucker*: "(The government's) purpose cannot be pursued by means that broadly stifle fundamental personal liberties when the end can be more narrowly achieved. The breadth of legislative abridgement must be . . . the least drastic means for achieving the same basic purpose."

real living contact with *both* parents has been severely undermined. A "legal" right without a living parental bond is no right at all for the child. Nor does it enable parents to carry on real parental relationships. The law should maximize the parental bond, not some legal abstractions of adult rights. This can be one result of a focus on the parent-child covenant in legal reform.

Some parents, to be sure, may want to waive some of these rights, and the court is helpless to force them to exercise them. However, it can and should seek to enforce their critical responsibilities, such as financial support. Moreover, along with public and voluntary organizations, the court can encourage parents to effect the more personal responsibilities of contact, guidance, and care. It can also safeguard the parental bond from arbitrary dismemberment.

The question then becomes how this bundle of rights and duties shall be exercised in the breakup of the marriage and household. How should the family, that is, the parental and sibling bonds, be restructured?

Working from our ethical framework, we alter or remove those relationships least important to the parent-child bond. A divorce is the death or failure of a marital bond. This condition can exist psychologically within an "intact" family. A legal divorce or separation alters a household, with its attached property rights. This is the essential restructuring involved in legal divorce.

How then should the law provide for establishing a flexible framework which observes the hierarchy of rights, responsibilities, and interests, as well as changing developmental needs? First, it should provide for the establishment of a parental plan covering the basic considerations of parenting under these circumstances. The plan involves both "legal" and "physical" aspects of the parental bond. Both must be part of the child-parent relationship. We have given great attention to the division of material assets in divorce but almost no attention to the rights of children to the preservation of their bonds with both their parents. A plan reflecting the unique parenting pattern of each family can minimize misunderstandings concerning the exercise of parental rights and duties. It can be as detailed or as spare as the parents need. A clear plan gives children a dependable structure to rely on.

Second, the law should provide for an equitable sharing of the financial burdens of parenting in two households. While this is a thorny issue involving both the means for establishing fair apportionments and the provisions for paying them, we should not let this deter us from our efforts to honor the need for an ongoing parenting structure.

Third, the law should provide for ways of changing the plan by common consent as the children mature and circumstances change. The plan should contain provisions for mediation in case of disputes over its implementation or modification.

Finally, the court should have the power to order a shared custodial plan that seems best for the child in light of the circumstances. Deviation from a maximal preservation of the children's bonds with both parents should be done only for good cause.

The beneficial effects of such an approach will be numerous. Much of the trauma of divorce concerns parental fears that they will be cut off from children. These fears have driven people to suicide, murder, various forms of vengeance, kidnapping, and abandonment. With a framework that assures parents of their continued parental status, these incidents should lessen.

In addition, by rehabilitating the equal partnership of fathers in parenting, mothers would have increased time for pursuing their own work and public life. Fathers would be more likely to make their fair contribution to the children's financial needs. Both parents will put their money into their children rather than exhausting "you win–I lose" battles for custody.

Even with all these positive possibilities, many people still resist such an approach. Given the woeful state of fathering in our culture, many women feel that men will somehow use their equal status to reduce their financial contribution. Moreover, women fear losing one of their advantages in divorce negotiations—their status as mothers. These are inevitable risks in our struggle to be more adequate parents. What must be kept uppermost in our minds is the rights of our children to the parental bonds which they need to mature into whole human beings. The justice of the goal should motivate us to find the resources to achieve it in an equitable fashion. To this end we need to devote the resources of church, government, and voluntary association.

8

OIKOS
Convergence in Business Ethics

Work, family, and faith are the core of most people's lives. The way we put together our commitments to these three facets of life affects how we work as well as marry, raise children, and believe. These patterns have immense importance for the way work is organized and managed. However, because they are considered private matters, no forum exists where we can talk about how they are in fact related in people's lives. Moreover, the secularist bias of scientific management precludes such an engagement lest it raise unmanageable conflicts.

However, developments in a number of fields point to a new convergence of these perennial human concerns. It is time to discuss these matters in a public, organized way as we struggle to understand how people put their lives together around these commitments and what difference these structures make for the workplace.

Convergent thrusts arise from the areas of business ethics, management theory, and family studies. I want to highlight them and then present a basic model for examining the interrelationships of work, family, and faith. In the process I will expose some implications of these patterns for business life and ethics.

CONVERGENCE I: BUSINESS ETHICS

Discussions of business ethics have greatly multiplied in the past decade. According to a 1979 study sponsored by the Ethics Resource Center over 70 percent of large US corporations have codes of ethics.[1] Workshops, seminars, conferences and numerous books and articles have been devoted to this theme. Ethics has moved from being a luxury curiosity for the soft-minded to a necessity for survival in an awakened marketplace.

This little seedling has grown to flower in the midst of changing winds and fickle climates. However, it has also grown within a small pot. It began with a limited focus on personal and corporate honesty, fairness, and integrity. It was an effort to reclaim the original values of the Anglo-American economic order. Now, because of new conditions as well as the dynamic of its own genes, it needs to be transplanted to a bigger pot. If it is to realize its potential, business ethics has to be placed within a wider context. To understand this need, we have to look at the typical shape of business ethics discussions.

A number of fine books have appeared in recent years written by ethicists, business leaders, philosophers, economists, and theologians.[2] Typically, they deal with cases such as conflict of interest, pricing, marketing, bribery, and corporate responsibility for the environment and wider public. These ethical discussions arose because of burning policy issues and even specific decisions, beginning with the General Electric price-fixing cases of the late 1950s to the issues of bribery at Lockheed and industrial safety and pollution at Allied in the seventies. They were concerned with how managers were to make policy decisions within a frame of values wider than short-term marketplace considerations.

It soon became apparent that ethical discussion could not be simply a component of immediate decisions. It should shape the character of managers. It should form new horizons, new imaginations, and new sensitivities within managers so that their decision-making evidences a new

---

1. See the study Ethics Resource Center, *Codes of Ethics*. Also, see Bank of America Corporation, *Bibliography*. For an overview of business ethics teaching see Jones, "Teaching Business Ethics" and, with a fine bibliography, Jones, *Business, Religion, and Ethics*. Robert V. Krikorian provides a realistic argument for business ethics in Krikorian, "Self-Regulation."

2. To name only a sample: Walton, *Ethos and the Executive*; Barry, *Moral Issues*; Beauchamp and Bowie, *Ethical Theory and Business*; Stevens, *Business Ethics*; and DeGeorge, *Business Ethics*. Unusual in its consideration of the relation of work and family life is Solomon and Hanson, *Above the Bottom Line*, ch. 14.

way of approaching conflicts, planning, and policies.³ Business ethics had to focus on the development of business leaders. Here we find the importance of ethics courses in business schools, programs of managerial development, and periodic review of corporate codes of conduct.⁴ Ethics had to develop a character base underneath its concern for specific decisions and policies.

This wider horizon concerns not only matters of cognition—of perspective, vision, and conceptual tools. It also touches on the motivation to act ethically. Earlier efforts to broaden the ethical horizon have been important and necessary but not always very fruitful. They lacked attention to the proper integration of knowledge and motivation. The recent attention to the roots of ethical vision and action must go beyond them but also take seriously their attention to the ethical importance of societal structures.

For instance, business ethics has always been augmented by the wider discussion of economic ethics, in which the values of alternative economic systems have been analyzed.⁵ However, this choice of alternatives is quite beyond the range of any individual manager or corporation, so it has been hard to connect the two, except in special cases.⁶ The new thrust in business ethics now poses the need for fresh perspectives on the linkage between faith, vision, organization, and personal motivation.

Because most ethical discussion has focused on the action of individual managers, it has had great difficulty moving to these wider connections between corporate life and environing institutions such as those of family and religion. It has been much clearer about relationships with governments, for these concern specific decisions and policies. Moreover, there is a long if somewhat atrophied tradition regarding the relationship of business and government in our society. It is a well-rehearsed argument, regardless of one's opinion about the level of sophistication it achieves. More lamentably, while Max Weber, in his classic study, *The*

---

3. Oliver Williams has brought considerable focus to the discussion of character in business ethics. See Williams and Houck, *Full Value*. See also Solomon and Hanson, *Above the Bottom Line*, 418–24.

4. See Jones, *Doing Ethics in Business*, for reports on several programs, and Powers, *Ethics*.

5. For example, Wogaman, *Great Economics Debate*.

6. Occasionally the macro issues do get raised in managerial journals, as with Wuthnow, "Moral Crisis." For a business ethics text discussion see Davis and Frederick, *Business and Society*, ch. 6.

Protestant Ethic and the Spirit of Capitalism, gave us concepts for linking religion and economics, we have not moved much beyond them. We have been buried in his concepts without pursuing the spirit of his inquiry.[7]

On the practical side, in most management experience the junction of religion and business has been forged in encounters with religious and community stockholder groups which have engaged firms and top management on specific issues, policies, and decisions. In accepting the stock-democracy of corporate life these groups moved within the ambit of most business ethics discussions and within the horizon of business leaders, but the combative, often single-issue focus tended to seal the debate off from the ethical discussion within business circles.[8] The common agreement to work within the existing structures of corporate decision-making has held both management and religious groups from joining the issue of overall organization and motivation for economic life. The current debate over church pronouncements on economic life could offer the forum for a broader engagement if it does not vaporize in volatile fulminations.[9]

From the domestic side, business ethics has only been tangentially affected by profound changes in family relations. The separation of home and office has also compartmentalized our thinking. However, since World War II people's action, role, and status in the marketplace and public forum have been increasingly separated from the conditions of their birth—gender, race, ethnicity, and physical condition. The person at work has emerged as an individual shorn of all other associations.

Strong cultural perspectives have encouraged this development. Laws against discrimination have reinforced it so that women and ethnic minorities have made some advances in the workplace. Corporate commitment to these new values of equality has been a deeply ethical matter. Ethical codes reflect this cultural decision to treat people only in terms of their work role and performance. However, the actual web of emotionally

---

7. The standard English translation is by Talcott Parsons as Weber, *Protestant Ethic.* The basic parameters of the argument are laid out in Green, *Protestantism, Capitalism and Social Science.* Gordon Marshall's history of the Weber thesis arguments confirms, though approvingly, its static character. See Marshall, *In Search of the Spirit.* For some efforts to go beyond it see Winter, "Elective Affinities."

8. Williams, "Who Cast the First Stone?"

9. Among recent church statements see World Council of Churches Central Committee, "Statement"; Baum, *Priority of Labor*; Lutheran Church in America, "Stewardship"; Canadian Conference of Catholic Bishops, "Ethical Reflections"; National Conference of Catholic Bishops, "Catholic Social Teaching."

charged relationships that energize and orient their lives are hidden under a shroud of privacy. This veil may be laudable as a protection from corporate manipulation of the springs of people's lives, but it prevents us from seeing the full ethical context of corporate policy and decisions. It may be an easy parallel to the protections offered by the first amendment, but it creates the same peculiar blindness to the intricate relations of spiritual roots and public action.

Moreover, this strict separation also guides popular understanding of what ethics is all about. Ethical reflection and codes of conduct tend to focus on conflicts of interests between private finances and public office. The wider ethical meaning of the complex interplay of these private and public spheres is lost in the hunt to sniff out any possible financial bridges between the two.

Business ethics must take account of this wider context if it is to move from an analysis of how people think about action to how they are motivated and structured for action. The stress on virtue, character formation, and faith vision has begun to move the discussion in this direction. The discussion of character now seeks to analyze the structures of family, community and public that shape the way people approach their work.[10] This thrust is assisted and reinforced by developments in management theory as well as family studies. First, let us look at management.

CONVERGENCE II: MANAGEMENT THEORY

While business ethics has moved to a deeper and wider interest in the character of decision-makers, managerial theory has raised the question of corporate ethos and culture to a central position.[11] Students of business culture hold that corporations succeed when they can develop a coherent and articulate value structure to guide and motivate every member of the organization.

There are sound organizational reasons why this is a necessary approach in our time. Corporate life is too complex to wait for decisions, policies, and solutions from the top. Conditions change so quickly that we need on-the-spot responses that are also in line with the wider thrust

---

10. Stanley Hauerwas begins to move the theological discussion of character into a social framework in Hauerwas, *Community of Character*, where the family takes a central role.

11. Peters and Waterman, Jr., *In Search of Excellence*; Deal and A. Kennedy, *Corporate Cultures*.

of the organization. Moreover, pure economic incentives are not enough to motivate people to contribute their abilities, loyalties, and ideas to the organization. They need a corporate culture that can resonate with deep dimensions of their own character.

Quite clearly, the corporate culture thrust of contemporary management theory coincides nicely with the ethical concern for the character of the decision-maker. It provides a wider value matrix to nourish and guide the character-building process and relate it to organizational needs. A vital corporate culture taps into the motivational dimension of ethics by cultivating people's loyalties as well as their muscles and brains. It appeals to their faith as well as their reason. Management theory is beginning to go beneath the abstract world of detached "values" to the earth that nourishes them. It can begin to talk about a personal and corporate "faith" that is more than a set of eccentric beliefs—a faith that is the total construct of values and loyalties shaping people's outlook and action.

However, there is more at work here than a new approach to the relationship of individuals to organizations. It is more than merely the movement from "values" to "culture" and "faith." There is a new paradigm of organization, a new icon in managerial vision. Whereas classical bureaucracy drew its models from the military, and small business from the family, we see now a new, perhaps unexpected influence here.

When I read a book like *In Search of Excellence*, the most celebrated expression of this trend, I am acutely aware that the authors are asking businesses to act more like churches. They are to have a "faith," that is deep conviction about the overarching commitments of the group. This faith not only needs clear articulation, it needs to be ritualized continuously and colorfully in the activities of the organization. It isn't far from the "hoopla" they applaud to the revivals of my Baptist youth, though perhaps a more established corporation would prefer the gentle sobriety of Anglican liturgy. Their model of a good corporation resounds with themes from the life of evangelical, somewhat sectarian churches. One wonders whether the success of their book results from its attention to the bottom-line concerns of most managerial theory or from its resonance with the dominant religious piety and organization of the country.

In their companion book, *Corporate Cultures*, Terrence Deal and Allan Kennedy draw even more explicitly on ecclesiological references. Each corporation's culture demands a "priest" or "priestess" to transmit and maintain it. When corporate culture is strong, it can bind its members together so tightly in common purpose that it can be radically decentralized without fragmenting. Their examples? McDonald's and the Roman Catholic Church!

What this means is that the relation of management to religion lies not merely at the juncture of ethics and the ordering of values. This ethical analysis is just as easily a task for philosophers as well. Perhaps a more important juncture is that of ecclesiology, the study of church organization. It is in the various ecclesiologies of the denominations that we see living experiments in giving structural form to faith. Ecclesiology is a fertile ground for those seeking to study the conjunction of emotional loyalty, faith perspectives, and organizational form.[12] Both the trend in business ethics toward the importance of virtue and faith as well as the managerial concern for corporate culture leads to the ecclesiological shore.

What this emerging convergence points to, however, is the importance of people's wider cultural formation for the way businesses operate. It means that management theory as well as business ethics must take account of the two primary arenas for the formation of emotional loyalties and conduct—the family and the religious community, whether church, temple or synagogue.

Most of us are aware to some extent that what divides churches is disagreements over right organization. Intellectual questions of belief or personal devotion do little to help us understand these differences. Some churches are clearly patriarchal institutions honoring God the Father and Christ the Son. Others are shaped by assemblies which give expression to the consciences of each believer. Still others place central power and authority in the hands of the clergy as a college of professionals. Churches that resonate with Peters' and Waterman's excellent companies consist of spontaneous small groups led by the spirit and committed to a distinctive ethical discipline. The more Catholic ecclesiology of Deal and Kennedy demands a more formalized cult presided over by a tradition-oriented priest. All of these forms cultivate their own kind of spirituality, ethos, theology, and orientation toward the social environment. Both business ethics and management theory would be rewarded by a careful examination of these relationships and dynamics.

It is clear, however, that religious institutions do not descend solely from heaven. Just as management theories reflect their military ancestry,

---

12. An excellent introduction to the relationship between ecclesiology and social structure is Stackhouse, *Ethics and the Urban Ethos*. For a specific focus on organizational theory and ecclesiology see Rudge, *Ministry and Management*.

so churches, even more distinctly, evince their deep ties with family forms. The parental care of the pastor, the appeal to love and family-like care, as well as the permeating residuum of family metaphors in religious language, all work to draw on the emotional bonds of family to shape and motivate church activity. An examination of ecclesiologies leads us to a greater appreciation of the way public organizations are linked to the patterns of the family.

Ever since Charles Dickens wrote his searing portrayals of the domestic effects of industrial work we have been well aware of the enormous and usually deleterious impact of work on the family life of wage earners. With William H. Whyte we were forcefully confronted with the impact of managerial life on family relations. With increasing frequency contemporary studies of managers report on the deep tensions managers feel between family, community, and work commitments.[13]

However, this is not merely a matter of time allocation. Studies of the impact of female employment and careers have opened up the discussion of the way domestic life and values may affect work roles, authority, and the exercise of power. Family values and organization are seen as active participants in a reciprocal exchange with work life. Changes in family styles and organization toward more equality are being felt very strongly in the workplace. An adequate appreciation of management dynamics has to take into account this crucial arena of human development and nurture. It is in this area of family studies that we find a third converging trend significant for managerial theory as well as business ethics.

CONVERGENCE III: FAMILY STUDIES

The study of the family, like that of management, has always been closely linked with efforts to improve it. For decades inquiry into the family was closely tied to psychotherapies which sought to heal people deranged by patriarchal oppression and sexual repression. Out of this psychoanalytic ground emerged fuller developmental perspectives on both individuals and the family.[14] The theory that families themselves have a "develop-

---

13. Management studies are increasingly aware of the impact of organizational life on the family, as in Margolis, *The Managers,* and Kanter, *Men and Women.* Other studies reveal acute managerial awareness of the tensions between home and work, as in Schmidt and Posner, *Managerial Values,* and Bartolome, "Work Alibi." The reciprocal relation of work and family is more evident in books like Ehrenreich, *Hearts of Men.*

14. Erik Erikson set in motion a long string of research with Erikson, *Childhood*

mental cycle" soon led to an inquiry into the dynamics of the family as a system itself.[15] Families came to be seen as organic systems whose internal relationships need to be changed if their members are to improve their well-being.

All of these approaches, however, still tended to see the family as a somewhat autonomous unit. Such a perspective disguised the family's interaction with other systems in the society—especially the economy. They reflected the extreme separation of family and work typical of urban industrial economies. With increasing participation of women and mothers in public life and the work force, family studies has turned to the connection between family and economy. The problem of dual-job and dual-career spouses has come to the center of attention.[16] Social workers and other family professionals are looking for new ways they can engage businesses to take a role in providing the support that used to be furnished by relatives and neighborhood groups.

Moreover, new patterns of authority and communication are being exchanged between management and the family. For instance, communication techniques developed for family enrichment prove equally interesting to management training.[17] Communication patterns of command-response give way to those of negotiation, contract, and problem-solving as people explore more egalitarian relationships in marriage and project-oriented relations at work.

From the management side corporations have moved from efforts to draw on the energies of wives to advance corporate goals to a concern for helping spouses handle two career paths and parents to balance work with child care.[18] Not only are corporations seeking new relations to spousal and parental roles, they are also seeking more professional, non-paternalistic ways to deal with the personal problems of employees. Various employee assistance programs have arisen to provide counseling

and Society. The most popular example of the family development approach is Duvall, *Marriage and Family Development*.

15. For an overview of systems theories of the family see Foley, "Family Therapy," and Minuchin, *Families and Family Therapy*.

16. The rising tide of research in this area is spearheaded by Hood, *Becoming a Two-Job Family*; Kanter, *Work and Family*; Kamerman, *Parenting*; Portner with Etkin, *Work and Family*; Staines and Pleck, *Impact of Work*; to mention only a few. See Baden, *Work and Family*.

17. For an example see Miller, Wachman, et al., *Working Together*.

18. An important resource here is offered by the Catalyst Career and Family Center, New York, NY.

and therapeutic services to employees and their families.[19] Large firms manage in-house programs while smaller firms are increasingly contracting this service out to family service organizations. While the gains in employee protection and therapeutic effectiveness are laudable, the effect of this move may be to seal off this dimension of life from managerial awareness and in turn from business ethics.

We are increasingly aware that people actually do not live in compartments like that anymore. Most people are engaged in a continual struggle to hold together their work life, families, and faith, whether that faith be personal or institutional, whether it be expressed as "beliefs" or as "basic values." They are trying to maintain some kind of personal integrity while playing roles of parent, manager, employee, spouse, citizen, and believer.[20] Moreover, they strain toward some kind of consistent pattern by which to exercise these roles. They look for a common script to knit together the drama of their life. People are not isolated individuals who can be approached simply in terms of one role. This is the myopia of most managerial practices and theories as well as of the business ethics that attends them. However, it is a myopia that is blind to the wider context which determines how people will in fact act, regardless of official codes of conduct, role definitions, managerial structures, and elaborate governmental regulations.

Corporations and families are seeking new patterns of relationship amidst these fundamental changes in domestic life. Neither institution can any longer ignore the other. The century of effort to isolate a "rational" economy from the "irrational" world of mothers and children has ended. Not only are we more aware of the way the economy in general influences the quality of family life. We are also more aware of how patterns of power, authority, communication, and cooperation in one sphere become transmitted to the other as people seek a consistent pattern of action and values for their lives.

It is in this triangle of people's work, family, and basic values that a convergence of concerns is now occurring. It is in this nexus that people live out some of their most important values and aspirations. Yet,

---

19. For an overview of employee assistance programs see Wrich, *Employee Assistance Program*, and Masi, *Human Services*. Vincent Barry probes the question of personal privacy and the corporation from an ethical angle in Barry, *Moral Issues*, ch. 5.

20. The classic portrayal of this is in Fustel de Coulanges, *Ancient City*. For a philosophical approach see Simon, *Work, Society and Culture*, and for an approach in terms of the tight fusion of gender, household and work see Illich, *Gender*.

ironically, it is these very matters which we are least able to discuss. We are freer to discuss sex and corruption than faith and family in the typical business or academic setting. They are too important to talk about. Family and religion have been off limits for business concern. Churches and synagogues have seen their main mission in terms of families and consider economic organization foreign to their task. Family services have tended to focus exclusively on the psychological dynamics of family systems and bracketed off people's religious roots or their economic context. This is the constricted pot that is stunting the flowering of business ethics as well as management theory.

The kind of broadened perspective we need is not a deepening of the rutted paths of our usual conversation but an awareness of the depth energies and structures that integrate people's actual lives. It is in the profound interplay of workplace, family, and faith that we have enormous difficulty coping with the profound changes that are upon us.

This timidity is understandable, for we are dealing with the roots of human action and meaning. Institutions have always become tyrannical when they could dominate this central sphere of life, whether they were governmental, ecclesiastical, or economic. However, in our fervor for personal liberty we have excluded the center of our lives from the public realm and public discussion. As a result, we have lost a sense for its character and more importantly we have lost a language for addressing its complexities. We are left with fervent rhetoric about "faith," "God," "the family," and "values," but little substance to guide us in the real complexities of our life. No wonder serious students of business or precise academics don't want to tangle with it.

At this point then we need not only an orientation to this crucial nexus of work, family, and faith, but also a language for talking about it. We need some symbols for getting a grasp on this integrated reality. One way to get to the root of this integration is to recover its ancient origins.

RE-DISCOVERING THE *OIKOS*

In all traditional cultures the family is a productive, usually agricultural, unit. Moreover, its basic religiosity, or faith, is bound up in the honoring of ancestral customs and beliefs.[21] Economics, family, and religion are tied together in an integrated bundle. In the Greek this was called the

---

21. James S. Bowman articulates this point in Bowman, "Altering the Fabric."

oikos, that is, habitat, or household. The *oikos* embraced not only a house, but also a lineage, a hearth, ancestral claims, and the land of the fathers.

In the course of Western development these aspects became differentiated from each other. Production and distribution became "economics" (*oikonomia*). *Oikonomia* is quite literally the laws of resource management—the interrelationhip of people, buildings, and land. Today, when we wish to speak of the dynamics of the household we have to define it carefully as home economics. The broader sense of resource management has moved out of the domestic sphere but has failed to stay in touch with the religious and familial roots of people's ongoing lives.

Religion became pluralized into numerous houses of worship, though the strain toward a world-embracing unity in faith is still called "ecumenics," from another *oikos* word—*oikoumene*, meaning the whole inhabited world. While it is mostly used of Christian efforts at unity, it also means any effort at religious cooperation. *Oikos* was also the symbol underlying the way classical theologians spoke of God's "economy" as the whole process by which God saves the world. The "economy of salvation" is the way God claims people into a perfected household of love and peace. *Oikos* and its progeny, then, are redolent with religious associations.

Finally, the concern for how everything fits together on the land like a well-ordered household now has its own special name—ecology, from "*oikos-logos*", or "household theory." Our language tells us that all these dimensions of the effective management of people and resources were once integrated with each other. It is an integration deeply embedded in archaic memory and pressing for new consciousness in our own time.

Now it is quite clear that we cannot recover the same pattern of integration as that in ancient Greece or in traditional agrarian culture. However, the concept of the *oikos* gives us an orientation toward the process of integration. It symbolizes the human desire to relate these crucial dimensions of existence. *Oikos* is a heuristic concept for analyzing a set of relationships that shapes the way people relate to work, family, and faith. It can be applied to individuals as well as to organizations. It is both psychological and structural.

A person's *oikos* is the way he or she relates these three dimensions to each other. Each of us has an *oikos* of some kind. Some are more explicit than others. Some are more internally congruent and coherent than others. Recognition of our own *oikos* and the way we are struggling toward congruence and integration is a fundamental life task. The vision of an ideal *oikos* shapes our judgments about personal relationships as well as institutions, whether they be businesses, families, or churches.

Likewise, each organization has its *oikos*—its model of the way it wants its members to organize their various *oikoi* (the plural of *oikos*). Moreover, each organization, no matter how "rationally" it has been separated from religion and family, tends to develop its own *oikos* to embrace these perennial human dynamics. Firms in the past have given considerable attention to the role and character of the employee's wife (to say spouse in this context would be disingenuous indeed). Her role in the corporate *oikos* was to provide a safe haven and launching pad for her husband. Faith and values were construed in individual terms of personal morals and conformity to the ethos of marketplace dynamics—honesty, promise-keeping, obedience, and rational even-handedness. The function of religion was to keep home life stable so that men could be raised and supported in their economic pursuits.

The religious concept of vocation was transformed into the commitment to a life-long career—at least for professionals. For the entrepreneurs and managers it was a vehicle for expressing their vocation or peculiar talents. For workers, the workplace was to be a source of money to support their family. From a religious standpoint it was a discipline for selfish and sexual inclinations. This pattern dominated the industrial *oikos*. Both the firm and the family were to be run by a lofty patriarch guiding and disciplining weaker and more unruly subordinates. Both poles were united by a religious vision of a divine Father, obedient Son, and dutiful, devoted and inspiring mother—the social vision of the traditional Christian Trinity.

Now, however, there has been a shift in the official *oikos*. Religion is no longer content with supporting the family alone. Or, more precisely, many religious groups are no longer committed to the patriarchalism and male individualism that permeated the industrial oikos. Not only do they seek to make direct impact on the marketplace or on corporations. They also lift up other models for human relationships—of collegiality, friendship, equality, or nurture.

Correlatively, wives are no longer so content to hitch their wagons to their husband's corporate star. And women don't have to be wives at all to get a piece of the managerial cake. Moreover, sharing of parental tasks has meant that people want flexible hours, job and career patterns, allowances for child-care, home health emergencies, and even paternity leave. Many men are no longer solely committed to their occupational career but see their life's vocation in broader terms that encompass occupational changes, community service, family life, and ongoing personal reassessment.

In short, corporations cannot assume the existence of a single type of *oikos* among their members. Moreover, the *oikos* pattern the corporation projects, whether explicitly or implicitly, greatly affects how employees will relate to it, if they choose to at all. We are therefore confronted not only by a change in our dominant *oikos*, but the presence of a variety of *oikoi*. Families, corporations, and religious bodies all want a say in how this is to develop, because each has a primordial claim on the integrated reality they all participate in.

OIKOS: NEW DIRECTIONS

It only takes a few examples to raise to consciousness the problem of the *oikos* in our lives. The concept of the *oikos* directs our attention to an ensemble of relationships rather than to its compartmentalized sectors. It prepares us to reflect on the structural as well as motivational aspects of these relationships. It leads us to talk about how our lives are energized and oriented as well as how they are ordered.

In this respect the concept of the *oikos* brings together the concerns of developmental studies such as *Seasons of a Man's Life* and *The Corporate Steeplechase* with the structural concerns of most organizational theories.[22] It provides a way for studies of character and personal spirituality to be linked to the emotionally charged relationships that order most of our lives.

The concept of the *oikos* also resonates with archaic memory. It calls us to a deeper awareness of the historic roots, developments, and variations of the dominant *oikoi* of our civilization. This historical sensitivity can help us see that these *oikoi* lie on a spectrum as well as consisting of many alternative patterns.

The *oikos* spectrum ranges from the fused *oikos* of traditional agrarian life to the fragmented *oikos* of a highly individualistic, specialized,

---

22. Psychoanalytically inclined students of management have most often seen the connection of work and family life. In addition to Maccoby, *Gamesman*, see Levinson, *Executive Stress*. The relation between male adult life stages and work only partially surfaces in the Jungian orientations of Levinson, et al., *Seasons of a Man's Life*. Srully Blotnick provides a career-focused study of adult development in Blotnick, *Corporate Steeplechase*.

and pluralistic society. In between we find degrees of differentiation and integration which follow a number of different patterns. Some people feel too great a fragmentation and yearn for the fused *oikos* of primordial memory. Others feel suffocated and oppressed by the fused *oikos* and long for the untasted freedom of a more differentiated *oikos* usually found in urban areas. Most of us struggle with the various patterns of relative integration possible in a differentiated, though not fragmented pattern.

The spectrum from fusion to fragmentation is not hard to grasp and can help us typify the range of *oikoi* people participate in. The buzzing confusion arises when we become more aware of the alternative entrees provided from each relatively differentiated sector of the urban *oikos*—business, family, and faith. It is the exploration of these dynamics which I have been increasingly engaged in over the past few years. In this introductory essay I can only delineate some of the factors entering into the complex equations of people's lives. Over the next few years I want to develop a more coherent theory about these relationships through the ongoing program of education and research we call the OIKOS Project.

Each sector of the contemporary oikos provides some powerful symbols or models for organizing our own *oikoi*. They are emotionally charged reference points for establishing our identity and relationships. In the world of work we find the Agrarian, Artisan, Entrepreneur, Manager, Worker, and Professional. In addition, we see families, couples, and individuals engaged in family businesses, with their own peculiar *oikos*. Each of these contains a highly valued self-image as well as implications for a pattern of work life.

The agrarian, or farmer, is bonded with the land and the plants and animals nourished directly by it. Within this bond and its uncontrollable weather the farmer exercises a husbandry of resources with all the resonance of fertilization, control, and care that word evokes.

Similarly, the artisan exhibits a typical pattern of life, central to which are tools, materials, and workplace. Action according to a plan, a commitment to precision and control, and a personal identification with the product all characterize the artisan life and permeate its family and religious outlook.

Numerous studies have set forth typifications of these as well as the entrepreneur, manager, worker, and professional.[23] The directors of a family business have been less adequately treated, probably because

---

23. Maccoby, *Gamesman*; Deal and Kennedy, *Corporate Cultures*; Poster, *Critical Theory*.

they begin to blur the lines between the two spheres. Here again, however, we see some typical patterns—parental care over the enterprise and its employees, an identification of family needs and business interests, preference for stability over growth, and the like.[24] Moreover, the family business type has its own variations. It can be run by the couple, by one of the spouses, by siblings, or collections of relatives, each with its own pattern of organization.

The family business, like the agrarian household, raises to consciousness but does not exhaust the question of the relation of work to marriage and family. Each work type implies or presumes a family configuration for its most effective realization.

Family Types

Family life exhibits a variety of emotionally charged models for our sense of identity and relationship. We are familiar with patriarchal and perhaps matriarchal models. Egalitarian models are being developed more assiduously in our own time. The organic type of relationship, in which relationships are ruled by functional needs rather than personal dominance, is less familiar to us, though widespread in traditional cultures.

In our own time we are increasingly aware of what I would call "nodal" family models, in which the relationships once confined to a single household are spread out among several, whether through divorce, death, remarriage, adoption, fostering, or single motherhood within a network of relatives. Nodal families generate their own patterns of expectations, interdependency, self-image, and authority.

All of these patterns make a contribution to the total *oikos* people build up as workers and family members. The agrarian, for instance, typically has demanded a co-working wife and a large brood of laborers and heirs for what used to be a labor-intensive operation. The autonomous, market-oriented, and competitive entrepreneur or manager has historically depended on a patriarchal family structure, a matriarchal household, and a clear distinction between the sphere of free enterprise and the domestic or social concerns of raising children, maintaining neighborhoods, or developing public institutions.

---

24. Stein, "Company Family." Danco, *Inside the Family Business,* and Danco, *From the Other Side,* provide consultation, seminars, and a wealth of practical advice for family businesses through the Center for Family Business.

Similar preferences could be identified for the other work types. Each has taken on notable alternative forms with their own peculiar logic and dynamic. Each of us, whether as children, parents, or spouses, is caught up in the struggle to find the patterns most congruent with each other and with our basic values. That is to say, the decision over these basic alternatives takes place in the garden of faith. It is an encounter with thorns as well as roses. For some it is Eden, for others Gethsemane.

Faith Patterns

The sector of faith contains both cultural traditions and institutions. It produces powerful symbols of sacrifice and longing as well as the Baptist church on the corner and the Conservative synagogue up the street. I have already pointed out how these traditions and institutions bear ecclesiologies which can mold our preferences for one pattern of organization over another. Here, however, let me point to some basic symbols which shape the way we approach the task of relating work and family.

Management theorists are often aware that the religious concept of vocation lies behind the work commitments of many people as well as the way work is organized in our society.[25]

They are less aware that the career commitment we now associate with it is an exact reversal of its original religious meaning as a call away from the world's present order to enter, as a community, a new and heavenly order. However, regardless of the way this symbol has been interpreted, it still bears with it a sense of wholehearted commitment to a task, apart from family allegiances.

People whose faith is permeated by some sense of vocation develop a single-minded commitment to a specific task which becomes the meaning of their life. In the *oikos* it has meant a strict separation of home and work, usually of male work and female householding. Moreover, it has usually meant a life-long, total commitment requiring stability and self-denial. It has been a powerful spring of action, with its own family preferences, but can also be very destructive for people trying to put together an *oikos* grounded in other faith commitments.

Covenant is another powerful religious symbol in our heritage. A covenant is a set of promises ordering human relationships, especially as they are related to the ultimate order of things. It conveys a sense of

---

25. See Blotnick, *Corporate Steeplechase*, esp. Introduction and ch. 16.

freedom to make, keep, and even break promises. It has more of a relational focus than does vocation. It opens people out to a complex set of obligations established by their own choice.

Contract is its offspring. In that guise it leads us to a multiplicity of discrete, often unrelated contracts governing various aspects of our lives—at home, at work, at church, or in public affairs. However, in that form it tends to legitimate the fragmented *oikos* rather than the integrated one implied by covenant. In either form, however, people are drawn to make specific commitments to each other, negotiated in freedom. Because of the centrality of communication to this process, covenant tends to legitimate egalitarian domestic relationships and team project approaches in the workplace.

These two examples should give some sense of the role played by religious symbols in the *oikos*.[26] This is not to overlook the importance of other symbols, such as that of sacrament, which puts cultic action and organization to the center, or communion, which emphasizes ecstatic union with the divine source of inspiration. Each of these bears important values for ordering our *oikos*, whether through an emphasis on cultic rehearsal of memorable events or a stress on the spirit's free movement in all the members of the organization. My purpose here is only to provide a vista for our exploration. It is to describe some of the parameters of the *oikos* dynamic.

THE *OIKOS* AT WORK

Some of the implications of the *oikos* concept for business and business ethics should already be evident. Let me highlight only a few. The rest would require a longer treatment.

Sensitivity to the *oikos* dynamic in people's lives should help us become more aware of the way the three spheres interpenetrate and affect each other. Models of authority in the family provide emotional energy for the models we seek to pursue at work. Religious symbols, concepts, and organizational patterns serve to legitimate or undermine the patterns we may pursue in family life and business—whether they be patriarchal, egalitarian, organic, or collegial.

In highly integrated cultures the patterns in these three areas tend to fuse together, focusing people's loyalties and energies along common

---

26. A fuller exposition of these themes is available in Everett, *Blessed Be the Bond*.

lines. In a more pluralistic society, we can expect that they will tend to diverge. Work organizations have to balance a number of patterns even as they seek to draw on the emotional commitments grounded in people's family and faith life.

This plurality of *oikoi* in our society already is forcing work organizations to create more flexible schedules, especially for parents of young children. They have to make room for changing career patterns that arise from the different stages people reach in family life —whether they be childbirth, parenting, divorce, or post-parenting marriage. Moreover, businesses must be much more sensitive to ways to respond to people's religious faith, as in respect for the integrity of people's values.

In addition, as Oliver Williams recently pointed out, they need to respond more creatively to the concerns raised by religious organizations—organizations that energize as well as represent a side of people often hidden at work but still deeply felt.[27] Through the confrontation of religious and business institutions issues get raised that individual employees cannot raise at work. The confrontation is not only a mark of our fragmented *oikos* but also an opportunity for greater integration.

In the midst of this clash of *oikoi*, businesses are being driven to define their own *oikos* more clearly—their expectations about relationships at work, patterns of authority, balance between work demands and family life, and relationships with religious and community organizations. All of these facets must be attended to in an adequate articulation of the corporate culture. Inattention to any aspect of the *oikos* leads to a truncated and ineffectual stance.

Just as we officially seek greater public clarity about our guiding values, we also know that we often fudge, obscure, and ignore them in order to maneuver through the shifting currents posed by the variety of competing claims from the various aspects of our fragmented *oikos*, not to mention the plethora of organizations in the wider public realm. Even in these murky movements, however, awareness of the *oikos* as a reality can help us discern more clearly what is going on in the fog of human motives, goals, and emotional energies.

Finally, awareness of the *oikos* can help order and guide our ethical reflection as well as its public expression in codes, cultures, and corporations. We must attend not only to values in the abstract but to the *oikos* which energizes people to seek them. We must be sensitive not only to

---

27. See Oliver Williams, "Who Cast the First Stone?"

the psychological depths of motivation welling up in conjugal love, parental care, and religious worship, but also to the structures in which we rehearse them as husband, wife, child, and congregant. The concept of the *oikos* leads us to dimensions of a life which seeks ethical and emotional integrity as well as productivity.

OIKOS: AN INVITATION

This brief prospectus on the convergence between business ethics, management theory, and family studies has set forth an organizing concept—the *oikos*—for ordering our understanding of some central relationships in human life. I have tried to highlight a few components of the *oikos* and some of the ways they interrelate. At this point we have only an aerial observation of the terrain and a few rough maps. This article stands as an invitation to us to begin mapping this land more specifically, linking together valleys long separated and tracing rivers of inspiration back to their source. It is a project that can give us new appreciation for the intricate interplay of ethics, management, faith, and family life.

9

Transformation *at* Work

INTRODUCTION

Religious education seeks to transform people. It leads us out of our old selves into new selves. It leads us out of old social structures into new ones. It is not merely a communication of ideas but a transformation in faith. It seeks to transform institutions as well as persons and groups. Key figures in our history of religious education—Bushnell and Coe—have all emphasized this manifold character of religious education as a work of social as well as personal transformation.[1] In this chapter I want to focus this concern on the world of work.

As soon as we examine the process by which people are transformed we see that this conversion results not only through the impact of religious ideas and institutions but also through the impact of all the institutions in which we invest our loyalties and lives. Not only are we called to transform our workplaces, we are also deeply formed and transformed by the organization and conditions of work itself. How we work shapes our faith and ethics. Thus, the relation of religious education to work is reciprocal. Work shapes religious education even as religious education seeks to shape work. Work is not only a target of conversion, it is also a partner in the process of human transformation. To explore this complex dynamic I shall first expose the theory of human transformation

---

1. Coe, *Social Theory*. For a history, discussion, and proposals for recovering this thrust see Seymour et al., *Education of the Public*.

underlying my own approach. This theory will already orient us to the important relationships we need to examine in the relation of religion and work.

We shall then examine the relationship between these two partner institutions. In this case we find a history marked by professed concern and practical estrangement. The relations between the two spheres have been hidden and indirect. We need to lift them up so we can face new patterns of relationship more intentionally.

In seeking a more dynamic and intentional relation between these two arenas of transformation we must then proceed to an identification of the central changes that are going on in the world of work. This analysis should lift up not only key partners for the process of human transformation but also identify targets of the church's concern. We can then explore the impact of these changes in the workplace on the process of religious transformation, especially as this is advanced by churches themselves.

In turn, we need to identify important alternative strategies pursued by Christians in shaping the workplace. Here we need to exercise our critical judgment to evaluate approaches showing more promise in light of the changes going on around us.

The need to choose among alternative responses forces us to clarify our underlying theological commitments. I shall therefore direct us to some theological issues requiring reformulation in our present circumstances. I shall sketch some elements in the move from individual asceticism to covenant publicity as the guiding concept for moving us forward in the dialectic between work and religious education. On this basis we can then conclude by highlighting some emerging church responses to the changing workplace.

THE PROCESS OF HUMAN TRANSFORMATION

In this examination of the relation between work and faith I presuppose a particular theory about human transformation; namely, that we are transformed through the ways we are emotionally bonded to others and to significant beings and objects in our world. We can only hear information along the charged lines of our relationships. Information that comes in to us along lines that are emotionally dead never reaches us.

These emotional bonds are established not only in families, as Martin Lang and others have pointed out, but also in our work.² Sometimes work and family are very closely associated and resonate with each other, as in the family farm, family firm, or in homemaking and child rearing. Other times work and family are so distant from one another that they compete, even to the extent that one sphere loses all emotional significance to one's life, as with monotonous factory drudgery or the isolated home life of the overtime executive. In any case, no effort at education and transformation, whether in church, school, or clinic, can overlook the power of the emotional structures we gain in our work and family life.

The emotional bonds of marriage, parenthood, kinship, and work create a charged grid or filter through which we can assimilate new relationships, new ways of acting, and new objects of loyalty. In computer jargon, our basic emotional patterns are an operating program which receives some commands but not others. Our emotional bonds are patterns of love as well as faith. They are patterns of love in that they dispose us to embrace certain kinds of commands and demands but not others. They are patterns of faith in that they establish an adherence to a net of relationships we take as ultimately trustworthy, whether it is our relationships with our parents in childhood or our relationships in school, church, work, or even sports. Our basic faith is rooted in an internalized pattern of relationships we can take into all areas of life. Faith is a network of fidelity. Our life is a continual effort to give public form to our covenant of trustworthy relationships. It always contains an effort to make this faith public enough that it can be shared with others. Here we find some roots of the concept of covenant publicity which shall guide this analysis.

Transformation in our faith structures, that is, conversion, occurs not merely because of some ideas we have, but because the structures of our world change and force us to respond. Changes in our work, whether personal or institutional, dissolve old faith patterns and open us up to new ones. We can see this clearly in the typical life path of men and women: first career entry, reentry after child rearing, shift in career, and retirement. We can also see it in the collapse of old industries and the rise of new ones, the death of farms and the birth of new firms. We also know it in the devastating impact of unemployment. These are social changes which also entail personal transformation. They are transformations in

---

2. Lang, *Acquiring our Image*, and Everett, *Blessed Be the Bond*. Parker Palmer develops a relational approach to learning in Palmer, *To Know as We Are Known*.

work, but also transformations in faith. It is the church's task to enter into this dynamic in its own way. Engagement with transformations in the workplace is central to the task of religious education.

This chapter is too brief to spell out a refined theory of educative transformation along these lines. My main concern is to lift up the powerful way our basic emotional bonds shape learning and faith. Because of its primacy in our daily life, work is one of the most powerful workshops of the heart. It is one of the forges of faith.

THE ESTRANGEMENT OF WORK AND RELIGION

If work is so central to faith, why has it not occupied a central place in religious education? Religious educators have traditionally looked to the family and the school as the key partner institutions in the overall work of religious education. They are the most important shapers of religious education as well as the key institutions to be transformed by it. After that, some attention has been paid to political institutions, primarily in their legislative functions. The workplace, now often displaced by leisure institutions, comes in a distant third or fourth if it is mentioned at all.

Why has this pattern existed? First of all, religious education has typically focused on childhood education. Religious education was seen as preparation for work rather than reflection on work. This focus on the child can be seen as one effect of the separation between work and family life that became normative in the nineteenth century. In that situation, the home no longer effectively mediated work skills to the next generation. The child, independent of kinship bonds, went to school to gain the requisite skills and certification for work. This reliance on one's own self and on value-free knowledge gained from experts in the school nourished the rise of "personality" as a value. Or, to put it more precisely, these economic conditions cultivated a particular kind of acquisitive individualism whose "positive" side was a deep belief in the importance of "personality." Individuals each had to have and find their special talent (how many people were raised on that parable?) in order to negotiate the many decisions and take the many risks necessary to enter into their own occupation.

In this situation the school emerged as a partner to the factory and tried to cultivate virtues compatible with industrial organization— punctuality, efficiency, deference to scientific experts, and a mechanical division of knowledge and work.[3] This ascetic individualism was nurtured in self-control through moral injunctions voiced in family, church, and home—sexual restraint, teetotalism, self-denial, thrift, and self-examination. The churches supported this model with the development of the Sunday school—a little factory of religious values to fit the new structure of work and family life. Schooling, that is a controlled progression ("grades") in the acquisition of discrete specialized skills and concepts, came to dominate our approach to education just as the factory dominated the world of work.

In the culture of schooling, occupation was seen as a lifelong commitment—a secularized version of monastic profession. The occupations people were schooled for were not seen as flexible and changing. Once educated for that work, one no longer engaged in education. Education was not a lifelong pattern of reflection for the sake of transformation and change—both in the self and in the workplace. It was the achievement of a fixed vocation in which God's eternal call crystallized in the occupational order.

This emphasis on cultivating the individual personality characterized the strong idealistic strand in religion generally. Personality was a spiritual matter emanating from God. Work and the economic sphere were materialistic. The church stressed personality over materialistic determinism, voluntary association over the involuntary bonds of work, democracy over the hierarchies characteristic of the factory and business. Within this approach it was all too easy to contrast and oppose the realm of work and that of the spirit, in spite of the many strands of Christian utopias, communalism, and socialism that wove through the fabric of the times. Factory work was characteristically dull, repetitive, brutalizing, and mechanical. Being so opposed to the spiritual aims of religious education it could not even be examined for its reciprocal interplay with the church.

Of course, the personalism and democratic idealism embedded in the thought of people like George Albert Coe could fuel a zeal for reform of work. Coe wanted to pursue some kind of industrial democracy but

3. For a prominent example of the theological centrality of the idea of personality at the turn of the century see Troeltsch, *Social Teachings*, 51–68. Max Weber set the agenda for this approach to industrial virtues in Weber, *Protestant Ethic*. For an analysis of the American history see Gilbert, *Work Without Salvation*, and Rodgers, *Work Ethic*. C.B. Macpherson traces the Puritan roots of acquisitive individualism in Macpherson, *Possessive Individualism*. Samuel Bowles and Herbert Gintis present a radical critique in Bowles and Gintis, *Schooling*.

was unable to engage that task with any rigor. We need to ask, then, why religious educators did not pursue this challenge more vigorously?

Responses to this question lead us to other factors in the separation of work from religious education. In the nineteenth century we see the domestication of religion under the impact of its institutional separation from government and the development of an autonomous economic sphere. Both politics and economics were seen as spheres primarily of conflict rather than cooperation. Only the church and home could lift up alternative, spiritual ideals of love, cooperation, and emotional commitment.[4] The public sphere was seen as an arena of combat among free citizens, free entrepreneurs, and, most ironically, "free" laborers.

According to this "two-spheres doctrine" work and politics were reserved for men, the raising of children for women. While the church continued to be governed by men, its actual work was carried on by women as one facet of child-rearing and homemaking. Religious education, likewise, became women's work—an activity rigidly cut off from public life and the realm of work.

However, this rigid separation of work from religion and the home did not mean the absence of reciprocal influences. Inspection reveals a powerful and contradictory dynamic between work and religion, especially in Protestantism. The idealistic belief in free personality could provide a powerful critique of work life, but it also reinforced the individualism characteristic of its ideology. Free personality was also an ideal underpinning the freedom of the entrepreneur to do what he wished with his property and of laborers to enter repressive contracts even if they violated traditional claims of home, religion, and the common good. In short, the Protestant ethic of enterprising individualism legitimated the very economy it could also critique. In cultivating honesty, sincerity, promise-keeping, punctuality, literacy (and often literalism), efficiency, and thrift among the emerging middle classes it reinforced an economy rapidly sundering the common bonds of humanity between owner and worker, between the corporation and its environment, and among work, family, and faith.

---

4. Carl Degler offers a description of the two-spheres doctrine in *At Odds*. Christopher Lasch provides a contemporary psychoanalytical perspective on the dissolution of this structure in Lasch, *Haven*. Ann Douglas portrays the turning away from practical reform to literary endeavor in Douglas, *Feminization*. For a historical examination of the interaction between Protestantism and industrial era labor movements see Gutman, *Work, Culture and Society*, ch. 2.

In addition to cultivating these ideals of the Protestant ethic outlined by Max Weber in his classic essay, the churches also reinforced patriarchal order through their repristination of the home as a kind of castle ruled by the male and managed by the female. This emotionally grounded image of right order was reinforced in worship as well as church order, further legitimating hierarchies of obedience in other spheres.

The historical scene was of course richer than this brief sketch of major elements can convey. There were always countercurrents and lone voices with singular insight. However, these four factors operated generally to divorce the church and religious educators from attention to the workplace as an institution of central concern for the transformation sought through religious education.

TRANSFORMATIONS IN WORK

With this preliminary grasp of the dialectic between work and religion in our own culture we turn to an analysis of the workplace itself. First, what is being meant by work? My primary concern, in light of my approach to transformation as a process of covenant publicity, is the human relationships in the workplace; that is, how work is organized. Within the field of economics I am concerned first of all with the firm. This is the basic structure of work in our society, whether it is a small family business or IBM. It is the point of identification for people. It is their emotionally significant frame of activity.[5]

The way our national economy is shaped both internally and in its profound connections with global markets and industries also conditions the shape of individual firms but is not my central focus here.[6] While concerned with these wider forces I want to focus here on the network of human relationships they shape in the workplace. Moreover, it is this web of organization, rather than its statistical face, which primarily concerns me. The bottom line of a work organization is not found in its quarterly statement but in the relationships of love, respect, cooperation, faith, and trust that enable people to work together in the first place.

---

5. To begin an inquiry into the literature on work see Richardson, *Work in America*; Ginzberg, *Good Jobs, Bad Jobs*; and Kanter and Stein, *Life in Organizations*. For religious perspectives see Raines and Day-Lower, *Modern Work*, and Gillett, *Human Enterprise*.

6. Contemporary discussion on the globalization of economic life begins with Barnett and Mueller, *Global Reach*; Turner, *Multinational Companies*, and Vernon, *Storm*.

Work in this perspective is the public expression of a certain set of covenants defining our life in relation to people, natural resources, land, and to God. It is a complex pattern of expectations among actors playing such roles as consumers, producers, organizers, clients, and coordinators. The entire network of work relations articulates a web of public trust. In this sense it expresses a particular faith binding people together in a process of transformation—not only of their environment but of themselves. It is an arena of covenant and public interchange.

Since World War II dramatic changes in the workplace have ruptured the industrial era's interplay of work and religion. Not only has the pattern of organization shifted, but the two-spheres division of labor has dissolved. The workplace has been undergoing great transformations which demand changes in our patterns of religious education. These changes also demand new challenges from our faith traditions. First, let us review the salient changes that have occurred in work before moving to the religious impact and response.

One of the most heralded developments is the rise of women in the workplace at increasing levels of authority. By 1986, 60 percent of all women between sixteen and sixty-five held jobs outside the home. Though most of these were not stepping into a lifelong career, they still have produced enormous changes in the way we raise children (at day care centers), feed the family (at the drive thru), and maintain households (call the service professionals). In short, we have moved from the doctrine of the two spheres to a life in two jobs if not two careers.[7]

Two-earner and two-career couples have begun to bring about some changes in the way work and career are envisaged. While senior management is still permeated by expectations of unlimited devotion to the firm, people are beginning to accept the importance of family and community within a sense of vocation broader than occupation. People almost expect to change occupation, often drastically, once or twice in their lifetime. Education is a continuing expectation in developing one's work life.

This expectation of continuing re-education is augmented by the rapid technological changes in work. Computerization and robotization are still working their way through the economy, with the outcome profound but still unclear. In addition, the global expansion of economic organization introduces greater complexity into these processes of change.

---

7. Hood, *Two-Job Family*; and Kanter, *Men and Women*.

Work requires increasing skills in communication, problem solving, adaptation to organizational change, and cultural (not to mention gender) pluralism. Technological change is joined by rapid financial movements to restructure organizations constantly. Not only does this require that people repeatedly change the way they work, but also increases the chance they will be out of work, not only in their robust years but in the form of retirement. Men and women are faced with many years of active life beyond their expected devotion to work and rearing children. We have hardly begun to assimilate the impact of these changes in our institutions and in church life.

As a result of this constantly changing work environment businesses have become major educators. The corporate classroom reaches as many people each year as are enrolled in higher education—at an expense of over $50 billion.[8] This education is both preparative and reflective. It has become a dimension of work itself.

All of these changes bring to sharper contrast the tension between the political and the technical-scientific character of work. While many people try to reduce work and work relationships to "objective" rules, formulas, and statistics, the actual fact of organizational complexity and change requires a heightened attention to its political character. Not only must people in an organization counsel together more frequently and openly, they must also recognize the rights and claims of co-workers and the many outside "stakeholders"—neighborhoods, consumers, suppliers, governments, natural environments, and future generations.[9] Thus work becomes a crucial arena for people's efforts to make their lives public—to find expression, recognition, and confirmation through the formation of dependable covenants with each other. It becomes even more clearly a place where people can forge and remold covenants of public trust. This is not only the religious theme I am pursuing here; it is also a crucial development in work life today.

---

8. Eurich, *Corporate Classroom*, 6–12.

9. Bradshaw and Vogel, *Corporations and their Critics*; and Kirby, *Prophecy vs. Profits*. For a good business text survey see Farmer and Hogue, *Corporate Social Responsibility*.

TRANSFORMATIONS IN RELIGION

What impacts are these changes having on people's basic patterns of relationships, that is, their faith—both in work and in the church? The movement of women into the workplace demands that people transform their basic instincts about the relation of men and women. People have to move from relations of master and mistress to those of colleague and collaborator. The intimacy of the home as the structure for relations of men and women yields to the civility of the public as the model of interaction.

Both men and women begin operating out of patterns of rational problem solving and civility learned in the workplace. We start living in terms of contracts for domestic as well as church relationships. Discrete and specified obligations and rights replace the diffuse and totalistic embrace of the traditional home and parish.

It is already clear that lifelong learning patterns have become general expectations in our society. This has always been a strand in the Sunday school movement and the CCD. To this is added an awareness that the focus of this process lies in the crises and junctures where people must deal with the end of old relationships and the beginning of new ones—divorce, job loss, job change, work force reentry, retirement, and the like. Religious education, whether conducted through classes, workshops, retreats, or programs of counseling and spiritual direction, engages people where they face changes in the way they put together their work and family life. Here is where the structure of fidelity upholding our lives is tested, broken, and converted. Here is where integrity must be fed by the communion of the saints.

Work changes have also drastically affected the availability of women for traditional volunteer work in the church. While most of us may not lament the passing of women's circles and luncheons, we all feel the pinch when no one can devote extra time and energy to Sunday school, outreach, and maintenance. We have yet to figure out whether and how to replace these women volunteers with older retired people, for our society has not conditioned us to expect their contributions. In any event they would differ from those previously tendered by mothers and housewives.

This loss of volunteers in the face of changed work patterns leads to a renewed professionalism in church staffs. While great lip service is paid to the volunteer aspect of church life, mainline churches have become increasingly specialized and professionalized. Work relationships at church also have become more contractual, specific, and monetary. While some churches, especially those trying to resist these changes, use this professionalism to construct more schools based on the old model, there is considerable question whether a schooling model can help people cope with the actual challenges of a new economy and family life.

At the same time that church work and religious education in particular loses its volunteer character, it loses its label as "women's work." It becomes first of all everyone's work and then the work of the professional. Indeed, to sum up one strand of this argument here we could say that religious education is shifting from being an expression of women's work to being an engagement with the quality of work. The heart of our transformation moves from the old alliance of home and church to include a new nexus of work and church. Religious education is now challenged to take seriously this new center of gravity in our lives.

All of these transformations in work have therefore effected changes in family life and church, the two primordial molders of our deepest emotional bonds. This change presents challenges and dilemmas to a church that would seek to be about the work of human transformation.

The church's dilemma lies in the fact that the workplace has become increasingly important as a separate arena of deep human loyalty, commitment, suffering, and transformation. Work life is profoundly important for both men and women. It must be lived out religiously through its many calls, transitions, and corruptions. It has enormous impact on the way people organize their family, church, and community life.

However, in the face of this the churches remain deeply estranged from work. Even entertainment is more closely allied to the church than are the ordinary processes of work. The place needing religious response the most is furthest from the church's ambit.

Moreover, the church's traditional way of affecting work through family life and the school has been greatly attenuated. The loss of the family farm and the family firm as dominant patterns even within middle class Protestantism cut one nerve. The gap between work and ordinary schooling, whether in public, private, Sunday, or parochial schools, severs another. Since people can no longer be prepared for a lifetime of work, the kind of schooling based on lifetime specializations is no longer relevant for the student. Not only do the emotional bonds of the family not carry over into school, but those of the school do not carry over into the workplace.

RELIGIOUS RESPONSES TO WORK ISSUES

These are some of the reverberations from the workplace revolutions we are undergoing. They confront the church with two questions: How will it bridge the gap between religious and workplace processes of transformation? Moreover, with what critical edge, with what definition of its faith, will it respond to workplace changes in a faithful manner? Contemporary church responses generally take three forms: *reactive, adaptive,* and *reformist.* Each of these offers handholds for entering the work-based dialectic of faith transformation.

Reactive responses seek to recover a nineteenth-century (so-called "traditional") pattern of work, with its attendant connections with family and church, typical of an earlier time. Some Christians seek to foster and preserve Christian businesses whose ethos resembles that of the Christian home and church community. The Fellowship of Companies for Christ (Atlanta), for example, comprised almost wholly of family-owned private companies, typifies this response. Here we find an effort to maintain a close relation of church, family, and work that was widespread in the last century but is very hard to preserve in publicly owned corporations in a pluralistic world.

A very different type of reactive response is the Christian commune or cooperative. These have been a regular though marginal phenomenon throughout American history, from the Puritan experiment to the Shakers and Owenites and the counterculture Christians of the 1960s.[10] These experiments do not accept the estrangement among work, family, and church. Communes seek to rejoin them, usually on an agrarian or artisan base. Cooperatives generally tolerate greater differentiation of these three spheres. In all of them, however, there is a tight relationship among the domestic, religious, and economic arenas.

In this context it is also important to remind ourselves of the radical Christian responses to industrial capitalism. In addition to Christian utopias and communes, we have also seen the way Christian faith informed the populism of people like William Jennings Bryan and the great twentieth-century socialist leader Norman Thomas. From Henry Demarest Lloyd we have a pithy summary of evangelical populism: "The proof that an individual has been regenerated is that he proceeds

---

10. For classic histories of the commune movement in America see Hinds, *American Communities,* and Holloway, *Heavens on Earth.* Contemporary developments are discussed in TeSelle, *Family, Communes, and Utopian Societies.*

to regenerate things around him—and that's Democracy and that's the Religion of Labor."[11] More recently groups such as American Christians Toward Socialism (ACTS) carried this banner in response to the emergence of United States globalism and multinational corporations.

While these quite varied responses lift up alternatives to our present fragmentation or hostility, they both require a pattern of work life increasingly marginalized in our own society. They are for the few. They remind us of the critical edge necessary in every Christian stance, but make that stance dependent on a particular form of economic organization. This selection of a specific organizational form over against the mainstream of economic developments helps us remember that faith transformation demands larger structural as well as interpersonal expression.

Adaptive responses generally accept the new economic forms but help people cope better with them. Here we find first of all the many programs in lay ministry, faith and life, and Christian leadership. Drawing on religious themes, theories, and values, they assist people with business decisions, cultivation of religiously grounded leadership styles, and problems in work life transitions. For instance, the Institute for Servant Leadership at Emory University draws on the writings of Robert Greenleaf to try to reshape executive leadership along religious lines.[12] The Center for the Ministry of the Laity at Andover Newton Theological School seeks to help people in all areas of organizational life exert a relevant Christian influence through their work. There are also many workshop seminars sponsored by business organizations, independent consultants, churches, and universities seeking to bridge the gap between faith and work, though often in disguised ways.

The positive contribution of these efforts is that they probe deeply into the many interrelationships of work and faith, sometimes also picking up some family themes. Their weakness is twofold: They tend to miss union employees and working-class people, especially Blacks and Hispanics, and they tend to skirt issues of structural change. Many union employees and most hourly wage people have a very different orientation to work than do middle class and managerial Christians. For them work

---

11. Quoted in Gilbert, *Work Without Salvation*, 104. For Thomas see Johnpoll, *Pacifist's Progress*.

12. Greenleaf, *Servant Leadership*. Other centers include the Trinity Center for Ethics and Corporate Policy (New York), the Center for Ethics and Corporate Policy (Chicago), and the Center for Ethics and Social Policy (Berkeley). For a critical reflection on earlier efforts see Clark, *Ministries of Dialogue*.

often does not have vocational importance in the religious sense. It doesn't carry the same kind of emotional weight. It is a means for maintaining the family rather than securing the central worth of the individual. In short, these people demand different responses. The quandary here is that the church and labor union connections once forged primarily by the Roman Catholic church have lost their momentum. We are confronted by a very different economic situation requiring a recasting of those old alliances.

The second kind of adaptive response is more practical in an immediate sense. It involves the de facto cooperation between churches and businesses through church-sponsored day care centers for children.[13] While the relationship may not be a formal one, churches and businesses have a symbiotic relationship in which businesses need quality day care for their employees and churches need a creative use for their space. Many of these programs seek to enhance child development and even religious growth. In turn, the work day creates the structure governing the interaction if not the very culture being transmitted to the child. While the day care center may provide for an encounter of parent, child, and church, it is not at all clear that churches have found ways to go beyond the facade to deal with deeper issues. Moreover, churches need to ask more critically whether they should even be filling this role in the economy.

Churches also seek to assist individuals and families in dealing with problems which often have origins in the organization of work.[14] They provide or support counseling centers dealing with work-related problems or support pastoral counselors active in industrial chaplaincies and Employee Assistance Programs. EAPs began as industry efforts to deal with alcoholism and addiction but have now developed a whole range of services from financial advice to family life. Here again there exists an opportunity to help people shape more adequately the transformations they are undergoing at work in the light of Christian faith. Here too we find the tension between making a profound impact on the lives of individuals and attending to transformation of the structures that create family problems, personal self-destruction, and spiritual alienation.

---

13. The church involvement in day care is surveyed by Lindner, Mattis, and Rogers, *When Churches Mind the Children*. Sheila Kamerman and Alfred Kahn investigate numerous workplace developments in Kamerman and Kahn, *Responsive Workplace*. See also Rosabeth Moss Kanter's penetrating analysis of nursery dynamics in Kanter, "Organization Child."

14. For a practical entree see Rightor, *Pastoral Counseling*.

For reformist responses to the need for structural changes we turn to the many church statements on economic matters that have emerged from the Roman Catholic, Methodist, Lutheran, Presbyterian, and other mainline denominations.[15] While rarely calling for wholesale overhaul of the economy or an outright rejection of the trends I have lifted up, they often focus in a critical manner on the ruthless exploitation of land and natural resources, the mistreatment of part-time, seasonal, and hourly wage workers, the misallocation of resources through the market system and the link between militarism and economic growth.

These statements can and do serve to bring people into argument and even mutual education. Ministers, business people, union leaders, and academics are forced to come to the table together because they all share moral concerns but disagree on the best structures for advancing them. These pronouncements have generated a good deal of publicity about economic issues and have become a part of the educational diet in churches themselves. They are a hopeful sign that we are beginning to redress the estrangement that has characterized church/workplace relations for some time.

For more activist manifestations of reform we can look not only to church support for migrant farm worker movements but also to efforts to resist the massive dislocations in the steel industry through boycotts and employee buyouts in Pennsylvania and Ohio.[16] Churches have also maintained numerous quiet engagements to reform economic structures. They have been involved in creating urban and rural land trusts, producer and consumer cooperatives, credit unions, low-income housing, and community development organizations. Religious organizations have also been leaders in shareholder actions to reshape corporate policies toward greater justice and environmental responsibility.

Here we see a focus not only on corporate change but also on the very conditions necessary for work to exist at all—basic necessities, land, housing, and community infrastructure. Sometimes these may operate with a rather radical stance toward our present economic organization. Other times they are simply trying to help people survive an economy that has no room for them.

---

15. National Conference of Catholic Bishops, "Catholic Social Teaching"; Baum, *Priority of Labor* (including the Encyclical Letter of Pope John Paul II, "Laborem Exercens"); Smock, *Christian Faith and Economic Life*; General Assembly of the Presbyterian Church in the United States, "International Economic Justice."

16. Wolcott, "Church and Social Action."

The human transformation needed in the face of the deformation of land, communities, and people by our economic revolutions requires concerted attention not only from religious educators but from the entire church. Religious educators, however, have to lead in helping people understand and think through how faith should speak in the midst of these transformations. This demands not only appropriate analyses of economic change but also a clear articulation of a faith vision. We move now to the final part of this chapter—clarification of the challenges before us.

REWORKING RELIGION

As a first step we need to refashion key theological concepts in the light of a changed economy. Without theological clarity we will simply imitate the language, thought patterns, and practices of the workplace without a prophetic edge. Moreover, many traditional Christian concepts have, often unwittingly, become embedded in reinforcing an older pattern of work. We need now to disengage them from those older uses and recast them in the light of contemporary theological argument as well as social circumstance.

Every theory of religious education, and therefore of its relation to work, operates under some master concept. George Albert Coe and Walter Rauschenbusch cast religious transformation in terms of democracy and urged us to democratize the workplace.[17] Paulo Freire and Ivan Illich called for conscientization, humanization, and de-schooling. More recently, Thomas Groome has advanced themes of liberation and freedom. My own thrust uses the concept of covenant publicity.

I have already pointed out how the long heritage of Christian asceticism came to reinforce the industrial obedience and entrepreneurial acquisitiveness central to the industrial revolution of the eighteenth and nineteenth centuries. It is a revolution still reverberating throughout the world. In our own time, however, another religious theme reemerges with new force. I call it the theme of covenant publicity.[18] It has the possibility of being the contemporary theological bridge between church and work.

17. In addition to Coe, *Social Theory*, see Rauschenbusch, *Christianity and the Social Crisis*, and Rauschenbusch, *A Theology for the Social Gospel*; Freire, *Pedagogy of the Oppressed*; Illich, *De-Schooling*; and Groome, *Christian Religious Education*.

18. For a much fuller discussion of the implications of covenant publicity see Everett, *God's Federal Republic*. M. Douglas Meeks presents a theology based in an expansive notion of economy in Meeks, *God the Economist*. See also the essays in Krueger

Briefly stated, covenant publicity points to people's need and capacity for joining others in public expression of their life. Publicity highlights people's need to participate in the decisions affecting their lives. Covenant lifts up the need that people be joined in enduring bonds of trust that bring together their work, family, and faith in relationship to the land. This expansive embrace of covenant making is rooted in ancient scripture and finds renewed validity in the light of our modern ecological awareness of human life.

Both of these themes find points of resonance in contemporary work developments, though they also find much challenge and resistance. In nurturing persons who seek to live out lives of covenant publicity, churches can foster an appropriate spirituality for the workplace. In developing and supporting pilot projects exemplifying these ideals, whether in cooperatives, community development, or in internal church life, churches can sharpen their dialogue with the world of work.

Moving to a new religious orientation enables us to reinterpret numerous aspects of our faith tradition. For instance, we can reexamine the concept of vocation so that we can move from the individualistic view of vocation as job and occupation to one which captures the original biblical view that vocation is God's call to a whole people. To respond to God's call was originally not an individual selection of a career but a response by a whole community to reshape its common life around a higher standard of justice.

Similarly, we can reexamine the notion of stewardship to overcome a fixation on its monetary aspects.[19] Taking the parable of the talents (Matt 25:13–45) as an example, we can see that the "talents" given to the various stewards were not personal properties, as our culture typically views it, but representations of the gospel itself. The "talent" Jesus was talking about symbolized the good news of God's transformation of creation. We are called to be stewards of that good news as members of the church, multiplying it by our very sharing of it. What does this original meaning of the parable mean for an economic culture that prizes talents as individual abilities to be marketed to the highest bidder? What does it mean to our society to reinterpret this parable, which has been used to legitimate its occupational structure and market economy?

---

and Grelle, *Christianity and Capitalism*.

19. Max L. Stackhouse offers a critical theological view of stewardship in Stackhouse, *Public Theology*. See also Hall, *Imaging God*.

Moreover, we need to recover the lasting meaning of Sabbath in a time when it no longer sets limits to the dynamics of work. Sabbath means that there are and must be boundaries to the demands people can place on each other whether in the exploitation of workers by managers and owners or the exploitation of self by the workaholic. The idea of the Sabbath as a day set apart also means that it is not simply a day of rest for an individual. It is a time when all can refrain from work together and remember their covenant as a people dedicated to higher purposes. Work must honor these other relations of life and the transcendence of a God who speaks and listens to all people as citizens of a divine republic. We are not simply producers and consumers. We are not simply workers. We are actors in the theater of God's drama, players in a wider story of redemption.

The Sabbath can thus be seen as that moment when we remember that we are partners in covenant, a concept which goes far beyond the individual contractual relations typical of the modern marketplace. A life in covenant involves the whole range of biblical concerns: God's relation with our history, with all of creation, with other peoples, and with each person as a partner in world transformation.

In drawing on such theological resources we can develop further the concepts of covenant, which bespeak the deep relational character of existence, and publicity, which bespeaks our yearning for a life in which God is fully revealed to us and we to each other. These themes, present in our work lives, need to be drawn out by the church to help transform the workplace as an arena of faithfulness.

These themes must also find practical expression in our society. What must be done in light of this transformed faith? First, churches need to attend to the creation and improvement of the bridge structures between churches and workplaces. Adult education centers, day care centers, employee counseling programs, alternative economic structures—all these are places where faith concerns can be nurtured. They are as important as the home in shaping our patterns of faithfulness. Religious education must reflect this importance, and resources for dealing with the issues of work need to be developed for those points of intersection.

Second, the schooling paradigm in religious education needs to be subordinated to the kinds of educational processes present in workplaces that have moved beyond a slavish devotion to the old factory model. In these contexts group relationships of open consultation and negotiated goal commitments replace a bureaucratic command structure which isolates individuals in rigid job descriptions. The mentoring and consultation emerging in those arenas need appropriate reinforcement in churches. The move toward spiritual guides and directors in Christian circles needs to be evaluated as a model for religious education generally. Relationships based on a common trust and commitment can then move to the center of educative transformation. The priority of reflective process reminds us that no curricular foundation of information, whether it be Bible stories or catechisms, is ever secure but must be redone continually in the light of new experience—especially transition experiences in our working lives.

Third, to grasp these transitions we need to investigate the way work is organized and how it affects people, families, churches, community organizations, and politics. Work is transforming people—enabling more women to have a public life, power in family relations, and greater citizenship in the church. Workplace changes often train people with new skills at communication and problem solving that enhance the rest of their lives.

Work also continues to deform people. It can subordinate them to market dynamics and imprison them in the hierarchies of firms unable to develop more pliable structures. It often narrows human vision and aspiration. It deprives people of energy, time, and access necessary for public as well as private life. Religious educators need to know both effects of the workplace so they can fashion appropriate responses to enable people to be active agents of the transformations their lives need before God.

Fourth, to fashion this critical knowledge of the workplace, churches constantly need to be engaged in actions affecting the workplace, whether they are constructing alternative housing markets, pushing for corporate responsibility as shareholders, or conducting seminars in business ethics. Only within this engagement can we experience the processes of transformation where God is calling us to exercise our capacities for public profession and covenant. In the struggle to understand what we are doing and to recast it in terms of faith, religious educators can play a crucial role.

Fifth, to augment these external engagements religious educators need to take serious account of the way the church itself is a work institution. As a workplace it educates its members just like any other business. Many times church organizations compare very poorly with businesses in their attention to the dignity of employees, their rights, compensation packages, and environment, not to mention management methods. They smother with love and good intentions the deformations produced by patriarchal command or simple disorganization. Not only the staff but the members of the church internalize a model of human relationships through participating in the church. Religious education must lead in the task of helping us become more aware of what we are doing in church administration and organization. Management educates and shapes faith. It should be a central concern of educators.

In becoming more aware of the transformative impact of management methods churches need to be critical of the patterns they take up from the workplace, whether it is in achieving goals through contracts, programing with marketing models, or computerizing office procedures. Just as every baptism must be preceded by a catechumenate, so must organizational patterns and methods be subjected to theological scrutiny and refinement. In this way the church is always a critic as well as a partner in the process of human transformation.

10

Sunday Monarchists
and Monday Citizens?

On Sunday morning Christians across America sing praises to their "king," ascribe all glory and power to Him. pray for His "enthronement" and speak of a God whose Son has inherited the divine "kingdom." During the week, however, they engage in political campaigns, vote in elections, decry the emergence of hereditary dynasties, and resist the centralization of power. They are monarchists in worship and democratic republicans in daily life.

I used to find this incongruous, Now I find it painful and even dangerous. How can we celebrate kingship in church and still function as committed citizens in a republic? There are very few genuine monarchies left, but there are federal republics, democratic republics, and socialist republics. How has this fervent Christian anomaly survived two centuries of republican and democratic development? Isn't it past time that we reconstructed our Christian thought and worship in order to engage the world in which we actually live? Though the visions of worship cannot coincide with the patterns of politics, we need to worship and live in some kind of common language if we are to shape a coherent life of faith.

We can't set aside the language of monarchy without replacing it, however, for governance language is intrinsic to Hebrew and Christian faith. It affirms that we are struggling toward God's perfect order where we will be related to each other in justice. But if we are to be faithful to a biblical perspective. we need to re-examine the language of worship when it no longer speaks our language of governance. Moreover, we need to critique models of governance in the light of Jesus's ministry and the work of the Holy Spirit. The question of governance language is one of authenticity as well as relevance.

Feminists have begun to dismantle the exclusively male language of worship. Their effort to show how our religious language can reinforce injustice has led to far-ranging changes in music, prayer, and Scripture reading. However, the move to inclusive language can merely reinforce the domestic language of mother and father, householder and parent, sister and brother. If we simply eliminate governance language, we may reinforce the restriction of religion to the private sphere where women have been confined in the past. Inclusive language doesn't help us negotiate the second step: reconstructing the language of political governance itself. The issue here is not maleness but monarchy.

We face the challenge of being not only an inclusive church but also a public church—a church that is a public, an *ecclesia,* a genuine republic of Christ. I call this the task of *covenant publicity,* for which we can draw on the ancient traditions of covenant that underlie modern federalism as well as on the ideas of council, republic, and *ecclesia.*

Israel itself was not always a kingdom, and it struggled mightily with the introduction of kingship. As a confederation of tribes living under the law, it governed through assemblies rather than monarchs. Covenant meant first of all banding together under God's published orders, the Torah. With David and his court singers, worship as kingship adoration arose. Covenant and worship focused no longer on the tribal assembly but on God's relationship with the king. By painting God as a great king, the Psalms provided Israel's monarch with a crucial though subordinate legitimation. Conversely, the language of earthly rule became the language of divine governance.

The kingship tradition persisted, intertwined with the confederal traditions of ancient Israel, through the rest of Israel's history. And while the echoes of confederation, covenant, and council can be heard in the background of the Gospels, it is the fanfare of Davidic monarchy that greeted Jesus. He was received as king rather than as prophet, priest, elder or president. Though the church drew its primary name from the *ecclesia* of Greek public assemblies, this theme was gradually submerged under the monarchy of Christ, of the Father, of the bishop, and later of the pope.

For 1,500 years Jesus and his promised new creation were embellished with layers of monarchical and imperial symbolism. The emperor's basilica became the model for church architecture, and his throne became the bishop's chair. The feudal gestures of homage to king and lord were mirrored in the clasped hands of prayer, the kneeling at the rail, and the prostrate body of the ordinand. The trappings of the court became the framework of worship.

When this close alliance of throne and altar was broken in the revolutions of the modern world, the churches generally held onto their kingship forms by taking them into the home, safe from the emergent public square. The churches could survive the revolution from monarchical to republican government if they kept their worship private. The monarchy of Christ invoked in Christian liturgies retreated from the governmental sphere to the heart and hearth. Thus, emerging republics were spared the conflicts of old religious differences while the church was spared the loss of its monarchical worship forms.

Similarly, the ideal nineteenth-century Christian home was a castle where the man was king, the woman queen, and the children silent subjects. Gothic homes and Gothic churches joined to praise a divine monarchy of the heart. The feudal kingdom was translated into the hierarchy of marriage and parenthood, where women and children were subject to the domestic king. The ensemble of marriage, family, home, and church preserved Christian kingship in a world of democratic republics.

This relocation of kingship is nowhere more evident than in the monarchical language of Christmas. At Christmas we turn Matthew's wise men into kings and celebrate the birth of Jesus as the only begotten heir of David's throne—a throne residing solely in our hearts. Later in the church year, on Christ the King Sunday, we pray that Christ will reign as king in our hearts—since we can hardly pray for a return to monarchy in our republic. The struggle over public order is reduced to a struggle over the inner conflicts of the heart. Only a divine monarchy in our psyches can ultimately create peace for the world. In fact, Sigmund Freud developed a similar analysis of a little kingdom of superego, ego, and id warring for possession of our psychic throne.

Yet a proposal to reconstruct worship practices raises two critical questions. First, doesn't kingship symbolism attest to God's transcendence over any and all forms of government? Isn't God's sovereignty actually enhanced if we speak of it in a language no longer spoken by the people of our world?

To respond to these questions. we must first distinguish transcendence from irrelevance. The issue is not mere transcendence but critical engagement. Our task is not only to relativize the powers of this world but to transform them in accord with a faithful vision. It is not enough to overthrow an oppressive monarchy if we do not revolutionize the images of governance that legitimated it. We must choose the gods that will illumine our conceptions of right order.

The language of biblical worship must be both incarnational and transcendent. Just as kingship symbolism functioned for 1,500 years both as a language of cultural engagement and as a critical tool against all earthly kings, so republican symbolism can provide us with a language of transcendence as well as immanence for our own time. Not only does it pick up our actual memories of public testimony, election, and governance by law. but it draws us toward the perfection of our governance, helping us judge our present efforts, Jesus's exercise of his presidency through persuasion and open argument can be a plumbline for judging all presidents, and God's republic can be the ideal for our efforts at participation and public debate.

Symbols of monarchy make the church a nursery of reaction to republican life altogether. Monarchical kingship treats us like members of a household represented solely by its head, while republican order assumes the equal headship of all, each professing her or his own conviction about the common good in an arena of debate. That is, our new life in Christ is a longing for "publicity" in our lives, not for the comforting subordination of children. Moreover, we are searching for a "covenant publicity," one which prompts us to listen to others and enter into new bonds of relationships with them. We are drawn to form covenants that go beyond conditions of our birth, gender, race or nationality to a wider republic of justice.

We still face another key question. Doesn't real worship have to be embedded in archaic memories and language? Isn't worship essentially born of tradition and required to carry it on? Isn't it an illusion to think we can make radical changes in it? Won't the dew of mystery dissipate under the hot glare of analysis and manipulation?

Christian worship must honor not only our memories but our anticipation, the work of the Spirit as well as the divine founder of creation and the church.

Moreover, kingship is only one tradition in the history of Israel and the church. Even more archaic is that of covenant, Torah, council, and *ecclesia*. These traditions actually claim our fullest loyalties. They reflect increasing numbers of people—in the US, the Soviet Union, India or South Africa—who long for a more perfect republic of participation and genuine debate. The language of election, federation, congress, and council has also become the language of our operative faith. This is now the nature which needs perfection by God's grace. This is the nature to be purified in the worship where Christ presides.

What might this new language look like in worship? Our first challenge is to become more aware of the way our present worship is shaped by kingship symbolism. King, lord, son, throne, kingdom, court, crown, and glory only begin the list of terms we need to reconsider. We need to see the way the patriarchal formula for the Trinity—Father, Son and Holy Ghost—was rooted in the transfer of royal rule through household inheritance. We also need to see how the architecture of sanctuary and church shapes our images of divine governance. Do we worship in a throne room of the king or in an assembly of the people? In prayer and communion do we kneel before a feudal lord or do we share as equals around the table of Christ?

Second, we need to become conscious of the themes, language, and gestures of governance that have emerged in the past 200 years in the savagery of extinguished publics and broken covenants, whether in Nazi Germany, Chile, Czechoslovakia or the Trail of Tears. We can also explore stories of personal efforts to find a more expansive covenant with others, such as in biographies of alcoholics, homosexuals, divorcees, or victims of abuse. In short, we need to lift up our public and private stories of longing for covenant publicity, for a transformed life in God's republic.

We then must reclaim the biblical narratives that can shape these memories into a language of devotion articulated in the light of Jesus, whose presidency was powered by listening, whose republic was founded in covenant bonding, and whose election was rooted in self-sacrifice. The giving of Torah to Israel can reshape our grasp of constitutions. The theme of the wanderer or exile so central to the biblical story can amplify the cry for global citizenship by today's refugee. The demand to preach the gospel under any regime can sharpen our support of publicity for all peoples. Synagogue and *ecclesia* can become places where the rehearsal of covenant-making and public witness prepares us to confront a world of secrecy, lies, and coercion.

More practically, this transformation of worship will require changes in architecture, music, ritual movement, dress, language, and the shape of the church year. For instance, many churches have already moved the focus of attention away from throne, pulpit, and altar to the people assembled in response to God. Rather than subjecting people to the tyranny of one voice, every voice in worship could be amplified by the microphone. Many of our hymns are still deeply infused with the monarchical language of King George's England. The best of them need adaptation, the worst must be replaced.

Similarly, we need to assess the costumes of our worship leaders. Do they create a separate world of mysterious nostalgia or do they orient us to a new engagement with the powers and authorities of our world? Perhaps it is time to reassess the role of the business suit as an appropriate symbol for presidency in worship.

Even the language of the church year must be re-examined. Christ the King Sunday is an obvious relic of monarchy. Should it be lifted higher than an annual recognition of Human Rights or of the United Nations? Should the language of ascension be used after Easter? Or should we speak the language of inauguration, with our own covenantal swearing in to God's emerging republic?

These questions are only preludes to creating a worship that reformulates the way we address God, ally ourselves with Jesus, and invoke the Holy Spirit—not merely for the sake of being good citizens, but because of our responses to the nature of a divinely inspired assembly. This reconstruction is a matter not only of the integrity of our faith, but of our action in and toward God's world.

11

COUPLES *at* WORK
A Study of Patterns of Work, Family, and Faith[1]

> I don't know, we enjoy being with each other. It's fun to work together. It has been for us. Now that might be an old-fashioned answer or something, but we enjoy being together. You know, I couldn't have done this by myself, and she couldn't have done it by herself. And it is something, you look back and see where you came from, where you're at now and where you're planning on going. You know, there's kind of a surprise in that. We've accomplished a lot.
>
> —A Farm Couple

> SHE: To me it is real important to work together. It is part of what I believe in. If you're not really working with your partner, then are you really sharing your life? If you don't have common goals, especially that are outside the personal or romantic, then I think that the relationship has a hollowness to it, whereas a commitment of our lives to a common purpose, is very valuable . . . I would feel the same way if we were farmers.

> HE: Even though we said we were going to start a theater and work together—I didn't understand what that meant, and it was only through doing it that I came to understand the real implications of what that means . . . I didn't know it was that unusual, it just seemed like the thing we were supposed to do, because

---

1. Researched and written in collaboration with Sylvia Johnson Everett

we were together, and we're soul mates, and part of being soul mates, I think, means continuing to try to come to an understanding of that. What does it mean when souls commingle? I don't think it is something to be considered lightly or to be sloughed off. I think we are supposed to be active in trying to understand that, just as we feel it.

—Arts Directors

Work and family are the center of most people's lives. Work and family are not only the sources of our livelihood, but also the source and expression of some of our deepest values. They are crucibles as well as manifestations of our faith. The various forms of this faith are the subject of this chapter.

By "faith" we mean the ultimately trustworthy relationships by which we live. Faith is not simply a set of beliefs or propositions. It is not simply a set of values. It is more like an image of fundamental relationships. Faith is our pattern of deepest fidelity.[2] With that definition, many of the controversies of our day can be seen as matters of faith—the right order of parenthood and coupling, the right to abortion, the right way to organize the economy and to educate children. "Right religion" and "right family and work" are inextricably matters of faith in our sense.[3]

In examining changing patterns of work and family, we are therefore also examining changing patterns of faith. This study focuses on a particular pattern for combining work and family—that in which a couple work together. It is based on interview and questionnaire data collected in 1988. It explores the patterns of trustworthy relationship underlying this pattern of life, in an effort to see its possible implications for religious life, especially for its symbolism and organization.

THE *OIKOS* IMAGE

We use the ancient Greek word *oikos* to talk about the ensemble of faith, work, and family, including its bond to the land.[4] The *oikos*, which we

2. This view is worked out in considerable systematic detail in Everett, *Blessed Be the Bond*.

3. For exemplary discussions, see D'Antonio and Aldous, *Families and Religions*. For one example of changing family symbolism in religion, see Neitz, *Charisma and Community*, ch. 5.

4. We have developed this language over the past six years through the OIKOS Project on Work, Family and Faith, based in Atlanta. For one exploration of these

usually translate as "house," actually embraced all of these components. In the ancient world, and in most traditional societies today, these elements are fused together. The history of our own civilization, however, has seen them differentiate from one another, even to the point of fragmentation and hostility.[5] In the process, we have been left with no English word for talking about them as a system. To recapture that systemic sense, we use the term *oikos*. Our *oikos* is the way we put work, family, and faith together, whether as persons or as a society. It is also a powerful and emotionally rooted image of how things ought to be. People tend to act and to evaluate life in terms of images they have of how the *oikos* should be put together. A particular *oikos* pattern shapes the way people spend their time and energy. It expresses the sense of durable relationship and right order standing at the center of our faith image. This study therefore focuses on the *oikos* images informing the lives of couples who work together.

It is these images which legitimate how we arrange our lives and our institutions. They deeply affect the way institutions, especially religious institutions, justify themselves. Acceptable patterns of sacred order are closely tied to the emotionally resonant patterns of our *oikos*. In addition, the governing images grounded in work and family help legitimate patterns of organization in politics, education, and health care, as well as the church. Images of this strength endure across time. Churches, for instance, bear with them governing images based on the *oikos* patterns of previous eras. But churches can also—through events of conversion or deep spiritual formation—reshape the *oikos* images people bring to them. Sometimes people consciously try to reshape the rest of life according to religiously legitimated *oikos* images, just as images that come from other times and institutions may shape the life of family or church. This complex interplay of fundamental images shapes the way we put our lives together.

---

connections, see Everett, "*OIKOS*: Convergence."

5. This simplified picture of family history needs to be augmented by the resurgent work in this field by, among many others, Goldthorpe, *Family Life*, especially ch. 2, 4, and 10. See also Segalen, *Historical Anthropology*; Degler, *At Odds*; Lantz, Schultz, and O'Hara, "Changing American Family"; Sweet, *Minister's Wife*; Tufte and Myerhoff, *Changing Images*; Young and Willmott, *Symmetrical Family*.

OIKOS TYPES

Oikos images can be categorized in several ways, two of which are central for this study. First, *oikos* images can be placed along a spectrum between a fusion of components and fragmentation. We call this the spatial dimension of the *oikos*. Second, *oikos* images can be examined in terms of the way the participants arrange and govern their relationships. This is the relational dimension.

The Spatial Dimension

Oikos patterns vary in terms of the "distance" among the components of work, family, faith, and land. At one end we have the fused *oikos* typical of traditional agrarian society, in which the family is a productive economic unit bound together on the same land. Moreover, their faith is a set of ancestral loyalties tightly bound to working the land. In the tight *oikos*, family, work, and land are still tied together, while religion has become somewhat independent. Religious organizations are likely, however, to follow the pattern of family relations. This pattern, most familiar to us in the family farm, has been typical of most of our society until the end of the nineteenth century.

The open *oikos* represents a further differentiation, in which work and family are separated from each other. Schools usually emerge in order to help children pass from the family sphere to the work sphere. In the nineteenth century, this pattern took on a form which we call the split *oikos*, in which the separated spheres of work and family were divided between men and women. The husband worked outside the home, and the wife took care of the domestic sphere. Religion was split between them, with the male taking over most formal leadership, and the female supervising religious instruction and matters of the spiritual life.

As women have now moved into the workplace on an increasingly equal footing, more and more functions are moving out of the household—funerals, birth, food preparation, manufacture of personal items, entertainment, care of the elderly and sick, and so forth. We call this pattern of intense differentiation the fragmented *oikos*. This *oikos* pattern has become, at least in some sectors, a cultural norm in our society.

The Relational Dimension

The other distinctions to be made regarding these basic governing images have to do with relationship and authority patterns. There appear to be three basic types—hierarchical, organic, and egalitarian.

The hierarchical pattern is one in which one person is the final authority in the relationship. This can take a strong form, in which one party controls the others in every aspect, or a more limited form, in which one person controls only the key decisions affecting the group. Most of the hierarchies we know have been male-dominant.

The organic pattern, often neglected in research, is one in which the activity of the group is divided up by functions. Decision-making and command are not as salient, because the members of the group do what is demanded by the task of the moment. The work, not a person, controls the group. Within each function, individuals exercise considerable autonomy in fulfilling their tasks. In some forms, they all participate in performing each function, as when members of a farm family all turn out for the harvest. In others the work is divided up into interdependent but distinct functions, with individuals assigned to each. This pattern occurs not only in farm families but also in small family-run businesses.

In the egalitarian pattern the members stand on a roughly equal footing, and negotiate tasks and policies. They form agreements about their common life. They may reshape these agreements when one or more of the parties feels it necessary. Here the group is controlled by the outcomes of communication, persuasion, and negotiation. This image has become dominant in the open and fragmented *oikos* as it is experienced in American culture today.

THE *OIKOS* SHIFT AND THE INTIMACY DILEMMA

North Atlantic societies have gradually shifted from a fused or tight *oikos* to an open or fragmented one. When women began to be accommodated in the workplace, there was, however, no change in the split *oikos* image that operated with sharp distinctions between work and family. This economic base enabled women to press for greater gender equality and to pursue the ideals of equality, persuasion, public participation, and genuine citizenship; but it has also created a fundamental strain in our life.[6] This strain in the *oikos* can be identified as a conflict between the intimate communication assigned to marriage and the individualistic career expectations of our economy. People are told that their marriage depends on intense communication, but they are not allowed to place their common work at the center of that communication. Thus, marriage comes to be a leisure activity divorced from work. Moreover, parents are estranged from the process of passing on their values, usually expressed in their work, to their children. Religious institutions exist increasingly in the leisure sphere, with their values also detached from the work-family connection.

The couple who work together present an alternative to this pattern—one in which the values of marriage and those of work are combined in one operation. In one sense, this arrangement seeks to preserve the values of the tight or fused *oikos*. On the other hand, it may provide an opportunity for the expression of the egalitarian values of the open and fragmented economy. Couples who work together, then, present a peculiar crucible for the refinement of basic *oikos* images, one which might have critical significance for religious symbolism and organization, as well as for family and economic life.[7]

THE STUDY

Through "snowball" sampling, we have slowly accumulated a list of over 130 couples, located throughout the US, who work together. We sent each couple a questionnaire, and we received forty-nine responses. We also conducted twenty in-depth interviews with couples close enough

---

6. For key discussions of the general phenomenon of two-earner couples see Aldous, *Two Paychecks*; Heckman and Bryson, "Problems of Professional Couples"; Herman and Gyllstrom, "Working Men and Women"; Hall, *Two-Career Couple*; Hertz, *More Equal than Others*; Hood, *Becoming a Two-Job Family*; Hunt, "Dilemmas and Contradictions"; Kanter, *Work and Family*; Papanek, "Men, Women and Work"; Pleck, *Working Wives/Working Husbands*; Rowatt and Rowatt, *Two-Career Marriage*; Taylor and Hartley, "Two-Person Career"; Ulrich and Dunne, Jr., *To Love and Work*; Voydanoff, *Work and Family*.

7. For specific discussion of couples who work together, see Barnett and Barnett, *Working Together*; Bryson and Bryson, and Licht and Licht, "Professional Pair"; Epstein, "Law Partners"; Hoffmann and DeSole, *Career and Couples*; Krupa, *Couple-Power*; Lyson, "Husband and Wife"; McKiernan-Allen and Allen, "Colleagues"; Nelton, *In Love and in Business*; Pepitone-Rockwell, *Dual Career Couples*; Railings and Pratte, *Two-Clergy Marriages*; Rapaport and Rapoport, *Working Couples*; Sommer and Sommer, *Two-Boss Business*; von Lackum and Kemper von Lackum, *Clergy Couples*.

to Atlanta to visit in person. Both the questionnaires and the interviews sought to identify the governing *oikos* images arising in the lives of couples who work together. By "working together," we mean that the couple interact more than fifty percent of their work time and/or produce more than fifty percent of their income in this manner. We identified some initial couples through newspaper accounts, friends, and colleagues, and one advertisement in a trade magazine. We then let that list snowball through subsequent referrals from couples responding to our questionnaire. We sought out as great a variety of occupations and backgrounds as possible, especially in our interview group. Since no one knows how many couples actually work together or what defines them, it is not possible to say whether our sample is a truly representative one. The few students of this phenomenon estimate that between one-half and one and a half million couples fulfill these criteria.[8]

In particular, we wanted to explore three types of questions.

The Context. What are the economic, educational, familial, geographic and religious characteristics of these couples? What economic structures foster or impede them?

The Oikos Images. Why do they try to put their lives together this way? Are they building on inherited patterns? Are they value-driven and idealistic, or are they a devising a pragmatic response to their circumstances? What types of *oikos* images (spatial and relational) are present with these couples, and how do they structure the way the spouses interact? What are the sources or key supports for these images? What kind of explicit or implicit spirituality is allied with them? What are the advantages and disadvantages couples have found in this pattern?

Institutional Ramifications. What implications do these patterns and experiences have for the workplace and economy? What changes do they imply for marriage and family? What are the relationships between these patterns and religious symbolism and organization? What kinds of religious involvement and values do they foster or undermine? How might churches and other religious organizations deal with these alternative patterns?

---

8. See Barnet and Barnet, *Working Together*, xxvii-xxviii, and Nelton, *In Love and in Business*, 12.

EXPLORING THE CONTEXTS

Where Are the Couples at Work? The majority of our forty-nine questionnaire respondents were self-employed in some way. Thirty-one operated their own businesses—three couples in farming, eighteen had a business firm, three a franchise, six had professional practices, and one couple were artists. Fifteen were nonprofit entrepreneurs or professionals, generally employed by churches as pastors or by educational institutions. Four couples were corporate employees with professional training.

Most worked in small-scale organizations and settings. Half were in work organizations of less than seven people, with even fewer people in their immediate work area.

As we expected, we were unable to find any couples sharing standard jobs in large corporations. The exceptions are the few professionals employed by corporations. We do know that some corporate job-sharers exist, but they are rare indeed. The opposition between the corporation and couples is near universal—unless, of course, they own the corporation, as many couples do.

What Are Their Common Characteristics? These couples, on average, were quite well educated. Their educational levels ranged from high school to doctoral degrees, along with certificates in the arts and skilled trades. Twenty-seven of the men had graduate or professional degrees, as did fourteen of the women. All men had studied beyond high school, as had all but three of the women. On average, the males had a few years more schooling than the females. Though education may not in itself produce couples at work it may contribute greatly to the capacity for economic autonomy, which seems critical to the development of a couple-career path.

The couples we discovered ranged in age from people in their late twenties who had worked together only six months to couples in their late sixties, one of which had worked together thirty-five years. The couples had from zero to nine children. Their gross incomes ranged from twenty thousand to over a hundred thousand dollars, with an average of about sixty thousand dollars, placing them solidly in the upper middle class.

Ethnically, the vast majority were Euro-Americans (ninety-two percent), especially from the British Isles (sixty-seven percent)—not surprising, given the concentration of the study in the Southeast. The remainder included African-Americans and Hispanics.[9]

---

9. Unfortunately, when asked to describe their ethnic origins, couples frequently

The religious affiliations we discovered covered a broad range, from Methodists (eighteen percent) and Presbyterians (fourteen percent) through Episcopalians (eleven percent), Baptists (eight percent), "Evangelical Protestants" (seven percent), and Jews (three percent). Eighteen percent checked "none," with "other" counting for another fourteen percent of the sample. Compared to the general population, Baptists are under-represented, and Episcopalians, "nones," and "others" are somewhat overrepresented. We believe this denominational distribution is reflective of the entrepreneurial and professional class where these couples are found. It also appears to be the result of changes that have taken place since childhood. Several reported childhood religious affiliations, but said they had none now. And most mainline denominational categories were more strongly represented among childhood affiliations than among current ones. However, nonaffiliates, Episcopalians, and "other" (Unity, Meher Baba, "New Age," and the like) have increased over reported childhood affiliations. Thus, the religious dynamics reflect two central features of these couples—their preponderance in the professional and entrepreneurial classes, and their affinity with novel, freshly constructed frameworks of meaning and relationship.

EXPLORING THE *OIKOS* IMAGES

Why Are They Working Together? While in some cases couples could identify ancestors who worked together, all but one had simply invented their own work together rather than inheriting it. This ran contrary to our expectation that ancestral images, largely from farming, would find re-expression in modern form. Among those ancestors who had worked together, half were in small business, one fourth in farming, and one fourth in professional life. Thus, the farm experience was not as important as entrepreneurship in shaping these couples' contemporary *oikos* patterns and images.

People cited a number of different reasons for originally working together: to meet financial need or opportunity (thirty-five percent), to have mutual support in their work (twenty-nine percent), and to maintain their common interests (twenty-seven percent). Eighteen percent

---

responded "middle class" or "suburban mainline," or else had a mixed and forgotten ancestry, making it practically impossible to discern linkages between ethnicity and *oikos* patterns, though, on the basis of our interviews, we believe there are important connections here.

launched their common effort because they worked well together. Only two couples listed domestic flexibility (child care, household equity) as a prime motivator, although the advent of children precipitated the decision for a number of couples. Couples with a more hierarchical *oikos* image tended to concentrate on financial reasons. Those with organic images supplemented financial reasons with companionship values, while more egalitarian couples concentrated on common interest, synergy, and companionship. This seems to indicate that egalitarian couples emphasize the importance of the direct, interpersonal, couple relationship, while for the others the values of household and children are more important.

When asked why they were *still* working together, "synergy" rose from eighteen percent to thirty-six percent of the reasons, with common interest dropping sharply, and financial need dropping slightly. Thus, common interest and financial need seemed to form a kind of foundation or prerequisite. Once that was established, these couples found out, contrary to common cultural myths, that they *can* work together quite well. As a couple in real estate put it:

> HE: Well, we had the dream of being together, but it wasn't work related.
>
> SHE: Although I remember you telling me years before that you had always wanted to work in real estate with your wife. You don't remember that? I remember you saying that. And, of course, at that time I was working on the base. I never dreamed about really working with him, particularly in real estate, but I do remember him saying that.
>
> HE: She always said she couldn't work with me.
>
> SHE: That's right. I really . . . I'm surprised we can work together. . . . He is so bossy, really and truly. And so am I. And I thought that we probably would have great difficulty working together, although we get along well. I guess we have such a division of things, I never really thought about it that way until you had [us answer] the questionnaire, but we kind of separate things so that we get along better. But I thought that we would have great difficulty because I'd want things my way and he'd want his way, you know.

On balance these couples have found working together very rewarding. It is this discovery that sustains their work arrangement. The widespread belief that spouses cannot work together may well be a cultural myth to justify the split *oikos* that has little basis in people's actual capacities and tested experience.

How Are They Integrating Life? How are they integrating work with family and spirituality? We expected to discover that couples who work together have a tight or even fused *oikos* structure. Not only are they working and living together, but they tend to have an intense involvement in this work as an expression of their key values. We expected that couple careers would be a contemporary expression of the tight *oikos* of our agrarian past.

The findings, however, indicate that these arrangements today clearly bear the mark of the open and fragmented *oikos*. Taking together the responses to a number of questions, we found that their *oikos* structure was on average more open than tight (average of 1.63 on a scale of 1 to 2). What was striking in the survey and interviews is that our respondents tend generally to be very couple-centered, and less embedded in a wider set of relations. Instead of being "tight," their world is focused. Moreover, it is focused on the couple and their construction of an *oikos* suited to their needs, rather than on the family as a transmitter of the values and patterns of a wider community.

The *oikos* of these couples is a voluntarily constructed pattern to achieve particular personal and couple goals. It is a part of the pluralism of our society, and seems to thrive only in the most autonomous and entrepreneurial sectors of the culture. Because they felt they were making it up as they went along, many couples felt their common work had just "happened."

> HE: This business just happened. The way it happened is totally unexplainable. Martha started it just to have something to play with. It grew, and her mother came in, and it grew some more, and her mother got out. I think that one of our biggest pitfalls in the four and a half or five years that we've been doing this is that our growth has been totally unplanned for the most part. We have developed business plans and tried to do annual budgets like all good businesses are supposed to do. Even when we sit down and do those, it's like, "how do we get that?"

While this improvisational character of the *oikos* made life exciting, it also often created considerable strain and demanded frequent attention to the way they made decisions and carried out tasks.

How Are They Relating To Each Other? Three key questions were used to identify the relational dimension of a couple's *oikos* image—questions about their image of their work relationship, about their spirituality image, and about their formal position descriptions. We discovered that the organic *oikos* image was central for work relations, while egalitarian themes dominated their spirituality, and were slightly more prominent in their formal position descriptions. We expected to see the organic pattern of work relationships, because of its familiarity from agrarian life. However, the combination of organic and egalitarian themes was even stronger than we expected.

Since there were not significant variations among the couples in spiritual images and position descriptions, we used the image of their work relationship to distinguish them. Thus, we found four couples with hierarchical images, thirteen with egalitarian images and thirty-two with organic images of their life together. The organic patterns have, however, maintained some of the traditional hierarchical elements. This was especially apparent in the tendency of husbands to control big decisions and represent the couple's enterprise to the outside world. This pattern persisted regardless of whether the couple professed hierarchical values or egalitarian ones. One wife in a largely egalitarian structure put it this way:

> I don't know, it's just that we've always, I mean I've always felt—now I know the young people don't look at it this way—but I've always felt, like, that he should make the decisions. You know, I can make them, but I would rather that he make them.

In its more organic version, the man is the public face of sales, interpretation, external contracting, and the like. The wife takes care of the inside work of financial records, office management, and internal monitoring. As a wife in a complex entrepreneurial partnership put it:

> I'm a good inside person, he's a good outside person, and once he gets the customers then they get taken care of very well.

It is important to see what others have also suspected, that hierarchy is not so much a matter of command and submission as of public and private spheres of influence. This pattern persists in our sample, though against a more egalitarian value scheme.

There was a striking regional variation in the couples' work images. Almost none of the respondents from outside the Southeast had any hierarchical components. They had fewer organic images as well, and were heavily egalitarian in their composition. This marked difference clearly shows that regional cultures are still very powerful, especially those shaping basic relational images and patterns.

When asked to describe their pattern of coworking, there was considerable preference for "team" and "director/assistant," with "council" as an important additional image. When we asked them to speak to particular role images within this overall pattern, we found great complexity in the way some couples would share a great many roles, others dividing roles up more clearly. Some emphasized the importance of common goals in dividing up work:

> HE: Rather than setting out tasks—this is your task, and this is my task . . . We don't do much about "are you doing fifty percent or am I doing fifty percent." That's not really the issue. The issue is more around our goal.

Others depend on an immediate need and a common knowledge of their whole operation.

> That is a big advantage, the fact that she knows everything that goes on and I know everything that goes on . . . I mean I can go—I don't have to worry about anything here and she can do the same thing because there's nothing about it that she don't know . . . If she's taking care of it, I leave it alone. If I'm taking care of it, she leaves it alone.

Despite the preponderance of organic images and patterns in their day-to-day activity, the *formal positions* claimed by these couples were more likely egalitarian or hierarchical. They designated their formal positions as partners, co-professionals, and frequently in terms of corporate designations (president, vice president, secretary, treasurer). In these terms eighteen were egalitarian, eighteen were hierarchical, and thirteen were organic. Thus, the public face of their work tends not to reflect their felt working reality. Officially, our public culture and official language are generally either hierarchical or egalitarian. The bureaucratic corporation and political democracy dominate our imagery. The organic reality of the *oikos* known by many people is effaced, creating considerable tension between the self-understanding of these couples and other people's perceptions of them. What is experienced as an organic division of labor that fits their own individual gifts and needs is translated into hierarchical job titles. Indeed, they themselves may be at a loss for words to describe how they work together.

The values of equal dignity and recognition, also strong among these couples, can create feelings of resentment about this imposition of hierarchical assumptions. As one wife in a strongly egalitarian partnership put it:

> SHE: From my perspective, I think there are two things at work. One is, I think, the very male patriarchal world we live in. Like, I'm the cofounder of the theater, and people refer to "Robert Barnard's Theater" all the time. Depending on how emotionally stable I am feeling, that may hurt. Like, am I invisible here? On the other hand, if you look at me as a human being, I am not particularly aggressive or outspoken, I think I am those things privately, but in large groups, I am shy. And so if I am feeling strong and confident in myself, then I don't have a problem with that. I feel like it is a complementary thing. If we were both shy, what would happen? We would both be invisible. So I think part of it is the structure and part of it is my own personal personality. So sometimes I feel bad about that and other times I say well why not, that's perfectly okay. But also . . . he had a name for himself, before I even met him. So it was like a marketing or PR tool, and frankly, I don't enjoy that, that public exposure as much.

This tension of public and private roles dominated the list of dissatisfactions we shall examine later.

Not surprisingly, the "organic" division of labor between these husbands and wives includes about twice as many hours of housework for women as for men.[10] Egalitarian couples exceeded organic and hierarchical couples by only a small margin in their sharing of household work hours. Here, as with dual career couples, women have moved into public life, but men have not moved into much of the private world.

Age was more complexly related to *oikos* images than we expected. Older couples do tend to have hierarchical images and younger couples organic ones. However, among retired couples who had taken up a new work together, we found very strong egalitarian motifs. Their couple values shaped their work, rather than work shaping their relation. In this case, their relative freedom from job expectations and job-derived income enabled them to express their underlying sense of equality more directly in their work.

---

10. See, most recently, Hochschild, *Second Shift*.

This finding offsets the impression that younger people are more individualistic and egalitarian. In fact, they may be much more oriented to building a joint pattern rather than simply fulfilling their potential as individuals.[11] Younger couples often reported that they tried to do everything together as equals before settling on a more organic but nevertheless negotiated pattern. As a ministerial couple put it:

> For us, what has been crucial here is to get some separation in job description. But early on we were sort of trotting around together as a team, and we found that that really wasn't the best use of time. And so we tried to separate out some paths, both for some good use of time and also for some sort of accountability to the profession as well.

Each age group has been shaped not only by the values of their cohort culture, but also by their current stage in the life cycle. It seems that the more intensely they are involved in their common work, the more organic their image (or, when one is taking the lead, the more hierarchical). When their work demands less of them, as in the middle years and retirement, more egalitarian values can come into play.

What Is Their Spirituality? "Spirituality" signifies the root orientations behind people's images and values. Any *oikos* pattern has a spiritual dimension in this sense. The tight or fused *oikos*, for instance, lay at the heart of most of our traditional religious symbolism. In this symbolism, the patriarchal fused *oikos* was reinforced by the religious images of Christ's rule over the church (his bride), "her" children, and the church's lands and corporation. The church also exercised control over marriage and family, through which most economic and governance power was exercised. The open *oikos,* however, tends to emphasize the direct relation of believers to God, usually through the guidance of the Holy Spirit or an immediate grasp of God's Word. This kind of symbolism has usually reinforced personal autonomy in marital, economic, and political matters.

We asked people to consider a number of classic spiritual images of relationships—such as co-creator, friend, Abraham and Sarah, Mary and Joseph, and servants of God—as well as less obviously religious ones, such as soul mates. Some of our respondents chose explicitly Biblical or

---

11. In Daniel Levinson's terms, they are climbing the ladder at this point. See Levinson et al., *Seasons.*

Christian images. Others were more general. Some people were very explicit about their spirituality, while others were more indirect.

Overall, people chose much more egalitarian spiritual images than were present in their work. Of the forty-five responding to both questions, forty-one chose egalitarian spiritual images, three chose organic, and one chose a hierarchical spiritual image; whereas eleven chose egalitarian work images, thirty-one organic images and three hierarchical ones. Almost all respondents also reported that their domestic and intimate relationships were more egalitarian than their work relationship. We also found that the more ecstatic or mystical the spiritual image, the more egalitarian the couple. It is not too surprising that an egalitarian spirituality corresponds more closely to people's intimate life and their felt equality than to work, since the privatization of religion is a signal aspect of the fragmented *oikos* of modernized societies.

The strongest link between work images and spiritual images was in the relationship between a spirituality based on rational principles and ethics and a more hierarchical and/or task-oriented *oikos* pattern and image.[12] These differences were also related to gender. More men than women (twenty-eight compared to twenty-two) checked "ethical values" as crucial for their spirituality, whereas women more frequently checked "Holy Spirit" (nineteen women to twelve men). This in turn seems to be tied to gender-based work patterns. Women who were co-entrepreneurs and couples with hierarchical and organic *oikos* images were more likely to emphasize "ethical values" as the source of their spirituality. This may occur because men tend to have the public role of dealing with decisions affecting the general public. They are therefore trained to think abstractly about values. Women, with their traditional assignment to the inner and private sphere, have a spirituality of particular affections and less publicly rational dynamics. Women also chose more "other" patterns (twelve to six), reflecting, we think, the possibility that their creativity has been channeled into cultural and religious realms, since they were traditionally excluded from the world of public work. Thus, they tend to be more sophisticated and innovative in spiritual matters.

What Were their Satisfactions and Dissatisfactions? By far the biggest satisfaction in working together was the cultivation of greater mutuality (seventy-one percent). Couple values are decisive here. Working together, as some couples stressed in interviews, made them respect each

---

12. This connection between rational ethical principles and task-oriented images would surely not have surprised Max Weber. See Weber, *Protestant Ethic*.

other more and discipline themselves for better communication. This communication was also a source of professional support. In the words of a ministerial couple:

> HE: . . . this is a new form of ministry, and we're individuals, but we're more than the sum of each person separately . . . I've learned a lot about myself by being in partnership. So it may be that first we have gained a partnership and in that we have the space and the freedom to explore who we are.
>
> SHE: Well, it's [a big advantage] having a trained spouse that you could go [to] and either with a sermon or a pastoral care issue, who understands, who's had the same training and is a colleague and you could go and say, "I need help with this issue. Help me brainstorm about this part of my sermon." That's wonderful to be able to have that support and somebody who understands ministry . . .

The focus on the couple that we have already seen is also apparent in the satisfactions they report. Only two couples put high regard on creating something to pass on to their children. In fact, the more egalitarian the values of the couple, the fewer children they had. For a number of couples, however, involving their children in their work was an important way of passing on family values and skills for dealing with the public, even at an early age. Some also emphasized that their work pattern enabled them to integrate their lives more satisfactorily than was possible in the corporation.

This very focusing of the *oikos,* however, requires extra attention to differentiating work and family more consciously. Some couples reported that, though they were able to pursue the goal of having more time together, they also found themselves forced to have too much time together. They were often searching for an optimum balance between togetherness and apartness. Fourteen of the couples reported dissatisfaction over the relation of the private and public sectors of their lives. Being equally involved in everything tended to break down sectoral boundaries. In a tight or fused *oikos,* those boundaries tend to disappear. Many of these couples were actively engaged in redrawing them, establishing clear rules about work and family spheres. For many of them work problems easily seeped into every conversation and activity. Listen to how one couple developed this awareness:

SHE: I know our son used to fine us a quarter at the dinner table if we would bring up anything about our patients. I think that really helped us because . . . he wanted his time and he didn't want all these people floating into it. But I think even before that we were working pretty hard to get distance between the office and the home . . . I think that was more of the case when we were less sure of ourselves professionally, so we were seeking more assurance personally about it. I think that as we have become more confident and skilled professionally it is, like, "I don't need to deal with this." You can put those boundaries around it, and look the other way and not worry about it.

The advice books on working together spend a good deal of time emphasizing a self-imposed differentiation between the two spheres, something these couples had to develop for themselves. While this *focused oikos* centers on the priority of the couple relationship, it still experiences and needs considerable differentiation.

A number of couples found it difficult to move back and forth between the organic/hierarchical values of their work endeavors and the egalitarian values of their spiritual, emotional, intimate, and domestic life. This applied to their relationship as couples, as well as to their roles as parents. Many felt that women kept lower boundaries between these two spheres, and operated out of a single image in their various roles. As one husband put it:

> Mary has always been the love center of whatever we did. Whether or not she even worked there, she was always the one who understood the human dynamics that would play in the company. And we have always—and in each company we did it better—created a family environment. It was an extension of our family.

Separating out or integrating these dynamics is even more complex in the two-generation family business. In the words of one father in a family firm:

> This is probably the toughest part of a family operation. We have two sons here in the business, and the final responsibility of this operation lies with me. You can't have but one manager. You can't have but one person who is guiding the situation. Now at the same time, one has to try to keep a balance. I have a relationship with my wife here at the office. and I have a relationship with the boys. But here at the office, it's not a cold business relationship,

but in the end it does have to be a business relationship, and I can't let emotion or let my heart get too involved sometimes with dealing in a businesslike way with her or with the boys. Now whether this is inherent in a woman or a mother, she probably is a mother first. And that may not be exactly fair, because she is so objective when it comes to numbers. But it does get sort of wrapped up, mixed up together, it's hard to separate. We have two sons. We love them the same ... But I have to consider them as employees, at a certain point, rather than just sons. It comes to certain situations where I can't treat them equally.

Here we can clearly see the complex dynamics involved in balancing hierarchical work images, egalitarian couple dynamics, and parental care. Each couple in such a situation has to develop their own pattern for dealing with these points of potential strain as well as mutual reinforcement.

"Couple stress" was a reported dissatisfaction from nineteen of the respondents. Hierarchical patterns tended to increase this stress. This may be due to the stress of the more financially oriented pursuits of these couples. It may also arise from the stress of hierarchy itself, which typically leaves the dominant person feeling lonely and the subordinate person resentful. Some of this tension may well result from the enormous demands of their work, with some couples reporting sixty to sixty-five work hours per week. In other cases it seems to have occurred because they tend to become isolated from non-work friends and pursuits. The *oikos* patterns of these couples are often so demanding (especially for the entrepreneurs) that they seem to exclude many other engagements. Involvement of children has to be a very conscious part of the enterprise, or else the couple's work tends to cut heavily into parental time. While this tension was recognized by many couples, they were generally unwilling to adjust their *oikos* to overcome it.

IMPLICATIONS FOR THE WORKPLACE

The modern corporation required a radical separation of work from family. Rules against nepotism crystallized this separation. Work would be ruled by rational, nonfamilial norms. The family would be governed by love, emotion, and particular bonds. While increasing numbers of corporations have had to abandon explicit commitment to their anti-nepotism policies (they violate EEOC regulations against gender discrimination), actual changes in employment are minimal.[13]

The primary reason for inertia at this point is because work is still defined in rational units performed by individuals. Job-sharing in any form is still a marginal phenomenon, though its benefits have been clearly identified.[14] A number of couples spoke about the opportunity to combine their creative powers as a major reason for choosing to work together. In spite of such benefits, almost no corporations employ couples who are job-sharers.

Our inquiry reveals two possible avenues for change in this respect. One involves the professionalization of work. We did identify work-sharing in corporations employing professionals—the churches, journalism, the media. Their relative autonomy and the structure of the work make possible job-sharing by professional couples. Secondly, we found that franchise businesses not only attract couples, but in one case find them the best choice for franchise operation.[15] The more the large corporation devolves in the direction of subcontracting and franchising, the more opportunities for couples will be created.

Some of the couples developed joint work because the husband left his corporation in order to have more control over his life, especially his basic work values. Sometimes this dissatisfaction with the corporation was augmented by the couple's desire to have more flexibility in ordering their domestic and parental lives. In the face of values like these, corporations are beginning to adopt more flexible approaches to work time, not to mention career paths, as people seek a better balance in their lives.[16]

There are clearly certain niches in the economy where couples can pursue their desire to have a tighter or more focused integration of their life so that it can conform more closely to their own values. While there

---

13. For example, Pingree et al., "Anti-Nepotism's Ghost." More recently, many academic institutions have developed special contracts for couples. Other organizations are developing more refined rules to enable couples to work in the same corporation or department.

14. Best, *Work Sharing*; Meier, *Job-Sharing*, 134–147; McCarthy and Rosenberg, *Work Sharing*; Olmstead and Smith, *Job Sharing Handbook*.

15. For instance, the Norrell Corporation, a temporary employment franchise system, reported to us that, as of 1989, 28 percent of their franchises were owned by couples, with actual operation of franchises by couples approaching 50 percent. Couples are their primary source of owner-operators.

16. Best, *Flexible Life Scheduling*; Kamerman and Kahn, *Responsive Workplace*; Bureau of National Affairs, *Work and Family*; and Harriman, *Work/Leisure Trade Off*.

are ways that corporations could change to accommodate these desires and values, such change will come slowly, if at all. If it does, it will have to occur as part of a basic value shift in our culture, one which seeks to preserve personal autonomy but balance it with the needs for enduring bonds of mutual support and transmission of viable *oikos* patterns to the next generation.

IMPLICATIONS FOR MARRIAGE AND FAMILY

We have been exploring an *oikos* pattern which is focused on the couple who work together. It reflects not only the egalitarian commitments arising in the spouses' spiritual life and self-conception, but also the organic patterns arising in the way they work together. It is a couple held together not only by intimacy but also by work. This makes the marriage as much a public institution as it is a private refuge. In one woman's words:

> I think it is exciting that our relationship has its personal part, its romantic part, and all of that is also in the public and in the community. I just feel that that is good, rather than just sitting out in our little house in the suburbs someplace, and maybe we have a child or whatever, but there is no social responsibility involved in that partnership.

Thus, the civic qualities of respect and argument play a greater role here. This emphasis came from many couples in a variety of ways. In the words of a couple operating a consulting firm:

> SHE: We're very tolerant with each other. I think we're definitely two different types of people, but with great respect for the other.
>
> HE: First as individuals, you know in individual respect, human respect. And then professional respect as well.
>
> INTERVIEWER: Has that changed over the period of years?
>
> HE: I think it's matured a lot. I mean I feel like it's matured to the point where we can now disagree on something and discuss it, where in the early days if we disagreed it was too emotional and we couldn't discuss it.

In succinct form a farm couple put it this way:

> HE: Well, it's really easy. All it takes is that you love and respect one another and then you just don't have a problem.

> SHE: There is no problem. I mean I respect what he does, and he respects what I do. You know, we've been married thirty-six years and we just never have had no problems. I guess people think we're not telling the truth when we say that, but there is no problem, you know, what else can I say?

The respect of these couples is one earned through concrete achievements in the workplace. It is not the respect that derives from the status they occupy by nature of their sex or role as mother or father.

Inasmuch as the couple itself is the focus of this *oikos*, the spouses move away from the individualism of the dual-career couple. The couple's sense of "we-ness" is reinforced by their work, rather than threatened by it. This does not necessarily mean that these couples will be less prone to divorce. The values of dignity, respect, and fair play are necessary for intimate communication and tend to enhance marriage. But the enhancement gained through joint work might have only a marginal impact on the divorce rate, since workplace strains alone can also deeply aggravate weaknesses in the couple's relationship. An inquiry into these dynamics demands further research.

The couple focus we have seen here also means that this *oikos* arrangement is not primarily intergenerational. It is not a "mom and pop" operation, since parenthood is not its focus; coupling is. It is a "his and hers" operation in two senses. Each spouse is seeking fuller expression in socially and publicly recognized terms. That is its individual side. It is also "his and hers" in that the whole operation is shared. To a high degree the couple have at least psychological co-ownership in the operation. That is its couple side. They represent their enterprise to a public world. Here are just two versions of that widespread theme:

> There are two of us who are doing the same work. You don't have to pull the plow by yourself. It's like being in two places at once.
>
> So there is an advantage of working together, the fact that the two of us know what's going on and she can deal with these people as well as I can. In fact, a lot of these people call me "Mr. Ruth Burns!"

In this sense these couples evidence a significant variation on the individualism so much vaunted and maligned in American life. It is life lived in intimate relationship, not merely through direct emotional communication, but through the public instruments of work. Marriage here expresses itself not only in the bedroom, but in the board or work room as well. The couple shares a publicly recognized world, which in turn shapes their marriage. This is what gives the organic shape to their relationship. They are related to each other through their work, as well as through their personalities, emotions, and, frequently, children. In this sense, American culture may actually be a culture of couples rather than of individuals. Its purported individualism may actually mask a pervasive desire for coupling that is an even more important dynamic in American life.[17]

These couples also raise up for us a public vocation for couples that lasts longer than that of raising children. Since the historic separation of work and family, raising children and perhaps maintaining a home have been the chief means by which a couple could have what we call a marital vocation. While the home (and perhaps several pets) can last a marital lifetime, children grow up. What worldly expression and confirmation of marriage can hold people together in a world, and give them a basis for communication, struggle, and mutual appreciation? It is this need for some form of marital vocation which is being addressed by these couples. It is a need which our society does not address and indeed suppresses or trivializes. Meeting that need, it seems to us, can do much to strengthen the intimate bonds that give so much of life its vitality.

IMPLICATIONS FOR RELIGIOUS INSTITUTIONS

Christianity and Judaism continue to bear an ambiguous relationship to these developments. On the one hand, their traditional symbolism of male dominated hierarchy has reinforced the split *oikos* of the public male and the private female. On the other hand, they are also the bearers of egalitarian images of spiritual friendship and equal participation in the divine life. Both of these have been mediated by family life, and will continue to be so.

In our study we found that the more egalitarian the couple, the less involved they are with religious institutions and with traditional religious symbolism. Egalitarian couples were less likely even to respond to the religious questions. In substantial part, this seems to arise from the way churches are associated with symbols of hierarchical order. It may also

---

17. In Bellah et al, *Habits of the Heart*, ch. 4, Robert Bellah and his associates struggle with the problem of balancing the need for more individual equality in marriage with the desire for forms of commitment that might simply reinstate familial hierarchies.

arise from the perception that churches exist more to pass on traditional values to the next generation than to help today's adults, especially couples, invent new patterns for their lives. Moreover, these couples are very busy people. They have no time for the committees and volunteer work required for maintaining religious institutions. The church or synagogue does not play a big part in their life on either symbolic or organizational grounds.

The split *oikos* provided womanpower for religious life in nineteenth-century America. Today that pool of volunteers is rapidly evaporating. Today's couple needs private leisure time to cultivate the communication not possible during the workweek. Weekend religious activities compete for that time. In the words of one woman:

> I think it is really important, and I really like the quiet time at church, and the music and stimulation. But any time we tried to get involved in the church . . . you go two times and they want you to take on the youth group or you feel obligated with things like that. And we just don't have enough time. Sunday mornings is really about our only downtime. It is like we go, and just get what I want out of the situation, I don't have much left over to give back.

If churches are to find volunteers they will have to find them among retired couples rather than younger housewives. The organization of church work and activity has to adjust to people's dominant *oikos* patterns.

Moreover, these couples tend to want a more intense focus on the symbolic and worship life of the church. In worship they find a structured and insulated hour for personal reflection—regardless of what may be pronounced from the pulpit. That is, we got the impression that social relationships and community work are not what these couples want from religion. They want development and confirmation of their distinctive spiritual life. The church provides this opportunity, especially in public worship. However, few churches provide worship that draws on egalitarian or organic symbols that would be relevant to the *oikos* of these couples. They are at once more spiritual, then, but also more institutionally alienated.

Support for couples who work together offers a way that churches could meld their traditional concern for a more tightly integrated *oikos* with an equally compelling commitment to the equality arising in the call of men and women to a shared public life. By lifting up the option of combining work and family, the church can also counteract the historic tendency to confine its mission to the realm of female domesticity. By adjusting its symbolism and ritual as well as its program and personnel along these lines, it can be a place where men and women deal jointly with the interconnected dynamics of work and family. That twenty-one percent of our sample are clergy couples may indicate a skewed sample, but it also may indicate that the churches are one of the few places where this *oikos* pattern can be sustained. It may also mean that this is a group who will inevitably recast religious symbolism in their direction.

The question is whether religious leaders will shape religious symbolism to appeal to the organic and egalitarian patterns of the couples we have met, or to the hierarchical ones pervasive in the corporations maintaining the split or fragmented *oikos* of the last century. It may well be that when religious groups have sought to be more egalitarian, they have actually reinforced the individualism of separate spousal careers, with its attendant strains, rather than the organic bonds of couples seeking to integrate work and family more closely. Church-sponsored daycare centers, for instance, may simply support the dominant split between work and family, just as most marriage enrichment programs ignore the work-family interface. This is not to say that the *oikos* split between public men and private women is the only alternative to these arrangements. Here we have tried to explore the lives of people taking another route. At this point, religious institutions need to move beyond the individualism of the old, split *oikos* and its modern egalitarian counterpart, to legitimate other patterns that honor both the equality of people and their need for deep relationship in marriage and work.

12

Human Rights *in the* Church

The advancement of human rights in civil society has been fostered as well as blocked by Christian churches. This ambivalence in the civil arena is mirrored within the churches as well. In this essay I want to explore the relationship between legal processes within the churches and generally recognized human rights norms. From the outset it is clear that there is a variety of patterns in the life of the churches around the world. While we see clear efforts to suppress free speech in some churches, we also see agonized efforts to ensure representative governance and democratic decision-making in others. First, we need to understand how and why these various patterns arise. Second, we need to ask what we can learn from this encounter between the human rights movement and the churches.

FRAME OF ANALYSIS

In order to analyze this complex situation, we can begin with a rough ordering of the factors shaping the way churches do or do not embody human rights norms in their internal life. We can then review the actual situations in the churches with these factors in mind. After reviewing some of these key areas of controversy, we can then return to this preliminary framework and refine our understanding of how and why these human rights issues arise and what they portend for the future of human rights.

Four fundamental factors seem to shape the way churches deal with human rights issues in their internal life: (1) the degree to which they have a codified internal law; (2) their ecclesiology, that is, their understanding of their purpose and proper internal structure, (3) their relation to environing governments; and (4) their relation to their environing culture.

Church Codification of Law. Churches vary greatly in the degree to which they have developed a complex internal legal structure. The Roman Catholic Church has the most highly developed tradition of Canon Law, stretching back at least to the codification of Roman canon law in the eleventh and twelfth centuries.[1] All the European churches with a history of state sponsorship or close affiliation have an extensive corpus of law as well, either because of their internal size and complexity or the accumulation of legal agreements with their governments.[2] Even churches with no direct state ties have often developed very sophisticated bodies of internal law and legal procedure.[3] At the other end of the spectrum lies a multitude of smaller churches with little codified law, whether because of their theology, their charismatic mode of leadership, their parental pattern of governance, or their embeddedness in a subculture governed by tradition and custom.

Clearly, the way in which one might identify or evaluate human rights performance varies enormously depending on whether the church has a clear law and procedure. Informal or traditional churches may or may not conform to human rights norms. Highly legalized churches might have clear human rights codes on paper but fail to follow them in practice. In any case, since human rights norms generally have pushed for expression in covenants, constitutions, and codes, there is a tendency to assume that codification is good and that it reveals to a significant degree the extent of the church's conformity to human rights norms.

Type of Ecclesiology. The second factor is the type of ecclesiology shaping the church's internal life and external mission. A church's

---

1. For the current Code of Canon Law see Canon Law Society of America, *Code*. For a critical response see Provost and Walf, *Canon Law*.

2. For instance, for Germany see the whole history of *Staatskirchenrecht* (State Church Law) in Eichstaedt, *Kirchen als Körperschaften*; Dombois, *Recht der Gnade*; Quaritsch and Weber, *Staat und Kirchen*; Hesse, "Entwicklung"; Friesenhahn, Scheuner and Listl, *Handbuch*; and Listl, *Grundriss*.

3. See the United Methodist *Book of Discipline*, which receives some alteration every four years through actions of its General Conference. The Presbyterian *Book of Order* exhibits a similar pattern.

ecclesiology is shaped by theological principles, ethical values, and the cultural history of that church. Ecclesiology is the structural expression of that church's beliefs. The wide variety of ecclesiologies among Christians often surprises outsiders. That these ecclesiologies make a theological difference is often seen by insiders as a perversion of the Gospel, with its apocalyptic critique of the institutions of Jesus's time. For these Christians any institutionalization of Christian life is suspect, if not anathema. However, the primordial fact that the early church preserved four Gospels and was purportedly constituted as twelve apostolic churches embeds this plurality and at least primitive institutionalization in the church's very beginnings if not in the will of its founding Rabbi.

These plural ecclesiologies have been variously classified. Some theologians, working in the tradition of Ernst Troeltsch, root their typology in the relation of the Church to its culture. Others, such as Avery Dulles, seek it in terms of the church's understanding of its own highest value. Still others, such as Peter Rudge, anchor their typology in the affinity between theological categories and choices of organizational structure.[4] Each of these classification schemes illuminates different aspects of ecclesiology and helps explain differing facets.

In light of the issue of human rights in the church I would like to suggest that we ground our ecclesiological typology in the pattern of conflict over structure that goes back to Israelite origins and that was rehearsed in secular form by John Locke at the beginning of the modern movement for human rights. Namely, it is the conflict between patriarchal monarchy, on the one hand, and covenantal conciliarism on the other. In ancient Israel it was the conflict between the confederation of Israel gathered around the traveling Ark of the Mosaic covenant and the monarchy of David centered around the temple in Jerusalem. These very different conceptions of proper governance carried with them very divergent understandings of the rights of the people, differences which still reverberate in our debates.[5]

This conflict exploded again in the European recovery of this biblical dialectic in the sixteenth century. John Locke recast it in a partly secular form in his *Second Treatise on Government*, which was a trenchant attack on the whole ideology of patriarchy and a defense of the republican

---

4. Troeltsch, *Social Teachings*; Niebuhr, *Christ and Culture*; Rudge, *Ministry and Management*; Dulles, *Models*.

5. For a penetrating exploration of the Biblical roots see Elazar, *Covenant and Polity*. For further Protestant developments see Stackhouse, *Creeds*.

ideals rooted in Israel's covenantal and conciliar polity.[6] The individual liberties necessary for participation in these emerging republics became the core of what the American and French revolutionaries later called human rights.

Locke's *Treatise* was a brilliant polemic which condemned any polity modelled after the patriarchal family in order to make way for the emergent republican aspirations of his time. However, humanity's ongoing need for familial care, socialization, nurture, and security have gradually brought back a modern form of the patriarch (and in this case matriarch) in the form of the welfare state and the "rights of entitlement" to care and sustenance, if not through limited associations then through the state. In this situation the "second table" of social, economic, and cultural rights can be set alongside the "first table" of political and civil rights as modern expressions of this tension between parental care and adult liberty.[7]

In this climate it might be possible to construct a simple ecclesiological typology on the basis of this distinction between "parentalist" and monarchical ecclesiologies, on the one end, and covenantal conciliar types, on the other. How a church manages these two values in its ecclesiology shapes how it looks from the standpoint of a theory of human rights. But this theory of human rights is, in turn, critically illuminated in terms of how it relates the rights contained in these modern "two tables" of civil rights and socioeconomic rights.

Relation of Church and Government. The third essential factor affecting the relationship of the church and human rights is the relation between a particular church and the governments of its civil environment. Many churches are transnational and so have numerous environments and governments to deal with. The church's search for internal consistency on a global basis perforce collides with the plurality of national laws and the way they may or may not incorporate human rights norms. To some extent these relationships arise from the church's ecclesiology and patterns of internal governance. To a large extent they arise from

---

6. Locke, *Two Treatises*. For the theological dimension in Locke, see Dunn, *Political Thought*. Republican readings of the Bible emerge in Harrington, *Oceana*; Milton, "Readie and Easie Way"; and the famous "Putney debates" among Cromwell's soldiers. See Woodhouse, *Puritanism and Liberty*.

7. The United Nations Universal Declaration on Human Rights (1948) was followed in December 1966 by two U.N. Covenants spelling out political and civil rights (my "first table") and economic, social and cultural rights (my "second table"). The parallel with the two tables of covenantal law in the Bible draws me to this language. For the documents and a Reformed Christian response see Miller, *Christian Declaration*.

the constitutions, powers, and interests of governmental controls of their legal and political environment.

In some regions, the churches are essentially "private" associations entirely separate from the civil law. The whole construct of human rights, which presumes the relatively involuntary environment of the civil state, may not even be valid for such voluntary associations. The right to join or leave a voluntary church preempts all other human rights claims within that church.

In other regions and churches, the church may be under the control of the state to some degree, as is still the case in many parts of Europe. Here, human rights in the church quickly becomes a matter relating to state policy. Within these situations of close accommodation between states and churches, churches may be "established," as in England, Norway, or Sweden, "public corporations" with a privileged status in law, as in Germany, or be governed by concordats with the state, as is the case with the Roman Catholic Church and Germany, Italy, and other countries. In this last case, the church seeks to be seen essentially as another state rather than an entity in any way subordinate to national law. How we view human rights in these churches will thus have a different cast, since their legal contexts of enforcement and the degree to which membership is voluntary vary substantially.

Relation of Church and Culture. Lastly, both the churches' ecclesiologies and their relation to governments is shaped by the dominant cultures and circumstances of their regions. Some churches find themselves in an overwhelmingly dominant position in a culture. In this situation of close accommodation or assimilation, the values of the dominant culture permeate the churches, shaping their appropriation and implementation of human rights norms. In other cases, they may be in an acute minority, whether persecuted, as in China of recent years, or preferred because of governmental interests in redressing past grievances, as with some Christian tribal peoples of India. The dominant culture shapes not only whether human rights are appropriated but also how the relation between the rights to liberty and the rights to care are balanced and integrated. The degree to which church membership brings social benefits or liabilities thus shapes the extent to which one may press for the exercise of certain rights within the church.

CONTEMPORARY CONTROVERSIES

These four factors—the degree of legal development within the church, the type of ecclesiology, the church's relation to the state, and the cultural context of the church—help to orient the investigation of certain key human rights controversies within the churches. My investigation of these controversies is perforce illustrative, not exhaustive. I have not examined all Christian churches with a checklist to see how far they have incorporated human rights norms into their internal life. Nor have I surveyed the ways churches may have incorporated these norms without much controversy. Instead, I focus on the principal controversies over church discipline, governance, and administration that have divided church authorities and subjects—with illustrations drawn principally from India, Germany, and the United States. All four of the factors introduced in the foregoing section bear on the formulation, and resolution, of these controversies.

Church Discipline. Inasmuch as Christians are disciples of Jesus, seeking to follow "the Way" he set out originally, they are committed to some sort of discipline. The question is: Who interprets the discipline, and who enforces it? Churches vary widely in this respect, with some ecclesiologies supporting a clear central teaching authority and others leaving this interpretation to councils, congregations, and individuals themselves. The more this authority is removed from the immediate control of individuals the more questions of human rights may arise.

Membership. First, most churches with any degree of institutionalization must establish the limits of membership. Thus, they must reserve the right to dismiss members and to rule on their admission. Similarly, they tend to develop methods of behavioral control short of outright excommunication in an effort to redirect wayward members back to the path of discipline espoused by that church.

Very few people question the churches' need to discipline their members. Questions first arise, however, when that discipline carries with it civil or social penalties, especially where the church is closely tied to the state and the culture. Church discipline can damage people's reputations, social standing, or even eligibility for state-funded offices, as in European churches supported in some way by state funds. But this is not yet a question of human rights, unless the procedures followed in bringing to bear disciplinary sanctions violate human rights norms.

Assessment of questions of "due process" vary from region to region. American law differs from that of England and France. India has restricted its use by Constitutional means[8] and other countries have other legal traditions which could still fall within general human rights norms. Some of the most famous cases in this regard have arisen in the Roman Catholic Church, whose central authorities have removed permission to teach in Catholic institutions from Stephan Pfürtner in Switzerland,[9] Hans Küng in Germany,[10] and Charles Curran in the United States.[11] The Dutch theologian Eduard Schillebeeckx also underwent disciplinary scrutiny.[12] Pfürtner, a Swiss Dominican, was deprived of his teaching duties at the University of Freiburg, because he raised in public profound questions about the Roman Catholic Church's approach to sexual ethics, including questions of birth control and homosexuality. Hans Küng, a leading Catholic theologian at the University of Tübingen in Germany, was subjected to numerous investigations by the Vatican offices during the decade following Vatican Council II. These resulted in the termination of his permission to teach Catholic theology in 1979, thus depriving him of his position at Tübingen. Charles E. Curran, a respected American theologian, was deprived of his permission to teach Catholic theology at Catholic University in Washington, after a seven-year investigation of his teachings, especially those concerning sexual ethics.

In all of these cases the procedures themselves were characterized by a high degree of secrecy, though documents from them later emerged in the general public. Moreover, since they were ideological in character, the normal legal procedures involving guilt and innocence did not or could not come into play. Thus, we have a question of whether questions of truth, which lie at the heart of theology and the ecclesiology growing out of it, can be dealt with in the same way as questions of guilt or innocence, which are the general concerns of a court of law.

Freedom of Speech in the Church. Do human rights norms require freedom of speech in the church, even by those who are employed to teach a received tradition? Some churches give wide latitude to freedom

---

8. Austin, *Indian Constitution*, 73, 101.

9. Kaufmann, *Ein ungelöster Kirchenkonflikt*; and Ringeling, "Fall Stephan Pfürtner."

10. Swidler, *Küng in Conflict*; and United States Catholic Conference, *Küng Dialogue*.

11. May, *Vatican Authority*.

12. Swidler and Fransen, *Authority in the Church*.

of speech, asking only that officials declare clearly whether they are speaking for the church or for themselves. For others, the whole person is absorbed into his or her office, leaving no room for such a distinction. Like soldiers, it is claimed, officers or teachers in the church have given up certain ordinary rights in order to carry out their assignments. Churches with a clear internal distinction between the public and private aspects of life can make a distinction between official and personal statements, but churches for whom a familial model is dominant have difficulty recognizing such a distinction, just as parents do not recognize it in dealing with their children. In such a household model, there is no internal public where personal opinion can be given voice.

Many churches have experienced controversy of this kind. In the most notorious recent American case, many professors and administrators in schools controlled by the Southern Baptist Convention have been disciplined or dismissed on doctrinal bases.[13] This has resulted in disciplinary action from educational accrediting agencies though not from governments. Here again it is clear that the claim that human rights norms have been violated depends on holding an ecclesiology that maintains that the church itself, including its teaching offices, should be a free public similar to that envisioned in most human rights doctrines. However, those very doctrines of human rights would also contravene any requirement that churches must have such an ecclesiology. This would be a denial of religious freedom for religious groups and institutions. Moreover, it does not follow that the religious rights of individuals should be enforced by authorizing governments to intervene in those procedures to conform them to civil procedures incorporating human rights norms. Such a policy of intervention would reduce religious freedom solely to the beliefs and actions of individuals.

Confidentiality. A third area of discipline involves matters of confidentiality and pastoral care. Here no one is asking that matters of counseling and confession be public. The sphere of personal confession and counseling is inherently a private matter, as both churches and governments widely recognize. The point of contention arises when people wish to bring civil actions against churches for harms they allege to have suffered or might suffer in this confidential arena. This involves both the question of revealing evidence of child abuse as well as suffering from counselling which departs from acceptable professional norms.

---

13. Ammerman, *Baptist Battles;* and Ashcraft, "Southeastern Seminary."

For Americans, this was epitomized in the case of Ken Nally, whose parents sought to sue his church for being criminally responsible for his suicide.[14] Nally received counselling from the church's professional staff, who operated under the church's theological insistence that psychological distress is rooted in the person's sinfulness. Therefore, repentance must precede healing. Nally, his parents argued, was driven to suicide by this insistence, which only increased his guilt and self-hatred. The church claimed that it could not be prosecuted under secular counselling norms, because its practices were protected aspects of its religious freedom. But the church's resistance to being sued could also be seen as an interference with the rights of Nally's parents to their day in court. American courts sided with the church, thus leaving a zone of freedom in which churches could both secure important human goods, such as spiritual growth, but also wreak unknown harms.

Most obvious among these harms are violations of ordinary norms of sexual behavior. Here many churches face increasing pressure to follow ordinary civil procedures in handling such cases or to face direct civil action regardless of any claims to immunity from state interference in these confidential arenas.[15] The Roman Catholic Church, the Episcopal Church, and other large historic American denominations have all taken steps recently to conform their laws and procedure to the civil models in the United States.

Church Marriage Law. The final contested arena in church discipline involves church governance of marital and family relations. For centuries, marital relations in Europe were governed by ecclesiastical law linking the marriage contract to the sacramental order of the church. While this legal arrangement may once have been an advance over earlier familial control of marriage, it eventually was attacked first in the Reformation and then in the republican revolutions of the eighteenth and nineteenth centuries. Gradually, the laws of marriage and family, including matters of inheritance, were removed from church control and placed in the hands of courts and governments, which consider them secular matters.

Today, any interference in marriage and family matters by churches seeking to maintain their own control can be seen as violations of rights that people have been guaranteed by the state. In secularized North Atlantic countries, human rights would imply the freedom to marry and

---

14. *Nally v. Grace Community Church of the Valley*. See Battin, *Ethics*.
15. Clark, Jr., "Sexual Abuse"; Fortune, "Imitate the Navy."

divorce apart from all church or familial control. However, in regions where the state recognizes church law pertaining to marriage, the state's interference could be seen as restrictions on the freedom of religion, that is on the church's freedom to control marriage as a religious matter.

In India, for instance, there exist separate "personal codes" for each religious group, with an optional state code for those who seek it because of religious intermarriage or for personal reasons.[16] Here the clash between the human rights applicable to marriage and family and church law becomes sharpest. Since all these church codes reflect an earlier patriarchal model of family life, they come into direct conflict with women's rights. In two cases, that of Shah Bano in Madhya Pradesh and Mary Roy in Kerala, Indian courts encroached upon discriminatory personal codes in favor of democratic values, only to produce a storm of protest.[17] In both cases the passionate support for these traditional personal codes reflects a situation in which the formal aspects of religion are embedded in a dense communal culture which finds its expression in terms of religious law. For the church or other religious organizations to introduce human rights values into their law requires that the whole sub-culture move with them.

In England, on a very different plane, the Anglican Church's law on divorce has deeply affected the life of the English monarchy. Should the church's law govern who is eligible to become the next English monarch or should the English people (and their monarchs) be entitled to follow a secular norm compatible with most human rights doctrines?

These cases illustrate that the way the church is or is not embedded in a wider community deeply affects how one approaches the role of human rights in the church. Either one presses to introduce human rights norms into a church which is deeply embedded in the society, or one seeks to separate the church from governance over family matters in

---

16. Devadason, *Christian Law*; Mathew et al., *Cases and Materials*; Pinto, *Law of Marriage*; Derrett, *Death of a Marriage Law*; Khodie, *Readings*; Khory, "Shah Bano Case"; and Mansfield, "Personal Laws."

17. *Mary Roy v. State of Kerala* and *Mohammed Ahmed Khan v. Shah Bano Begum and others*. The *Shah Bano* decision precipitated riots leading to a parliamentary act overriding the court's decision. The *Roy* case had less violent, though still disturbing consequences. For Shah Bano, see Engineer, *Shah Bano Controversy*. For general discussion see Everett, "Religion and Constitutional Development"; Lawrence, "Woman as Subject." On inheritance matters (the *Roy* case) see Kishwar and Vanita, "Inheritance Rights."

order to resolve the conflict. Both directions imply a general model of social order as well as an ecclesiology.

Governance. Churches must also control who is allowed to set forth authoritative church teaching and carry on the general leadership of the churches, whether in congregations, missions, or schools. Some churches, such as those in the United States, are completely autonomous in their selection of leadership. Others must seek permission or confirmation from civil authorities, as with the Anglican Church in England, the Church of Sweden, and the Church of Norway. Less formal interference from state bodies in other countries can also affect these choices. Independent free churches are being pressured to democratize their selection procedures and to open eligibility not only to women and members of minority groups but also to homosexual persons or others who deviate from social norms but are otherwise prepared intellectually and spiritually for these positions. Major movements within the Roman Catholic Church have sought for some years to effect such changes, albeit with little success.[18]

Church Administration. The final arena of controversy involves persons employed in the church in a non-authoritative position. These have traditionally been the secretaries, custodians, administrators, and workers who run the institution on a daily basis. More recently, especially in regions experiencing a decrease in clergy, they have also held teaching and high-level administrative positions. The thorny question is this: At what point should these employees lose rights they would have in secular employment simply because they work in a church trying to pursue a religious purpose? For instance, should they be able to form labor unions? American courts held in at least one case that church rights should take precedence over the right to form or join unions.[19] Absent labor unions, should employees still experience the same rights of liberty and care applicable to the general population? Here we reach difficult problems not only of how we construe human rights but of practical possibility, in terms of both resources and enforcement.

Efforts to extend ordinary interpretations of human rights to church employees have stressed the distinction between what is essentially religious in ecclesial institutions and what is secular. This implies that the "religious" aspects should be handled by clergy and the "secular" aspects

---

18. See Coriden, *We, The People*; Coriden, *Case for Freedom*; Coriden, *Sexism and Church Law*; Coriden, "Human Rights in the Church"; Swidler, "Demokratia"; McCann, *New Experiment*; Provost and Walf, *Tabu of Democracy*.

19. *National Labor Relations Board v. The Catholic Bishop of Chicago, et al.*

can be handled by lay employees. This effort to drive a wedge between religious and secular realms within the church and between clergy and laity in church employment runs against the church claim that, regardless of ordination status, all employees are essential to its mission, which is a religious matter. The effort to extend rights to lay employees implies an ecclesiology which distinguishes between "secular" and "religious" dimensions of the church as institution, while the rejection of such efforts implies an ecclesiology which sees them as indivisible parts of one cloth. In traditional theological terms, the first ecclesiology is a "docetic" theory that separates the divinity and humanity in Christ, whose body is the church. The latter ecclesiology implies a highly incarnational theology in which these two dimensions are thoroughly merged in the church. Once again, the two camps are divided according to their theology and ecclesiology as well as their weighing of individual and institutional rights.

While many churches around the world are pressured to adopt the human rights norms established by local civil governments, European churches, which are closely related to the state, face a rather different pressure. For instance, German pastors are by law entitled to the same level of compensation and care as state officials.[20] Moreover, military chaplains in many countries have a governmental status that often involves privileges they would not enjoy in a normal church setting. In this case, the desire to follow a higher Gospel ethic leads many people to want to eliminate these privileges—privileges which to others might look like ordinary rights to liberty for soldiers or entitlements to care by church employees.[21] Whether these conflicts between a Gospel ethic and individual rights can be called human rights questions leads us back once again to the question of ecclesiology and normative models of church-state relations.

TOWARD AN UNDERSTANDING OF THE DIFFERENCES

Having identified and ordered a number of cases of controversy, we can return to our preliminary list of factors shaping the question of human rights in the churches. We can explore their relative weight and how they interact to shape particular situations. Hopefully, we can isolate with

---

20. Bock, *Das für all geltende Gesetz*.
21. Martin, *Frieden Staat Sicherheit*.

more precision the key issues at stake in raising this fundamental question in the first place.

Degree of Internal Legal Development. The degree to which a church has a highly developed legal system does not seem to be a crucial variable. The Roman Catholic Church, with probably the most sophisticated canon law of all, is frequently involved in controversies over human rights within the church. Conversely, churches such as the Society of Friends have virtually no distinct legal structure, and yet they are almost never involved in these controversies.

What seems apparent is that legal development is shaped by theological convictions as well as size and length of history. The New Testament contains several passages attacking involved legal recourse, either inside or outside the church. A church's decision to have extensive legal forms involves theological decisions that require law, regardless of the claims of biblical origins. Neither course touches the question of whether human rights is protected or threatened in such a legal development.

What is clear is that to introduce more of a sense of individual rights into church life requires the elaboration of judicial processes to advance and protect such individual rights. If the church is to resist this legal approach to incorporating human rights, it has to discover some other credible way to guarantee human rights in the church. It may restrict the size of its organization, subject all decision-making to saintly judges or special councils or discover some other way. This choice returns us to questions of ecclesiology.

Ecclesiology. Ecclesiology is the bridge between fundamental theological values and practical organization. On the one hand, it is an expression of these values and, on the other hand, it is an expression of models of organization found adequate in the surrounding culture and adopted for church governance. In the case of human rights, Christian ecclesiologies can be seen on a spectrum between familial and conciliar models.

The familial model takes the parent-child relation as fundamental to human flourishing as well as to eventual perfection or redemption. Its dominant form not only in the West but in the East and South has been patriarchal and monarchical. The father rules and provides for the household. The wife and children obey and receive. When extended beyond the family as a model, it implies monarchy, as Robert Filmer and John Locke both knew.

This familial model stresses the values of care, control, nurture, transmission of heritage, and safeguarding of "the flock" (to draw on a related metaphor of care). When rights talk enters this model, it stresses the rights of the members to receive that care, to sustain their membership, to be fed spiritually, and to receive what they need for their spiritual growth from the church. Both Catholic and German Protestant formulations of "Rights in the Church" have stressed this set of values, though on distinctly different grounds.[22] Behind both of these traditions lie strong parental and familial models of the church. For the Roman Catholic Church, this entails rule by the Pope, the "Papa" of the church. For the German church tradition it lies in the symbol of the "Landesvater," who combines the elements of rule and paternal care. In this model, church membership has about the same degree of voluntariness as does membership in a family. In this kind of church we are claimed by Christ and can secure this relationship only within the church. Thus, claiming adult participation in the church takes on much of the urgency of human rights demands in the relatively involuntary confines of membership within a national state. The claim of human rights participation is not trumped by the freedom to leave that church and join another.

This focus of rights talk echoes what I have called the rights to care in the "second table" of human rights—the economic, social, and cultural rights. Here we find not only the rights of individuals to fundamental security but also the rights of natural groups normally associated with providing this care, above all the family. What is at issue in these models is how the rights to care will be squared with the rights of these natural groups to preserve themselves and their traditions, even at the expense of any particular individual's right. The rights of women are frequently held in the balance within churches as well as in civil orders reflecting this model.

What is important to recognize is that from the inside of such an ecclesiology the rights question is largely one of serving the right to be

---

22. For instance, although Wolfgang Huber and Heinz Eduard Tödt, in Huber and Tödt, *Menschenrechte*, 204–8, stand clearly on the conciliar side of this debate, their list of ecclesial human rights begins with "access to faith." Arguing out of a German Protestant tradition, they emphasize that human rights values of participation, freedom, and equality have to be transformed theologically to find appropriate articulation in the church. Thus, for instance, freedom and equality have to be construed within the theological frame of neighbor love. For a Roman Catholic approach, in which sacramental participation historically plays a more important role, see Coriden, *Case for Freedom*, 12–14, and Coriden, "Human Rights in the Church," 72–74.

nurtured by and within the group—the extended family of faith led by a father imbued with the spirit of the ancestral tradition. This is a particular view of the self, its rightful claims, and also of the nature of faith and of the church.

The conciliar model takes as its point of departure the relations between adults in the public. It is rooted in covenants and contracts among adults rather than in the parental care invoked in familial ecclesiologies. When it comes to matters of human rights, it naturally focuses on the table of liberties necessary to enable people to engage in public argument and agreement. The fundamentally religious claims of the church arise more from the agreement among the people than from the tradition conveyed by the fathers of the church.

In this conciliar ecclesiology, our fundamental "humanum" is our adult capacity to enter into public friendships, arguments, and agreements with an inherently universal coverage that exceeds by far the bonds of family, ethnic community, and race. Salvation is not so much a matter of parental care by those who are wiser and more powerful; it is the illumination of conscience and its expression in the midst of others who can confirm, challenge, and expand our understanding. Freedom of speech is central to this ecclesiology just as admission to the nurturing grace of the sacraments is central to the familial model.

Obviously, this latter model lies much closer to the rights of liberty stretching back to Locke. The readiness of churches shaped by this model to embrace the whole contemporary panoply of human rights seems largely determined by how much distance they want to maintain vis-a-vis their environing societies and governments. Their commitment to separation from the state may occlude their conformity to many human rights norms, as in the present battles over ordination of homosexual persons. This decision is shaped not only by other theological factors such as their eschatology but also by cultural factors such as their relative privilege or deprivation in the social order.

The Relation of Churches to Governments. Regardless of where churches find themselves on this ecclesiological spectrum, they must address the question of their relationships to governments. Here the crucial issues revolve around their degree of autonomy from governments and other powerful institutions, such as the business corporation or stock market.

First of all, who is accountable for securing these rights, or adjudicating their abridgement, however they might be construed? Is the church the sole judge? Are the civil courts or other governmental agencies? What would it mean to say that "humanity" or "God" is the judge? Churches differ sharply on this point. Some are willing to leave wide ranges of judgment to the civil courts, as when Protestant churches by and large leave matters of family law to them. German Lutheran churches have historically left much of church administration to the state, though that tradition has been substantially criticized since the Nazi era.[23]

By keeping some claim to self-determination in this judicial realm, the churches counter the often submerged tendency to make the nation state the guarantor of human rights. That is, they take common cause with the claim that no institution can be entrusted with securing these rights. Only a society with genuine institutional pluralism of power can do so. This discussion of the political conditions for human rights cannot be followed here, though it is important to see how it intersects the subject before us.

If the decision is internal to the church, where should it reside? Does that choice reflect an ecclesiology as well as a theory of human rights? A familial model places these judgments with the parental figures in the church—those who bear the spirit of the ancestral deposit of faith. Those at the conciliar end ground it in the public debate of the assembly, often reduced to some form of public court.

On these points the human rights doctrines themselves exhibit a certain openness (or opaqueness), for they cannot give us detailed answers to matters of legal due process or the relative weighing of rights either between or within the two tables. The question posed for churches with both ecclesiological models is how they order their life to deal with these questions.

The second way the autonomy question arises is in ascertaining the degree to which ecclesiological models should conform to civil or public models of decision-making. Controversies over selection of leadership or over unionization rights bring us back to this point. Here we reach a fundamental dilemma, for the effort to change the practices of these churches to fit general human rights norms runs directly against other human rights provisions which defend the rights of churches to preserve their own beliefs and traditions—even if they violate human rights norms. But this is the rub: Are human rights really universal, or are they restricted

---

23. See the Barmen Declaration of 1934 in Cochrane, *Church's Confession*, 237–42. For an example of recent reflections on the implications of Barmen, see Burgsmüller, *Kirche als "Gemeinde von Brüdern,"* vol. 1.

to certain institutional arenas, namely the civil public? If restricted, then they are better called civil rights. If universal, then they deny the rights of persons to form associations that may dissent, indeed radically dissent, from the accepted norms of society.

This inherent tension in human rights doctrine cannot be pursued here, but it does point to the way that human rights doctrine has its own social, historical, and cultural location. Human rights doctrine has arisen to contest the patriarchal, monarchical, and feudal exclusion of persons from public life. It therefore is biased in favor of individual rights rather than the rights of associations. When religion is strictly a matter of individual belief and conduct, it will protect it. When it is a matter of institutional autonomy, it will seek to democratize that institution to reflect the democratic model of governance assumed in human rights doctrines. The only way these doctrines can remain internally non-contradictory is to remain at the level of the individual, leaving the state, in fact, as the only institution in the society. But this step would eviscerate most of the group rights, especially religious rights, which people would like to attribute to human rights.

This "rightful autonomy" of churches takes on different forms within different patterns of church-government relations.[24] In the "free church" pattern of the United States, for instance, churches struggle with the degree to which they can be reduced to the form and status of non-profit corporations. In a celebrated case, a California court held that the United Methodist Church is a hierarchical non-profit corporation, responsible for the actions of all its local "agents" despite clear and long established ecclesial statements to the contrary.[25] But if this direction is pushed far enough, the churches are robbed of their quality of being real publics, in which human rights doctrines are most applicable. They become a corporate version of the familial household under a hierarchical head.

In order for the churches in such a legal milieu to retain their special position as more than non-profit corporations, they must be clearly "religious" in a very goal-specific and delimited way, usually confined to some kind of cultic activity. To do more is to be involved in activity also pursued by purely secular institutions, whether for-profit or non-profit. These "secular" categories must then conform to general civil law—the

---

24. Laycock, "Church Autonomy."

25. *Barr v. United Methodist Church*, known as "the Pacific Homes Case." For discussion of the ramifications of the case see Gaffney, Jr., Sorenson, and Griffin, *Ascending Liability*.

usual medium for human rights norms. But this dissection of religious from secular aspects of the church betrays the very holistic and communal spread of religious concerns in Christianity as well as Judaism and Islam. To retain their autonomy in this situation, and with it their control over the application of human rights within their jurisdictions, they must be less than the holistic bodies they claim to be.

In situations where the churches are closely accommodated to the state, as among the German "public corporation" churches, governments may be able to introduce more human rights norms if that is the direction of governmental policy or constitutional requirement. Churches may likewise introduce their own conceptions of public civil rights norms into government policy. In Germany, high standards for employee treatment are shaped by government policy, while new understandings of the role of the military in society are being shaped by church initiatives in the wake of the recent German unification. Here the churches are made increasingly public but at the peril of their own future autonomy.

The concordat pattern introduces even more complications, for here the church—in this case the Roman Catholic Church—appears as a sovereign state in its relations with governments, even to the extent of claiming its own inviolable territory. Here the claim for autonomy reaches a kind of worldly apex by conceiving the church's exclusive loyalty to God in terms familiar to other states. In claiming its autonomy, in contrast to the free churches, it must claim to be more than a church. On the one hand, as a sovereign state it reaches the highest plateau of accountability for the incorporation of human rights norms, as its internal critics have hastened to point out. On the other hand, the point of this autonomy was precisely to avoid falling under norms extrinsic to the church's theology and ecclesiology. If we drive this paradox further we would have to ask of the Vatican that it subscribe to the United Nations covenants and extend these practices throughout its worldwide organization. If, however, it wishes to retain its ecclesial autonomy in other ways, it must give up the claim to state sovereignty. The general tendency by many states today to defer to the Vatican's claim only masks this internal contradiction. It does not resolve it.

Cultural Contexts and Human Rights in the Church. We turn now from conceptions of the church as a kind of corporation or even a kind of state to one in which the church is primarily a cultural entity. In this sense it is more than merely a corporation but less than a state. It is a "people." In contemporary parlance, it is an "ethnic community." This has deep

Biblical roots as well as historic authenticity. Particular churches or denominations are often characterized by this ethnic homogeneity even if it is not part of their official self-understanding.[26] Sometimes, these ethnic churches have a dominant position in the society, as in northern Europe. At other times, they are despised minorities, as is the case to some degree all over the world. At still other times, they may be the object of solicitous "uplift" by the government, as in the Indian case cited above.

What is peculiar to all these contexts is that the customs of the ethnic community tend to take precedence over norms that might support a widened public in the church community. Women, who might have exercised a powerful de facto role in the community, must then be seen as equals in a church public, contravening historic cultural views of their status. Government by consensus might be threatened by introduction of voting, with its tendency to move to majority rule. Patterns of inherited or charismatic priesthood might be displaced by rational elections according to "merit."

Most strikingly, where the church is legally vested with rights over marriage and family, the ethnic character of the church has an overwhelming impact on the presence of human rights norms in people's lives. The Indian case may be one of the more extreme, but it does not take long to find manifestations of it not only in European and South American contexts but in highly secularized countries, especially in matters of divorce and remarriage.

An additional element is introduced when the church becomes in effect the legal face of the ethnic community, which has no other standing before the state or corporations. Here the way the government treats the church is in fact the way it treats a whole ethnic community. I have already indicated some features of this in India, where tribal communities and religious communities evoke very different responses from the state. A different twist is given this in the United States, where the original inhabitants — the "American Indians" — have sought to preserve their communal autonomy under the umbrella of protecting their all-embracing traditional religion, even to the point of "establishing" it.[27] This is yet another coalescence of religious autonomy with ethnic national autonomy,

---

26. See Greeley, *Denominational Society*, for a discussion of the denominations as ethnic associations. The Church of Jesus Christ of Latter Day Saints is practically a state unto itself, while functioning totally within a seemingly secular state and federal constitution.

27. See American Indian Civil Rights Bill (1976) and O'Brien, *Tribal Governments*.

with similar impacts on the way these groups might assimilate or reject human rights norms and practices.

This tension between "external" human rights norms and a people's customary integrity is not merely a matter of the autonomy of the ethnic church but also of the way norms of human rights are to be implemented within a jurisdiction. They involve not only the contrast between parental and conciliar ecclesiologies but also between informal and formal procedures, between tradition and "rational" choice.

SUMMARY AND CONCLUSIONS

The way human rights norms are implemented within churches is affected by their ecclesiologies and their relation to civil governments, by the degree to which they constitute an ethnic community and bear the traditions of that community, and by their internal degree of legalization. The central issue uniting these factors is their choice of ecclesiology, which brings with it a conception of the church's relation to government, family, and culture. This choice of ecclesiology rests in turn on a theological base which carries with it a particular conception of "the human."

The conception of the human has at least two issues that affect both ecclesiology and human rights theory. First, are human beings seen primarily as autonomous individuals who have rights prior to their communal relationships? This view is represented civilly by the liberal rights tradition and an ecclesiology rooted.in voluntary association and congregationalism. Or, are human beings born out of a communal body of some kind, who have rights in order to participate effectively in that body? This is the assumption behind both socialist and traditional social theories and is echoed in a conception of the church as the Body of Christ or as a Mother from whom we emerge in faith.

The individualist conception of the human, anchored in the "first table" of political and civil rights, pits the individual against the state, depriving the multitude of intermediary institutions—including churches—of their legitimation and of human rights protection. The corporatist conception, anchored in the "second table" of social-economic rights, gives more protection to these intermediary institutions and communities but often at the expense of less powerful persons, especially women, within them. The question of the nature of the human is both ecclesiological and legal. A closer dialogue between ecclesiology and law may be able to lead us to more nuanced ways of linking them in a critical manner.

The second theological struggle over the meaning of "the human" highlights the difference between a parental and a conciliar ecclesiology, mirrored once again in the "two tables" of human rights. Here the root question emerges: Are human beings essentially stunted or corrupted and in need of the parental care of the bearers of historic wisdom and transcendental revelation? Or are human beings naturally oriented to the good, thus needing to have barriers to their growth toward adult citizenship removed by the state? Or are they both fallen and yet also redeemed, needing both a clear public space to express their goodness but also a differentiation of powers to keep their evil tendencies in check? Each of these perspectives gives rise to different emphases in both the conception and securing of human rights.

The distinction between the rights to care and the rights to liberty, between the "second table" and "first table" rights, closely parallels the ecclesiological distinction between voluntarist and corporatist views of the church, between parental and conciliar ecclesiologies. An examination of the way churches appropriate or reject human rights norms challenges us to enter a more embracing dialogue between ecclesiology and human rights thought to see how these two "tables" of our modern human rights covenant might be conjoined for the sake of human beings within and outside the churches.

13

CONSTITUTIONAL ORDER *in* UNITED METHODISM *and* AMERICAN CULTURE [1]

UNITED METHODISTS ARE DEEPLY engaged with questions about their organizational forms and purposes. Amid calls for new paradigms and more effective organizational systems, the church's General Conference has taken up numerous proposals for structural reform. These address a wide range of fundamental elements of the church's polity, such as the nature and role of episcopacy, annual conferences, and congregations in the overall governance of the church.

Debates about reform and restructuring, however, have suffered from a lack of critical discussion of the ecclesiology underlying them. Like other contemporary institutions, the church needs to address with greater clarity the elements that constitute it as a body and the nature of the authority that enables those elements to work together toward a unified purpose. In particular, as United Methodism considers a covenant with other Protestant bodies to bring together catholic, evangelical, and reformed ecclesial traditions in a Church of Christ Uniting (COCU), it must clarify its expectations of the constitution of such an ecclesial covenant as well as the distinctives of Methodist tradition that are its unique contribution to emerging forms of church.[2]

1. Co-authored with Thomas E. Frank.
2. See Moede, COCU Consensus.

The ferment over denominational forms as well as conversations about ecumenical relationships present United Methodism with an opportunity, then, to address polity issues in a fresh way. However, ecclesiological proposals must be formulated in critical engagement with the organizational forms prevalent in the larger society as well as in the church's own traditions.

Methodism grew up within a new nation deeply imbued with the voluntary ethos of associationalism. The nation sought to become a democratic republic that would be fertile ground both for economic enterprise and for associational life. Methodism as it became fully a denomination was profoundly shaped by these commitments. Indeed, historian of American religion Nathan O. Hatch has argued that Methodism "invented the American denomination, making obsolete the European reality of church at the cultural center and sect at the periphery."[3] Neither established church nor sect set apart from society, denominations became America's unique form of ecclesial voluntary association for authorizing ministries and supporting missions.

Methodism's vaunted pragmatism enabled it to be remarkably effective in expanding with the nation. William Warren Sweet noted that the Methodist Episcopal Church was "the first religious body in America to work out an independent and national organization."[4] By 1850 over a third of all church members in the USA were Methodists, with a geographic coverage such that even today over ninety percent of American counties have a United Methodist Church.[5] Methodist forms spread rapidly among German-speaking people and African Americans as well as the predominantly Scotch-Irish and English population of the Methodist Episcopal Church.

The church's rapid growth in the context of the formation of American political institutions led to heated debates. For some, especially Methodist Protestants, a church in which bishops appointed pastors and in which conferences assembled clergy but excluded laity was out of keeping with the emerging democracy. For others the crucial issue was the balance of powers between the episcopacy and the General Conference in an organization of growing resources and influence. These debates

---

3. Hatch, "Puzzle of American Methodism," 187.

4. Sweet, *Methodism*, 100.

5. Finke and Stark, *Churching of America*; Bradley et al., *Churches and Church Membership*, ix.

revealed both the diversity of traditions within an emergent Methodism and their tensions with the developing republican order of the nation.

Eclecticism in ecclesiology and polity was native to the Methodist movement. In their episcopal greetings introducing the 1939 *Discipline*, the bishops of the newly formed Methodist Church explained that

> The Methodist *Discipline* is a growth rather than a purposive creation. The founders of Methodism did not work with a set plan, as to details. They dealt with conditions as they arose . . .[6]

This disposition to "deal with conditions" and to let the church's discipline grow from efforts to advance people's spiritual life is fundamental to the character of United Methodism and its predecessor bodies. Methodism's central historic purpose has been to invite people to the disciplines and practices of Christian life and action, whatever their circumstances. Consequently, United Methodist polity is an exercise in practical theology—the interplay of theological reflection upon contexts and situations of ministry and actions of witness and service to the Gospel. What has been lacking, however, is adequate theological reflection on the relationship of the church's institutional forms with the models of governance projected by other major institutions in American culture.

Since the heady days of nineteenth century growth the church has devoted little debate to the constructive task of defining more precisely its nature as an ecclesial body. In a twentieth century marked by world wars, economic depression, and the expansion of a global market, the church has had to address vast issues of social change. In its continuing response to new contexts and needs, the church has tended to borrow elements uncritically from diverse forms of ecclesiology, corporate business, and civil polity. These have been woven together in a largely oral tradition about its ways of getting things done.

Often the church's pragmatism has degenerated into an inarticulate indifference about the church's organizational body, as in the immortal words of Bishop Charles Henry Fowler (elected 1884), that "Methodism means always doing the best possible thing."[7] Even the various church unions of this century have provoked little debate about polity issues among church members and scholars. Methodist groups have focused

6. *Doctrines and Discipline. . .* 1939, 1–2.
7. Quoted by Short, *Chosen to be Consecrated*, 75.

more on how various interests would retain power in a united church than on the ecclesiological rationale for the structure itself.

In popular discourse Methodists have often expressed what they see as connections between church order and American institutions. For instance, parallels are drawn between United Methodism and the three branches of US government. The first generations of Methodists were especially eager to show that their polity was not incompatible with ideals of the new republic. Thus, they pointed to the church's division of powers, with the conferences as the legislative branch, the episcopacy as the executive branch, and a judiciary function lying in a special committee of the General Conference. After the 1939 union, of course, Methodists could argue that their legislative bodies were now fully and equally delegated by both clergy and laity, and that the judiciary was now located in a separate Judicial Council parallel to the US Supreme Court.

In fact, however, this is a strained and misleading analogy, not only because United Methodism is not solely an American denomination but also because federalism is only one of at least three principles of organization in the church. Models of public association and corporate organization have also played a major role in the formation of the church's polity. We need to pursue a much more critical analysis of the way various models of organization are brought together with theological commitments to shape the church's ecclesiology.

Critical engagement between ecclesial and civil constitutional forms must undergird the search for an appropriate ecclesiology for United Methodists today. The church has an opportunity to constitute itself anew in a polity that more fully incorporates congregations and conferences as constituted assemblies, and that advances a broader sense of episcopacy. We believe that reflection on how elements of public assembly, federalism, and corporatism are brought together in a coherent ecclesiology can also be indispensable for wider institutional reforms in the society.

We shall first explore the meaning of these three major principles of organization and their impact on the development of Methodist polity. We shall then turn to a case study of the Pacific Homes litigation in the late 1970s, which revealed fundamental problems of church order in relationship to American legal culture. Finally, we will take up a critical examination of the key issues illuminated by this case that require careful attention in the current ecclesiological debates.

ELEMENTS OF PUBLIC ASSEMBLY

The victorious American colonies constituted themselves as a federal republic. As a republic, they sought to govern themselves through congresses, assemblies, and conventions rather than through the paternal guidance of a monarch. Though the US Constitution of 1787 presumed this lush proliferation of councils and assemblies it did not articulate their presence constitutionally. Only the First Amendment ensured their vital place by protecting freedom of religion and public assembly and communication. The right of assembling in publics, whether religious or civil, became the core of the people's constitutional liberty.[8]

The idea of a public is closely tied to the church's origin as well, for it was the ancient Greek notion of the *ekklesia*, the civic assembly of the *polis* or city, which formed the key notion of church and from which we derive the term "ecclesiology." In constituting the republican heritage in the American context, then, the church was only coming home to its mother tongue.

What then are the key dimensions of this "publicity" intrinsic to American constitutionalism? First, public assembly is a form of *participation* in power and authority. It is a mode of self-governance. People have a right to participate in the structures of authority that hold them in common obligations with others. They have a right to be in direct communication with others who participate in the powers of governance.

The religious echoes of this right to public participation were already manifest in the open public preaching of revivals, in which itinerant Methodist preachers played a central role. In the ecstatic gatherings of camp meetings people found their own voices—slaves and free citizens, women and men, people of all ages. The revivalist temper found its way into the wider political culture, and into the societies and churches of Methodism in particular. Many United Methodist local churches have functioned primarily as indoor camp meetings—assemblies for preaching, singing, and conversion—with affiliated organizations for education and missions. They have also been a center for public life in many towns and cities across America.

This congregating in public assembly, fundamentally dependent on the unpredictable work of the Spirit, always has an ephemeral character. It cannot be sustained by law but only by will and inspiration. Both

---

8. For the problem of constitutionalizing public assemblies see Arendt, *On Revolution*, 241–44.

civilly and ecclesially, the maintenance of the public assembly so crucial to governance has been very difficult. Neither the congregational life of United Methodism or the popular assemblies of the republic have been able to find explicit constitutional recognition and legitimation. Meanwhile in contemporary market culture, invocations to participate fall on the deaf ears of hurrying commuters and exhausted weekend couch potatoes. The spirit of the assembly soon becomes the object of bureaucratic programming.

Secondly, public assemblies are modes of reasonable *persuasion*. While people act and agree on the basis of many sinister as well as altruistic motives, non-coercive persuasion according to reason is the benchmark that legitimates the decisions of the public. This is why non-violence is so crucial to the existence of a public. Moreover, the criteria of persuasion require a rough economic and social equality among the participants in order to strengthen their immunity from external coercion.

Here again, the emerging idea of a republic found common chord with the religious impulses of evangelical Christianity. In the conversations and arguments of meetings, societies, congregations and conventions, believers had to hammer out common agreements in order to build and sustain the often-isolated congregations of an expanding commonwealth. Without hierarchical guidance people had to live by persuasion in order to form a largely cooperative order in the church as well as in civil society.

The work of persuasion in both churches and general publics has had to swim against the tide of subliminal advertising and media sound bites that seek instant impact over sustained argument. One-way communication, whether by preachers or by mass media, stultifies the habits of dialogue. Decisions by majority vote cut off the requirement for persuasion and the creating of consensus in both church and civil assemblies. How to cultivate the arts of persuasion is a major challenge for any institutions committed to public life.

Thirdly, a public always appears as a *plurality* of strangers who need conversation and argument in order to create understanding and agreement. In a public of persuasion the multiplicity of opinions requires public testing and evaluation. The task of a constitution, as James Madison understood it, was to translate the many natural factions and interests of society into the measured opinions of a public argument for the sake of governance.[9]

Similarly, by placing such a high premium on personal experience, evangelical Christianity also created the conditions for religious pluralism—of congregations, movements, denominations, and associations. Sometimes all these religious interests could be held together under a common tent of conversation, but at other times people set out to build their own tent, only to find a plurality of groups emerging within its folds. United Methodism, as it has expanded among a diverse population, has struggled with how varied ethnic, cultural, racial, and interest groups can share fully in the various publics of the church. How to honor our commitments to the particular sources of our faith experience as well as maintain commitments to the welfare of the whole has remained a key issue in ecclesiology as well as civil governance.

Finally, a public can sustain itself only if it can find sufficient points of *commonality* in which to ground its efforts at persuasion. A cultural world of common reference offers the benchmarks people need in order to reach a conclusion accessible to public reason. In the civil order this is constituted by common history, language, geography, shared values and visions, mutual interests and the like. Sustaining this commonality is a critical cultural task for any republic that cannot rely on the single inheritance of a monarch who purports to be kin (and king) to all.

Much evangelical Christianity has sought to resist the fissiparous effects of ecstatic experientialism by appeal to Scriptures as the sole necessary commonality of the Christian people. Methodism's theological commonality by contrast has focused less on doctrinal propositions about Scripture than on a method and process for theological reflection. In order to resist the dogmatic schisms arising from different interpretations of Scripture, Methodism both held onto its Anglican ecclesial heritage and invoked the varied commonalities of creeds, Wesleyan roots and history, statements, resolutions, and its *Book of (Doctrines and) Discipline*. Around this cultural core a host of stories, customs, and traditions has gradually entwined itself.

These four elements of public assembly—participation, persuasion, plurality, and commonality—are only partially realized in any public. Yet they offer a framework for examining the principle of public assembly that has so deeply shaped both the civil and ecclesial forms of American

---

9. *Federalist Papers*, Nos. 10, 51.

constitutional culture. They can serve as a checklist with which to look at the character of any assembly, whether religious or civil.

Most Methodists know the church through the public gatherings of the local congregation. They soon discover, though, that their local congregation occupies a politically complex and often ambiguous place within the wider church. The question of denominational life is how these local assemblies are linked together in the wider church. Federalism is a major way this constitutional question has been answered in American government and society.

ELEMENTS OF FEDERALISM

Federalism, like public assembly, has direct affinity with concepts and practices rooted in our religious heritage as well as political experience. Historically it is rooted not only in classical political theory but even more deeply in the covenantal polity of ancient Israel.[10] Concepts of a promissory God, of life according to a higher law, of relationships of mutual consent in making covenants and mutual reconciliation in their violation, have deeply permeated our religious consciousness. Many people see the American Constitution as a covenant-like document indebted to this heritage, and conversely, many church members have viewed denominational polity as a form of federal government.[11]

An examination of six key elements in federalism shows how United Methodist polity accords with federalist order as well as ways it departs from classic federal conceptions. First, a federal order entails a *constitutional definition of mutual rights and responsibilities among its constituting entities*. United Methodism was in this sense brought into being by a formal constitution in 1968.[12] Predecessor bodies had also written constitutions, beginning with the Methodist Episcopal Church in 1900. The Commission on the Constitution of the MEC, which worked from 1888 to 1900 to formalize a constitution for the church for the first time, defined it this way:

---

10. For an examination of the Biblical sources from a political theory standpoint see Elazar, *Covenant and Polity*. For a constructive theological appropriation of the federalist tradition see Everett, *God's Federal Republic*.

11. See Lutz, *Documents*, and McCoy and Baker, *Fountainhead*.

12. The following discussion of the 1968 Constitution draws upon a fuller analysis in Frank, *Polity*, ch. 3.

> A constitution is an instrument containing a recital of principles of organization and of declarations of power, permissions, and limitations which cannot be taken from, added to, or changed in any particular without the consent of the power which originally created the instrument, or by the legal process determined by the body possessing original power.[13]

In Division One of the 1968 Constitution the general provisions for the church first name the originating bodies of this Constitution by declaring the union of The Evangelical United Brethren Church and The Methodist Church (Art. I). The articles go on to give the new entity a name, The United Methodist Church (Art. II), specify continuity of doctrine with the bodies now joining (Art. III), and claim the physical assets through which the church expresses its ministries (Art.VI). They also make explicit the fundamental nature of the church as an open community striving toward "unity at all levels of church life" (Art. IV, V). This nature resonates with the principle of a free society in which "all persons, without regard to race, color, national origin, or economic condition" are "eligible" to participate (Art. IV). (Notice it does not say "welcome" to participate; here the language is of human rights, not hospitality.)

In the subsequent two divisions come the elements that constitute the body. Division Two is titled "organization" but actually is entirely about conferences and how they are constituted. Division Three is titled "episcopal supervision" and describes how bishops are constituted. The last two divisions of the Constitution create a process through which it can be amended or disputes over its meaning can be adjudicated.

Absent as constituting entities are local church congregations, general agencies, or clergy and lay members of the churches. Even the annual conference holds an ambiguous status constitutionally, since it has no legislative function, though it is the sole body that can vote on the conference relations of clergy and diaconal ministers.

Thus, Divisions Two and Three of the 1968 Constitution clearly establish conferences and episcopacy as United Methodism's constitutive

---

13. Methodist Episcopal Church, General Conference *Journal*, 338. The Methodist Episcopal Church South was content with its historic understanding that the Restrictive Rules, including a rule on representation in General Conference, comprised the church's constitution (though it was not so titled). Prior to the 1968 union the Evangelical United Brethren Church had a considerably longer Disciplinary section titled "constitutional law," including the basic units and powers of all conferences, local churches, and forms of ministry.

elements, as they have been from the beginning of Methodism. As the constitutional historian of an earlier period, John J. Tigert, wrote,

> Through all this time two constant factors of Methodist polity, (1) a superintending and appointing power, and (2) a consulting body called the Conference, have been continuously operative. These two factors are constitutional or elemental in the government of Methodism . . . the former chiefly executive and the latter chiefly legislative.[14]

To state the case more broadly, United Methodism is comprised of an unsettled, often unreflective, yet remarkably creative blending of two constitutive principles. The principle of republican democracy essential to free societies and voluntary organizations has been constitutive of United Methodism most visibly through its conferences. While these assemblies have their origin in the organization of Methodism as a voluntary society in the Church of England, they took on a distinctly American form in parallel with the legislative assemblies and political conventions of the emerging republic in the nineteenth century.

Wesley gathered conferences of Methodist preachers to consult with him, but they were never decision-making bodies. In the American context, however, from the moment the "Christmas Conference" of 1784 voted to accept Francis Asbury and Thomas Coke as general superintendents, the preachers collectively became the fundamental legislative body of the new church.

In fact, one could argue, with the weight of constitutional opinion in the late nineteenth century, that "the body of traveling elders," that is, all the ordained preachers collectively, were the "original or primary constituency" that brought the Methodist Episcopal Church into being. In addition to electing their general superintendents, they voted on changes to the *Discipline* each year in annual conference, created a quadrennial general conference of all ordained preachers in 1792, and in 1808 made it a delegated General Conference with restrictions on its powers. In a genealogical sense, then, both the conference system and the episcopacy hold their charter from this original body—the "traveling preachers" of this one unbroken "traveling Connection."[15]

Nearly a hundred years passed before laity had any representation in the conferences at all, and fifty more years before they gained equal

14. Tigert, *Constitutional History*, 15.
15. *Discipline* 1992, para. 38.

participation. Yet it was still this primal body of ordained preachers who had to vote to amend the constitution to include lay representation. This expanded and continuous aggregation of all clergy and corresponding lay members of all annual conferences are the electors who must vote on any amendments to the Constitution.[16]

The Constitution names the annual conferences as "the basic body in the church." One way this is true is that they elect delegates to general conferences. On the other hand, the General Conference is the only body in the church with legislative powers over the connection as a whole. It defines and fixes the powers of all other units, including the bishops, clergy, laity, annual conferences, and "connectional enterprises of the Church" in missions, education, and other ministries.[17] It is given powers even over the annual conferences whose delegates constitute it.

These powers are not unlimited, however. Beginning in 1808 the General Conference of the Methodist Episcopal Church built in certain checks on the delegated body it was creating. The "restrictive rules" have continued in force in mainly the same form ever since.[18] Most importantly from a polity standpoint, the restrictive rules make it very difficult for the General Conference "to do away with episcopacy or destroy the plan of our itinerant general superintendency."[19]

Episcopacy is the second constitutive element of United Methodism. However, because United Methodism is a synthesis of varying polities, this episcopacy is, in the laconic words of Bishop John L. Nuelsen, "an episcopacy that the Christian church ha[s] not yet seen."[20] The bishops in council are charged by the Constitution with planning "for the general oversight and promotion of the temporal and spiritual interests of the entire Church."[21] Though bishops are elected to office by lay and clergy delegates in Jurisdictional or Central Conferences and their presidential duties are limited to the bounds of those conferences, they are to practice a general superintendency.

In some ways United Methodist episcopacy is kin to its Anglican and Roman Catholic forebears. The Constitution grants bishops the

16. *Discipline*, para. 62, and Tigert, *Constitutional History*, 323–24, 403.
17. *Discipline*, para. 15.
18. *Discipline*, para. 16–20.
19. *Discipline*, para. 17.
20. John L. Nuelsen, *Die Ordination in Methodismus* (1935) quoted in Moede, "Bishops in the Methodist Tradition," 59.
21. *Discipline*, para. 50.

power to appoint ministers to the charges in consultation with the district superintendents, who are extensions of the episcopal office.[22] The Constitution also gives bishops life tenure. These are both features associated ecclesiologically with a monarchical episcopacy, that is, an episcopacy understood as rule rooted in a hierarchy of personal statuses and apostolic succession.

Yet while United Methodist episcopacy was initiated by Wesley, Coke, and others steeped in the Anglican heritage, it is significantly different. In Anglican ecclesiology, the bishop is the unit of church and presides over a diocese. Parishes are local expressions of the episcopal diocese. Clergy are the representatives of the bishop in a local place. The bishop visits each parish to perform confirmations, because all parishioners are members of the bishop's extended congregation.

Little of this applies in United Methodist episcopacy. Oddly, the Constitution fails to specify even that the bishops have authority to ordain the clergy—this is listed as one of the "presidential duties" in a later legislative paragraph[23]—much less other aspects of an Anglican model such as the authority to confirm new members. The sacraments are not mentioned either, which means that in United Methodism bishops are not understood as a distinct order retaining sacramental authority which is then delegated to clergy as extensions of the bishop when they are ordained elder.[24] Though the Constitution does not state this explicitly, later legislative paragraphs make clear that bishops remain part of the body of elders as well as the itineracy, receiving their assignments from the jurisdictional conference.[25]

Thus, United Methodist episcopacy constitutes the church in a functional way through the bishops' legislatively defined roles of oversight and superintendency. Bishops are called to a broad mandate of oversight. In one of the *Discipline*'s simplest and most compelling phrases, they are asked "to lead"—"to lead and oversee the spiritual and temporal affairs of The United Methodist Church."[26] The generality of this duty is also its strength. Bishops are given wide latitude to speak, intervene, encourage, preach, evangelize, to put themselves on the line. As with their ancestor

22. *Discipline*, para. 57.

23. *Discipline*, para. 515.

24. Mathews, *Set Apart*, 47.

25. *Discipline*, para. 503-4. For further discussion see Buckley, *Constitutional and Parliamentary History*, 187-88.

26. *Discipline*, Para. 514.1.

Asbury, their constant travel and omni-presence makes the connection tangible. The bishops have also had wide influence in the missional activism of the church. They serve on and preside in general agencies and boards of trustees of church-related institutions. Their energy and experience are in such demand that they have had to be wary of becoming mere program promoters and "sales managers."²⁷ The conferences—fleeting moments of assembly—would never be enough to sustain the connection without the itinerant superintendency. Little wonder, then, that many voices in United Methodism today call for stronger episcopal leadership—not authoritarianism or autocracy, much less monarchy—but a leadership of presence and vision that will give the connection its coherence for the future.

The persistent ecclesiological question, however, is in what sense the tasks of teaching, evangelism, and mission constitute "church." The traditional Methodist privileging of preaching and mission activism over sacramental order results in enduring ambiguities for the *episkopoi* of this church. The Restrictive Rule protects "the *plan* of our itinerant general superintendency" [emphasis ours]—that is, its role in advancing the church's mission—more than any inherent authority.

The episcopal prerogatives of appointment-making and life tenure have been challenged repeatedly—in the 1820s, 1890s, and 1970s. The word "episcopal" was dropped from the denominational name when The Methodist Church was formed in 1939, in part to make the Methodist Protestant partners to the union more comfortable in accepting ecclesial life with bishops, whose appointive authority had been one source of their protest in the 1820s. Moreover, many twentieth century United Methodists have complained of the association of episcopacy with patriarchy and an outmoded culture of deference and paternalism. All this unrest indicates that for lack of a firm ecclesiological basis, the functions of bishops depend far more on "the consent of the governed" than on organic roots in the traditional understanding of episcopacy. At the same time, it must be said that the church continues to depend on the episcopacy for much of its coherence and continuity.

The second element in federalism is provision for a *balance or division of powers between the constituting units*. Federalist theory harbors suspicion of any concentration of power and authority. Instead of an organic theory for differentiating functions within a unitary whole, it posits

---

27. Leiffer, "The Episcopacy," 169.

a dynamic equipoise among equal powers bound together in an overarching common covenant. The constituting entities must consciously come to agreement about major decisions.

Thus, the Tenth Amendment of the US Constitution states that "the powers not delegated to the United States by the Constitution, nor prohibited by it to the States, are reserved to the States respectively, or to the people." In United Methodism annual conferences retain "such other rights as have not been delegated to the General Conference under the Constitution."[28] At first glance the breadth of General Conference powers would not seem to leave much room for "other rights." The Constitution grants General Conference "full legislative power over all matters distinctively connectional," which includes determining the boundaries, duties and powers of annual conferences.[29] But at the same time, the annual conferences constitutionally retain the sole right to vote "on all matters relating to the character and conference relations of [their] ministerial members"—their historic primary function. It should be noted, however, that this constitutional function belongs only to the clergy session of the conference, not to the lay members.[30] Each annual conference can also develop its own program initiatives and raise its own money to support them as long as it does not financially obligate any other organizational unit of the church.[31] Moreover, as we have already pointed out, annual conference members collectively vote on constitutional amendments as the final power over all such changes.

Another way in which the church has attempted to achieve some balance among the constituting powers has been through the creation of jurisdictional and central conferences. "Jurisdiction" is both a geographic and a legal term. As a name for a geographic area it carries with it the intention of adapting United Methodism to regional cultures. In this sense it served well to enable the Methodist Episcopal Church South to sustain many of its customs and relationships after the 1939 union with the larger, more centralized, and more national Methodist Episcopal Church (north). The jurisdictional system also allowed the denomination to preserve racial segregation in a Central Jurisdiction based not on region but on race. This accommodation to racial patterns had the

28. *Discipline*, para. 36.

29. *Discipline*, para. 15.

30. Since this writing lay representation has been added to the Boards of Ordained Ministry.

31. *Discipline*, para. 704.2.

effect of modifying the "plan" of general superintendency. Mainly in order to prevent black bishops from presiding at white conferences, the Constitution restricted bishops' powers of "residential and presidential supervision" to their own jurisdiction. The restriction remains, even in the absence today of the racial jurisdiction.

As a legal term "jurisdiction" indicates the limits of power of various agents. The designers of this system intended it to create a balance or distribution of powers throughout the connection. Membership of the general agencies would be elected by jurisdictions. The work of general agencies would be shared by jurisdictional auxiliaries. Jurisdictional conferences would be responsible for and in many cases actually own the educational and mission institutions within their boundaries. Moreover, they would have power to determine the boundaries of the annual conferences within their own territory.[32] A survey of current jurisdictional conference journals, however, shows clearly that these hopes were never realized.[33] Only the Southeastern Jurisdiction—the region of the old MECS—has any significant level of staff and budget for program. In fact, all the institutions that technically are jurisdictionally owned relate in various ways to the entire connection.

Thus, the Jurisdictional Conference remains a constitutional anomaly originally invented to accommodate a social and cultural difference. All the duties assigned to it by the Constitution historically belonged to the General Conference. In effect, Jurisdictions are simply regional derivatives of the General Conference and do not have a fundamental role in constituting the church.

This is not in any way to dismiss the primary political function of Jurisdictional Conferences—the electing and consecrating of bishops—only to make clear that the function derives from the General Conference. The episcopacy in the USA is constituted by the Jurisdictional Conferences; that is, the conference elects and the college of bishops consecrates by the laying on of hands. But the number of bishops is determined on "a

32. John M. Moore, the MECS bishop who worked tirelessly for this plan, put the argument best in 1943: "The Jurisdictional Conference is the essential, vital, and principal administrative and promotional unit of the Church, with legislative power limited to regulations on regional affairs . . . [it] is the key and core of the entire system . . . The Plan of Union [1939] sets up a commonwealth of balancing bodies wherein no one shall be supreme, except in its own field." Moore, *The Long Road*, 192–93.

33. Already in 1958 James H. Straughn was complaining that the Northeast in particular was not taking the jurisdiction seriously as a programmatic unit, in Straughn, *Inside Methodist Union*, 141.

uniform basis" set by General Conference, and once bishops are elected their membership resides in the Council of Bishops.[34]

Central conferences have powers and duties parallel to the jurisdictional conferences, but in a distinctive context that makes them in practice much more constitutive of the church in their regions. Since they exercise those powers in nations outside the USA, the Constitution grants them the power "to make such rules and regulations for the administration of the work within their boundaries"—this in common with jurisdictional conferences—but also the power to make "such changes and adaptations of the General Discipline as the conditions in the respective areas may require."[35] Central Conferences may not make changes "contrary to the Constitution and the General Rules of The United Methodist Church,"[36] but this still leaves wide range for encouraging local churches and annual conferences to organize in the most suitable fashion, and for churches of a region to adapt their ministry, worship, and other practices to the local or national context.

Third, federalism requires a *balance of powers within its governing units*. For instance, legislation passed by Congress must be signed by the President and may be reviewed by the Supreme Court, and major presidential proposals must be considered by Congress. In United Methodism this balance of powers is perhaps most explicit in the division of rights and duties between the conferences and the episcopacy. General Conference has sole legislative power over the *Discipline* while annual conferences have the sole right to admit clergy to membership. Unlike the President of the United States or the governor of a state the bishops individually or collectively do not make any legislative or budgetary proposals. They do not manage agencies charged with carrying out legislation and are in no sense executive administrators of the conferences. Bishops are members of no conference and have no voice except as presiding officers. In particular, they have no voice in the admission of clergy to an annual conference.

On the other hand, bishops retain the sole right to ordain and to appoint pastors to their charges. They are also individually and as a council charged with two broad duties: (1) "general oversight and promotion of the temporal and spiritual interests of the entire Church," and (2)

---

34. *Discipline*, para. 15.10.
35. *Discipline*, para. 29.5.
36. *Discipline*, para. 638.9.

"carrying into effect the rules, regulations, and responsibilities prescribed and enjoined by the General Conference."[37] The possible conflict between these two duties is further evidence of the delicate balance of power between bishops and the General Conference.

Fourth, federalism requires an *independent judiciary* to interpret the constitution. This prevents either legislative or executive powers from defining basic rights and responsibilities, and thus is a critical element in balancing their powers. In the Constitution of the UMC, Division Four constitutes a Judicial Council to make final determination of "the constitutionality of any act of the General Conference." This helps to limit the powers of General Conference, which cannot—as it once did in the Methodist Episcopal Church—judge the constitutionality of its own acts. The Council also has the authority to review and "pass upon decisions of law made by bishops in Annual Conferences."[38] Thus, while bishops are given the constitutional mandate of making decisions of law in their presiding role in conferences, thereby acting as a judiciary, the Judicial Council is equally mandated to review their rulings.

Judicial powers fully exercised become constructive interventions in the government of the body. The very act of interpreting church law also contributes to the church's self-understanding and certainly affects its practices. As in civil polity, though, the judiciary has no means for enforcing its own decisions. It depends entirely on the legitimacy granted it by the constitutive units.

Fifth, federalism requires that the *constituting units have representation in the wider governing assembly*. While the exact forms and proportions of representation may vary, the principle itself is critical to a republic pursuing democratic values. That is, all citizens must have access to some public, which then delegates some of its members to participate in more general publics. These delegates then represent their electors in this wider assembly, and in that assembly must also consider the interests of the whole federation.

The United Methodist Constitution guarantees these representative rights to the annual conferences, who elect by a formula of proportionality their delegates to jurisdictional, central, and general conferences.[39] Throughout Methodist history annual conferences have been the locus of

37. *Discipline*, para. 50.
38. *Discipline*, para. 58, 59.
39. *Discipline*, para. 12–14.

membership for the clergy. Because they alone have the power to admit clergy to the ordained ministry, the conference is the body to which "all the ordained ministers [are] bound in special covenant."[40]

Laity are now members of annual conferences as well, but not on the same basis. Their church membership remains in their local church, and they are excluded from the core function of voting on clergy relations. Their membership in annual conference is not based on proportional representation of the lay members of the local church, but is tied to the pastoral charge (one lay member per pastor appointed to a charge). Thus, the congregation as such is not represented in the wider polity of the church. Indeed, congregations are not in any way self-constituting within this system. They are governed not by annual meetings but by a charge conference called and presided over by a District Superintendent. In Methodist tradition they are known primarily as "local churches," that is, the manifestation of the "connection" in a particular place.

Thus, while all United Methodist conferences—general, jurisdictional, central, and annual—are equally lay and clergy, the delegated bodies can hardly be called fully representative, since the basis of representation is already skewed toward the clergy in the annual conferences. Moreover, the formula for General Conference is based equally on the number of clergy and the number of church members in each annual conference, which does not adequately account for the fact that in regions with many large congregations (currently the Sunbelt of the USA) clergy may serve on average far more members than in areas with smaller congregations.[41] This problem is addressed by the practice of bicameral legislatures in civil polity, but United Methodism lacks such a solution.

Finally, in federalism *each citizen stands within the jurisdiction of two or more of the governing partners in the federation.* In the USA a citizen votes, enjoys representation, pays taxes, and receives services from municipal, county, state, and federal governments. In the UMC, while there is no popular vote for representatives or referenda on conference actions, each church member belongs not only to the local church but to the whole connection.[42] Each member has a right to petition General Conference, to study and accept or reject conference resolutions, and of course, to "vote" with financial donations as well.

40. *Discipline*, para. 422.

41. Thus, some United Methodists argue that church membership numbers should be given more weight in figuring representation.

42. *Discipline*, para. 210.

A review of these six principles of federalism shows that they have indeed been a formative influence in United Methodist polity. The very fact of the church's having a formal constitution, combined with its concerns for defined and balanced powers and rights of representation, indicates a federal model. At the same time, however, because United Methodist polity embodies elements of several ecclesial traditions and has developed as a growth in varying circumstances, a closer examination of its constituting units exhibits many ways in which it varies from federal principles. This departure is due not only to internal ecclesiological considerations but also to the presence of yet a third mode of organization—the corporation.

ELEMENTS OF CORPORATISM

A third major influence on the formation of United Methodist polity is the understanding of corporations in American—and now global—society. Corporations provide a means of organizing large and complex enterprises around defined purposes. They rationalize authority, distinguish and specialize functions bureaucratically, and institute chains of decision and command that facilitate action. At the same time, they tend toward hierarchy and centralization of power in a single "head" of the corporate "body."

Corporatism, like public assembly and federalism, has deep theological roots. The very notion of the church as the body of Christ generated the idea of an immortal institution distinct from its transient human members. Paul's image of this body as an organic interdependency of functions (Romans 12:4-8; 1 Cor. 12:12-31; Eph. 1:22-23, 4:11-16) has exerted extraordinary power within ecclesiology as well as Western society. Without it, modern society probably would not have the cultural base to legitimate the corporation.[43]

Corporatism is most visible in United Methodism in the general and annual conference agencies of the denomination. Administrative and programmatic units tend to be organized bureaucratically and to reflect and reinforce trends in business management. While they have been effective in organizing large enterprises such as missions, publishing, and pensions, they have also drawn the church into structural forms in tension with United Methodism's wider ecclesial character.

---

43. See Stackhouse, *Public Theology*, 113-37.

Two basic principles constitute corporatism.[44] First, *corporations generally organize themselves bureaucratically through the differentiation and specialization of functions.* This "rational" form of organization enables them to coordinate a complex set of tasks toward a clearly defined goal. It encourages the development of specialized knowledge and employment of persons with expert skills in particular kinds of work. Bureaucracy usually divides the practice of these skills from the setting of general policy, thus creating a hierarchy of control. A central body or head, represented in a chief executive officer (CEO), takes responsibility for making policy which subsidiary offices then carry out. Internal controls systematize the acquisition and use of funds and channel initiatives into structures for evaluating personnel and program.

United Methodism has utilized this principle of corporatism in a variety of ways over the past hundred years. The denomination began to elaborate a bureaucracy in the 1870s.[45] Both the Methodist Episcopal Church and Methodist Episcopal Church South exerted increasing control over independent associations for education, evangelism, and mission. These were centralized, housed in specially built office buildings, and funded by coordinated budgets. By 1972 the plan for restructuring the agencies of the new church—promising greater efficiency and coordination—created more centralization by consolidating the church's efforts in everything from children's educational materials to agricultural missions into four program boards.

Moreover, the portfolios of these boards, along with other specialized initiatives expressing the church's commitments such as standing commissions on Christian unity, religion and race, and the status and role of women, were mirrored in all organizational "levels" of the church. The General Conference mandated that local church congregations as well as annual conferences carry out parallel functions preferably in similarly specialized units. This would provide for the most efficient and effective communication throughout the connection.

All program initiatives were to be coordinated by Councils on Ministries (COM), thus providing the general and annual conferences as well as local churches with means of central control. In many annual conferences the COM represented the first centralized bureaucracy

---

44. The modern theory of corporation begins with Max Weber. See Weber, *From Max Weber*, 196-244. For more recent developments in organizational theory see Morgan, *Images of Organization*.

45. King, "Denominational Modernization."

ever attempted, with the COM director acting like a CEO of conference program.

Many critics in church and management challenge classical hierarchical views of the corporation. At the same time new forms of corporatism such as total quality management, chaos theory, and other organizational development schemes drawn from corporate life continue to shape the reform of church organizations.[46]

Secondly, while corporations are collective enterprises involving any number of people, *for legal purposes they are single bodies or fictional persons*. Corporate entities can acquire property, assume debt, and risk liability. They can sue and be sued in courts of law. Moreover, the CEO is normally the public face of the fictional corporate person, embodying both the enterprise itself and its action in the public sphere.

United Methodism has donned the corporate mask in order to acquire property, employ personnel, and manage funds. Most local church congregations and all annual conferences and general agencies are legally incorporated with trustees or boards of directors. Particularly in the general agencies, the general secretaries function most clearly as the CEOs who represent the enterprise as a whole. In local churches and annual conferences, the church has continued to borrow the corporate legal fiction while trying to avoid the corporate consequence of the single executive head.

As legal issues surrounding liability and personnel actions have proliferated, bureaucratic controls have become more elaborate. Policies and procedures govern virtually every aspect of corporate action. The chief administrative units for finance and administration and for pensions have become indispensable to the church's life.

However, unlike the modern corporation, there is no single "head" of the United Methodist Church. The bishops in particular are not recognized as heads of the church, whether as CEOs or chairs of the board, even though they carry out very important functions within the ecclesial body. Neither are agency executives heads of anything more than their clearly defined organization. Theologically, only Christ is the head of the church and exercises this headship through many forms—individual disciples, assemblies, conferences, agencies and bishops. Such a conception of religious corporation, however, eludes the typical practices and

---

46. Wheatley, *Leadership*; Senge, *Fifth Discipline*; Jones, *Quest for Quality*.

conceptions of economic corporations and civil courts alike, as the Pacific Homes case revealed in painful particularity.

PRINCIPLES ON TRIAL: THE PACIFIC HOMES CASE

Any reforms in civil or ecclesial governance need to take into account these three primary modes of constitutional order in American culture. Principles of public assembly, federalism, and corporatism all play a necessary role in American political and organizational culture. In appropriating these organizational forms the church needs to be guided by its own deepest theological and ecclesiological commitments. An appropriate integration requires awareness of the three modes as well as attention to the actual historical dynamics of their implementation. To gain greater critical awareness of this historical context we turn now to the complex litigation which embroiled the United Methodist Church in the late 1970s and brought many of these issues to the surface.

Beginning in 1912 church organizations which later became part of the United Methodist Church sponsored a number of nursing and retirement homes in southern California as part of their charitable mission.[47] These homes, including seven residential centers and seven convalescent hospitals, had by 1975 come to serve about 1700 people under the corporate name of Pacific Homes, a non-profit corporation created in 1929. The board of directors of Pacific Homes was elected annually by the Pacific and Southwest Annual Conference of the United Methodist Church.

In the late 1960s Pacific Homes began to encounter severe financial difficulties because the costs of care steadily exceeded the funds generated by the "life care" payment system, according to which retirees prepaid their life-time care with a lump sum fixed on entry into the Homes. Increase in longevity of the population and sharp inflationary increases made this arrangement increasingly impossible. The Annual Conference arranged and guaranteed several loans for the Homes and finally appointed a "crisis management team" to reorganize the Homes and convert to a pay-as-you-go system over an extended period of time.

By 1976 the Conference agreed to subsidize the Homes in the amount of $5 million over a nine-year period to bring them through this

---

47. Information for this section draws on Gaffney, Jr., Sorenson, and Griffin, *Ascending Liability*; Lyles, "Methodist Litigations"; United Methodist Church, "Report to the 1980 General Conference"; and materials graciously provided by the General Council on Finance and Administration, United Methodist Church, Evanston, Illinois.

reorganization. However, the State of California, where the Homes are incorporated, refused to approve this conversion plan. As a result, Pacific Homes had to declare bankruptcy in 1977. Although 91 percent of the residents voted to accept a bankruptcy plan proposed by the management team to enable Pacific Homes to regain viability, two class action suits brought on behalf of approximately 150 of the residents blocked this arrangement. These two suits sought to hold not only the Pacific Homes corporation but various United Methodist entities and "the United Methodist Church" itself responsible for claims totaling $366 million. Subsequent suits against not only United Methodist-related entities but also individuals involved in the administration of Pacific Homes brought the total to over $600 million.

Even though there is no legally incorporated body called "The United Methodist Church," plaintiffs sought to sue it as an "unincorporated association." California law gives legal personality to groups who act like an incorporated association if they have a common purpose and function under a common name under circumstances where fairness requires the group be recognized as a legal entity. Fairness includes those situations where persons dealing with the association contend their legal rights have been violated.[48] By breaking down the distinction between formally incorporated and unincorporated groups, courts had enabled people to sue labor unions, political parties, clubs, lodges, and churches with greater ease. Indeed, under the California definition the mere allegation of legal injury seemed to be enough to establish this corporate persona for an association.

As noted in various documents related to the case, no religious denomination as such had ever been held to be a jural entity liable to suit apart from its various incorporated units. Not even the Roman Catholic Church, with its purported hierarchy of command from the Vatican to the local parish, had been so treated. Because of the landmark character of the case, the National Council of Churches filed a brief *amicus curiae* with the defendants and the case received extensive national coverage.

---

48. *Barr v. United Methodist Church.* For discussion of liability of unincorporated associations see Gaffney, Jr., Sorenson, and Griffin, *Ascending Liability*, 20–40.

The Trial Court: Free Exercise of Religion vs. Ecclesial Corporality

At the first stage of litigation, the California trial court held that the UMC was not a jural body that can be held liable for the acts of its purported agents. The Court accepted the church's argument that there is no central representative for "The United Methodist Church," much less a chief executive officer who could be a principal directing the activities of UMC "agents" around the world. The court held that the UMC was a "spiritual confederation" held together by shared beliefs. The many entities holding this common name and sharing adherence to the United Methodist *Book of Discipline* did not compose either a single jural body or an unincorporated association amenable to suit. Most importantly, treating churches like the UMC as simply another corporate body would abrogate First Amendment protections for religion. Any evaluation of the church's internal structure would violate this shield protecting the church's freedom to practice its faith through the design of its institutional life.

Arguing from the standpoint of religious free exercise, the trial court judge, Ross G. Tharp, wrote:

> A contrary ruling would effectively destroy Methodism in this country, and would have a chilling effect on all churches and religious movements by inhibiting the free association of persons of similar religious beliefs. If all members of a particular faith were to be held personally liable for the transgressions of their fellow churchmen, church pews would soon be empty and the pulpits of America silent.[49]

In short, Constitutional protection of religion took precedence over the ordinary claims of corporate liability.

The Appeals Court: Market Fairness vs. Freedom of Religion

The Fourth District Court of Appeals in California then reversed this ruling, holding that such First Amendment protection gave churches an "unfair" advantage in "commercial affairs" such as nursing homes.[50] This claim rested on the courts' previous development of a doctrine of "neutral principles" by which to treat the "secular" aspects of religious and non-religious corporations and contracts alike. In applying these

---

49. Superior Court of California, "Minute Order."
50. *Barr v. United Methodist Church*, 328, 333.

"neutral principles" to the UMC the court argued that the church did not differ significantly from other corporations carrying on the same activities. To give the church special consideration might even violate the First Amendment's prohibition of the establishment of religion. Upon removing this veil of protection for the church, the Court then went on to examine its internal organization to find that it was indeed a corporate hierarchy which could be sued as a "principal" responsible for the actions of its "agent," in this case Pacific Homes, Inc.

To make this case for hierarchical agency, the court argued that "the Council of Bishops is equivalent to the board of directors of UMC." Quoting from the 1976 *Book of Discipline,* the court singled out the statement: "[the] Council of Bishops is thus the corporate expression of episcopal leadership in the Church" overseeing "the spiritual and temporal affairs of the whole Church."[51] In line with this reasoning the plaintiffs had indeed served their court summons and complaint to a former president of the Council, who then claimed that he could not speak as a representative of the United Methodist Church—something only the quadrennial General Conference can do. Since this General Conference, as a legislative body, does not exist between sessions, the court looked to the Council of Bishops as its interim executive representative, in short, as its "board of directors," despite *The Book of Discipline*'s distribution of powers to a variety of bodies and agents.

Drawing on corporate analogies, the court said in addition:

> It [UMC] is hierarchical; the 43,000 local churches and 114 Annual Conferences are governed through the structure described by the Book of Discipline of the United Methodist Church (Discipline). In United Methodism 'the local church is a part of the whole body of the general church and is subject to the higher authority of the organization and its laws and regulations.'[52]

The Appeals Court's argument thus drew on corporate models of executive authority to construe both the conferences and the episcopacy. It could then assert that the various parts of the whole UMC were agents of each other and mutually liable. To do otherwise would be unfair to similar corporate actors. In the court's view, any preferential treatment to religion, especially to treat it differently in civil law because of its own

---

51. *Barr v. United Methodist Church,* 329.

52. *Barr v. United Methodist Church,* 328. (Internal quotation from *Carnes v. Smith.*)

religious self-understanding, would do more injury to the establishment clause than to the free exercise clause. Moreover, it would provide a shield under which unscrupulous and fraudulent operators could seek religious immunity from lawsuits. The Pacific Homes case did not present any religious activities to the court but only commercial activities conducted by a religious body. Such secular activities had to comply fully with ordinary civil law.

Church defendants appealed to the US Supreme Court to render a final verdict on its suability, but the Court refused to do so on technical grounds. Fearing an even more ruinous siege of litigation the Conference and related defendants settled with the plaintiffs out of court for $21 million, not to mention legal costs of over $4 million to the Conference and national agencies of the church also involved in the suits. Various conferences and agencies of the UMC loaned the Conference funds to make these settlement payments.

As a result, every entity within the United Methodist connection had to take steps to establish "firewalls" to limit the liability of the entities comprising the entire denomination for the actions of its constituent parts.[53] Such firewalls called into question the nature of "connection" among conferences, agencies, and local churches.

One of the most striking effects of the Pacific Homes case was to lead the church's General Conference in 1984 to change the *Book of Discipline* so that the Council of Bishops, rather than being the "corporate" expression of episcopal leadership in the church, as cited by the Barr court, would now be the "collegial" expression of that leadership.[54] The language of corporate unity was replaced by that of cooperation among independent superintendents overseeing the work of the ministers—something that may have been more congenial to Wesley's intention but nevertheless marked a change of course among United Methodists with regard to their conception of episcopacy.

The California Appeals Court ruling presently stands as an anomaly, and the US Supreme Court has never issued a ruling which would

53. See Weeks, "Methodist Approach," for some examples. Some Lutheran organizational adjustments are described in Draheim, "Lutheran Approach."

54. See *Discipline* (1980), para. 526. In addition, the 1980 *Discipline* added specific language denying that either "master-servant" or "principal-agent" relationships—both with legal precedent—defined the relationship of general agencies and the General Conference. General agencies are amenable to the General Conference, but General Conference—which has no continuing executive body between sessions—is not liable for the actions of general agencies.

settle the underlying Constitutional issues. While courts in at least nine states subsequently dismissed suits brought against the denomination as a whole, legal actions of this kind continue to plague religious organizations.[55]

THE ISSUES IN LIGHT OF THE THREE MODES OF CONSTITUTIONAL ORDER

This complex and costly legal battle serves as a stethoscope which can enable us to hear deep internal dynamics in the relationship between civil and ecclesial constitutional forms in American culture. The litigation itself revolved around questions of establishment and free exercise of religion, the suability of churches, and the degree to which civil courts can construe questions of church law and ecclesiology.

UMC authorities had argued the church's case not only on the grounds of sheer organizational structure but also as a case of autonomy and free exercise for religious groups whose organizational self-understanding should be protected from re-interpretation by a civil court. Such an insulation of internal church decisions was well grounded in Constitutional law. A long series of legal precedents had confirmed that civil courts cannot interfere in internal church disputes, even when it appears that church authorities have acted in violation of their own established procedures.[56] There simply is no "higher court" to adjudicate theological matters.

However, this legal tradition was carved out largely in the face of governmental intrusions on religion. Liability suits now force the court to use these traditions in a radically different direction—a new context of consumer protection and marketplace performance. A church's dealings with an external marketplace are quite different from internal relationships historically insulated from court interference. In external matters the Appeals Court and in a related case the US Supreme Court, held that

---

55. Gaffney, Jr., Sorenson, and Griffin, *Ascending Liability*, 103–32.

56. However, in two Georgia cases involving Presbyterian churches in 1969 and 1979 the Court developed the concept of "neutral principles" by which to approach church property disputes without entangling itself in doctrinal matters. In fact, however, "neutrality" simply means formulations familiar to civil law. See Gaffney, Jr., Sorenson, and Griffin, *Ascending Liability*, 60–75, and Laycock, "Church Autonomy."

its arm of adjudication can be long indeed.[57] This change demands not only legal reinterpretation but ecclesiological rethinking as well.

The churches now face a new legal context in which to approach questions of ecclesiology. This new context affects the way we think about questions of association and federal order within religious organizations, of the unique character of publics as distinct from markets within the society and the church, and the nature of corporate liability in an ecclesial setting. How then can our awareness of the three modes of constitutional order help us illuminate the key issues facing the church in this situation?

CHURCH AS CORPORATION OR CHURCH AS PUBLIC ASSEMBLY?

The Pacific Homes case revealed a deep tension between the perspectives of assembly and federalism of the one hand and those of corporation on the other, both in ecclesial and civil realms. The trial court argued from traditions of public assembly and saw the case in terms of religious freedom. The appeals court argued from the perspectives of corporations and markets and saw the case as a question of corporate liability for marketplace derelictions. The Church was caught in the middle because it engaged in many corporate practices without a clear ecclesiological understanding of the differences between federal and corporate order and of the role of public assembly in defining the nature of the church.

The Church as a Corporate Entity

Indeed, the church has a profound tradition which understands the church in a corporate way. The church is seen as "the body of Christ" in two ways—as a spiritual or mystical body, with its unity in the resurrected Christ, and as a temporal body in which it is institutionalized for ordinary life among other worldly institutions. Language attaching to its life as a mystical body—unity, catholicity, common baptism and faith,

---

57. A separate appeal by the General Council on Finance and Administration of the UMC was rebuffed by the US Supreme Court with Justice Rehnquist's claim that "... the First and Fourteenth Amendments [cannot] prevent a civil court from independently examining, and making the ultimate decision regarding, the structure and actual operation of a hierarchical church and its constituent units in an action such as this." 439 U.S. 1355 (1979). Cited with comment in Gaffney, Jr., Sorenson, and Griffin, *Ascending Liability*, 75.

and consummation at the end of history—reflects this dimension of the church's ultimate character and its source in Jesus Christ. The church as a temporal, or secular, body takes on various institutional forms depending on history, culture, and circumstances of law and economics. This is the continuing incarnation of Christ. The seeming fragmented character of this corporality is not necessarily a corruption of the mystical body but the form of its work in the world. The church's factual brokenness continues the cruciform brokenness of Jesus Christ. How these two ecclesial bodies—spiritual and temporal, visible and invisible, broken and consummated—are related has ignited a great deal of theological controversy over the centuries.

Despite these distinctions, however, the church's "temporal" body cannot be severed from its "spiritual" body. Its organizational form is an expression of its spiritual body. Both constitute the one "body of Christ." Within this tradition, the Methodist "connection" could not be merely otherworldly. It had to have real presence in the world of human action. But, the church argued, the real presence of this connection did not necessarily have an ordinary legal form. There is a real arena of human activity—indeed, the most important kind of public activity—that exists prior to and apart from civil law. The appeals court, however, followed the logic of a corporatist interpretation in another direction—toward the marketplace of hierarchical business corporations rather than the republic of church and state.

Indeed, many of the arguments in this case agreed on the use of language of body and corporation to describe the organizational issues at stake. The *Discipline* drew on this language in describing the Council of Bishops as "the corporate expression of episcopal leadership in the church."[58] In United Methodist theological tradition, this language refers to the way Christ is really and truly present through the teaching and general pastoral oversight (the root of the word "episcopal") of the bishops. It is the way the mind of Christ really has a body among us. But the control of this body through the episcopal representative is essentially a matter of persuasion seeking the assent of faith rather than a command seeking automatic obedience. However, the appeals court seized on this corporal language within the framework of contemporary business enterprise to argue that therefore the Council of Bishops, with its president, is the chief executive organ of the church.

---

58. *Discipline*, para. 526.

Drawing on corporatist language in the *Discipline* and seeing in the term "connectional" a hierarchical principal, the Appeals Court agreed with the plaintiffs' claim that the UMC was a hierarchical denomination with a central principal agent, to which all subordinate agents were responsible. To do so, it drew upon a long-standing legal precedent which reduces a church's ecclesiology to a choice between "hierarchical" and "congregational" forms. This simplistic distinction derives from an 1872 ruling by the US Supreme Court in *Watson v. Jones* concerning a dispute over church property.[59]

In that case the Court first identified three forms of church order relevant to legal cases. In the first form church property is held in a trust which specifies the intent of the original donor(s). Those who administer that trust most closely in accord with the trust's original intent then have rightful claim to the property. The problem with such an approach is that it draws the courts into theological examinations in order to assess the meaning of the original trust. Such a theological function for the court would eventually erode the distinction between civil and religious spheres. Republican governance, with its inherent plurality of all kinds of expression, ought not go down that road, argued the Supreme Court.

The United Methodist Church, however, in keeping with the English trust tradition and adapting it to the American civil environment, has explicitly construed its entire property arrangement in terms of trusts. All property under the United Methodist name is held "in trust for The United Methodist Church" to "be used, kept, and maintained as a place of divine worship of the United Methodist ministry and members of The United Methodist Church."[60] Such a web of trusts is what many Methodists mean by calling their church a "connectional" church. Because American civil law since *Watson v. Jones* largely has chosen not to pursue the trust arrangement for settling church disputes, the courts have not developed any understanding of the intricate meanings of this trust connectionalism.[61] What indeed does it mean to hold property "in trust" for the United Methodist Church if such a person, real or fictional, has no address and cannot appear in court? Furthermore, how can a court use "neutral principles" to separate out the theological and the "secular" meanings of the trust?

59. *Watson v. Jones*.

60. *Discipline*, para. 2501, 2503.

61. In the Georgia case of *Carnes v. Smith* (1976) the state courts did use trust doctrine, but the US Supreme Court never reviewed the case.

The second form cited in *Watson v. Jones* involves congregations which are "strictly independent of other ecclesiastical associations" and owe "no fealty or obligation to any higher authority." Such cases can be resolved "by the ordinary principles which govern voluntary associations"—majority vote of the members or other procedures previously defined by association. Because of the seemingly clean character of this ecclesial type within American law it actually functions as the normative case.[62] Just as entrepreneurial corporatism is the normative type of American business, so congregationalism is the norm in religious associations. To go beyond pure congregationalism for the sake of the spiritual unity of the church universal as the body of Christ is to risk unceasing lawsuits in a litigious world.

The third form is one in which the conflicted group "is but a subordinate member of some general church organization in which there are superior ecclesiastical tribunals with a general and ultimate power of control more or less complete in some supreme judicatory over the whole membership of that general organization." In this case, the decisions of the highest judicial organ must be respected by the court.

The court in *Watson v. Jones* did not in fact use the term "hierarchy." It focused on the judicial appeals process within a religious organization rather than its line of executive command. In this sense, it was working within the world of federalism more than corporatism. However, by the time we reach the Pacific Homes case we find that the term "hierarchy" has been used to define such third types, possibly because of the use of that term in the context of priestly governance in cases involving the Orthodox Church.[63] Moreover, this "hierarchy" concerns not merely judicial appeals within a church but control over employment as well as property. In short, it is the hierarchy of a bureaucratic corporation. If the church is "congregational" then liability stops with the congregation. However, if it is "hierarchical," then liability touches all "levels" of the organization, conceived in corporatist terms.

This presents the churches with a critical problem. When "hierarchy" becomes conflated with the corporate forms of organization typical of American business firms, and when "non-profit" organizations are interpreted in terms of for-profit organizations, then depiction of a religious organization as "hierarchical" leads the courts to treat churches

---

62. For a detailed examination of this point see White, "Constitutional Norms."

63. *Kedroff v. St. Nicholas Cathedral,* and *Serbian Orthodox Diocese v. Milivojevich* involved matters of ecclesiastical appointment and related control of church property.

the same as other organizations engaged in commercial affairs—all of this in the name of "fairness" and justice with regard to the claims of persons against church-related agencies. Here we have moved from disputes between factions within churches to those between church authorities and their "customers" or "clients," conceived in terms of business life. The power of this business or market model of religion has worked to overturn the historic immunity to suit enjoyed by religious and charitable organizations. No longer do they represent a public or transcendent purpose rooted in a people's common good. They are treated like one interested party among others, all competing for the allegiance of free individuals in the marketplace.

Within this context of claims by "customers" and "clients" against church agencies, the formulation of *Watson v. Jones* becomes a question of who possesses liability within a complex church organization rather than whether church disputes should be settled within the church or in the civil courts. This collision of two worldviews—an ancient spiritual tradition and a modern commercial-legal positivism—lay at the heart of the multi-million dollar Pacific Homes dispute.

Federal "Connection" or Episcopal Hierarchy?

Over against the courts' conceptions of corporate liability the church posed its own terminology of connection, which draws in a complex way on its roots in practices of assembly and federalism as well as its theological understanding of the church as the Body of Christ. However, on this central term in United Methodist ecclesiology church spokespersons could find little agreement. Neither the court nor the church could turn it into a useful organizational or legal concept. The reason for this confusion is embedded not only in Methodist origins, principally in its reconstrual of episcopacy, but in the way Methodism engaged the corporate and legal forms of trusteeship.

The "Connexion" originally referred to the personal relationship between Mr. Wesley and his preachers. It was not a connection of congregations or agencies but a relationship among preachers who were seeking the regeneration of individuals through class meetings and societies, and a spreading of holiness through the world. Wesley preferred to call his oversight of the Connexion a superintendency, not an episcopacy, since the term bishop referred to Anglican prelates of the established church.

The functional bias of superintendency stood in sharp contrast to the corporate notion of the Bishop as a representative of the church as the Body of Christ.

As Methodism developed in America it adopted the organizational patterns around it, and it proved difficult to maintain Wesley's original conception of connection within a world Wesley could neither imagine nor desire. Connectionalism had to be accommodated to the forms of association known to American business and law. From a legal standpoint connection became equated with trust deeds for holding property. As the itinerant preachers began to settle into stations their preaching "charges" began to become congregations. When Methodists adopted bureaucratic organization toward the end of the nineteenth century, they inevitably assumed the corporate forms that made this possible in law. This is the leading edge of hierarchy and corporate liability that the Pacific Homes court and plaintiffs engaged.

In the course of this development Methodists could not clearly articulate what is "connected" and how these parts are connected to each other. Since congregations, conferences, and councils were not the original units of connection, they could not fit the original spiritual and ministerial meaning of Wesley. Yet these were the connections the court and the plaintiffs were looking for. They therefore took the theological language of the Wesleyan tradition and attached it to places totally unforeseen and even unimaginable to Wesley—and indeed to his present heirs in the leadership of the United Methodist Church.

In the pleadings and affidavits presented to the Pacific Homes courts we find a variety of interpretations of the church as a "connection." The UMC, claimed one church brief, is a "spiritual 'connection'" whose units—"churches, conferences, boards, agencies or institutions"—operate "within a loose, non-authoritarian confederation pursuant to the *Book of Discipline*."[64] In an important affidavit to the court the distinguished United Methodist sociologist of religious organizations Murray Leiffer put it this way:

> The general United Methodist denomination is a voluntary religious movement and connectional network of millions of persons, known as 'members,' and literally thousands of units, denominated as local churches, conferences, boards . . . [This]

---

64. "Memorandum . . . to Dismiss . . . " by attorneys representing persons served as representing the United Methodist Church in *Trigg et al, vs. Pacific Methodist Investment Fund, et al.*, 5.

connectional religious denomination and movement [is] structured around two fundamental church units, namely, (a) the local Charge, with its local congregation or congregations and (b) the Annual Conference.

However, Leiffer noted, these fundamental units may not act as agents for other units nor "bind or obligate them in any way." He concluded that

> the UMC is not a single entity . . . It is not a unified corporate body, but in actuality is an international religious faith and denomination which constitutionally declares itself to be a branch of the Church universal and the ministry of Christ.[65]

On the one hand Leiffer wanted to stress the autonomy of the various "units" in order to resist court intrusion on liability issues, which require relationships of agency and mutuality. On the other hand he wanted to preserve the organic sense of the universal Body of Christ. The exigencies of the situation pressed him to a "loose confederation" of "autonomous units"—a kind of corporate congregationalism. In the process, almost unnoticed, the United Methodist connection moved, again under the impact of civil law, from being a relation of superintendents with ministers to being a relationship among "churches," "conferences," "units" and "members." Rather than the deep mutuality and accountability characterizing Wesley's relationship with his ministers, the meaning of "connection" brought before the court emphasized the autonomy of "spiritually confederated" groups.

Recovering the Church as Assemblies in Covenant

The problem with approaching a concept of church from the standpoint of business models of incorporation is that corporate models have little room for the independent initiative and extra-organizational activity that struggles to articulation in the confusions of the Methodist statements about "connection." While the corporate analogies draw on important strands of theological tradition, they cannot grasp the peculiar mix of effervescence and order which characterize not only Methodism but mainline American religion. This peculiar mix is also embedded in the rich

---

65. Leiffer, "Affadavit," 4, 20, 21.

tapestry of publics that generated the original American Constitutional order.

The spontaneous mass meetings, revivals, camp meetings, and conferences driving the engine of personal revival in the early American Republic did not fit the corporate understandings of established Christianity. Rather, they brought to mind the ideas of assembly, congregation, convention—ideas resonant with the root word *ekklesia* and the ideas of public in the republican theorists of the day.

While a small number of theologians were able to grasp this eruption of religious experience in the early nineteenth century as a fundamental ecclesiological change (or renewal, if you will), most people understood these "great awakenings" as matters of individual experience—a psychological revolution. Thus, it was difficult to interpret the ecclesiological shift as an issue both for theology and society. Individualism rather than a sense of the total system of organized relationships carried the day. The meaning of these new understandings of organizational relationships—a shift from the hierarchy of command to a covenant or compact of distinct but related publics—was overwhelmed by individualistic concepts of personal experience.

Like these religious assemblies, the public meetings, conventions and conferences typical of the early republican movement also have no ongoing corporate form. They are not "constituted" in the US Constitution and have an ambiguous status in American society despite the veneration they receive. Yet they are the ecstatic moments of inspiration without which a republic cannot be sustained. Without these "effervescent" moments[66] the ordinary barriers of family, race, class, geography, and even language cannot be bridged to create a wider republic. As a part of this republican development, American Methodism is an effort to sustain such peculiar publics within a culture which generally speaks the language of commercial individualism and legal positivism. Spontaneous publics and the "connections" that sustain them are neither real nor legally responsible. Whether this crucial reality can be sustained in such an environment is the challenge facing not only the United Methodism of the Pacific Homes case but republican orders in general.

In response to the Pacific Homes case we need to step back from the language of corporation and hierarchy and reclaim the language of

---

66. For Emile Durkheim's characterization of those moments of collective enthusiasm that create new forms of legitimate social order, see Durkheim, *Elementary Forms*, 474–75. See also Arendt, *On Revolution*, ch. 3 and 6.

assembly that energized the original Methodist movement. We then need to examine the language of "connection" more closely in the light of federalist theories. From that point we might move to an ecclesiology that can be in critical conversation with federal republican theories and practices.

How, then, might we reconstrue connection in a way that avoids misplaced corporatism as well as a fragmented congregationalism? The federalist language contains not only a heritage of political wisdom for safeguarding the vitality of republics but also leads us to its roots in the biblical notion of covenant, both of which may assist us here.

In the court case we have examined, all parties referred continuously to the *Discipline* as in fact the body of commitments constituting the connection. Indeed, it is the *Discipline* which defines the mutual accountability of the various units—including ministers, bishops, conferences and agencies—constituting United Methodism. In that sense it is the covenant of this federal ecclesial order. However, "covenant" is not part of Methodism's original vocabulary, largely because of Methodism's origins, even though its practices are thoroughly immersed in the republican effervescence and law-like mutual accountability typical of the covenantal tradition and its federalist expressions. If anything *is* the "connection" it is the institutional form constituted by this covenant-like document. The *Discipline* is the interim authority, rather than any bishop, president, or council, between the periodic conferences which guide the organization.

In these respects United Methodism rehearses, perhaps more explicitly, tensions and deficiencies within the American polity itself. The more one presses an emphasis on the centrality of the spirited public assembly as the practical center of the church or of a society, the more one moves to some notion of covenant as the connecting bonds within and among those publics. The more one stresses the bodily functionality of a group as a stable organism, the more one moves to some notion of headship as the connecting and representative bond. Methodism exhibits both this covenantal "connection" and this corporatism—tendencies which we see in American society generally. Its inability to give clear voice to its basic self-understanding within this tension reflects American disputes over federal order and public freedom as well as offering some illuminating points of entry into a critical engagement with them.

Such a federalist vision of United Methodist ecclesiology can save it from a reductionist view of connection as a corporate hierarchy and renew the role of public assembly in its ecclesiology. In this vision a vital federalism is not a pyramid of representation and command but a web of cross-cutting accountabilities among a variety of publics. But these constitutional relationships are always interrupted, as it were, by the overarching commitments of the people making up the total church. Bishops interrupt to teach. Evangelistic movements offend and even bypass congregations to affect conferences. Conferences establish agencies which make demands on local congregations. This vital mix of spontaneity and order—of assembly and covenant—continuously presses the participants to articulate anew the fundamental covenant informing their constituted federal order.

CONCLUSION

In the many legislative changes of the past century the United Methodist Church has largely ignored fundamental issues of ecclesiology and polity. In particular the connectional covenant by which United Methodists agree to live together needs major attention from the laity, clergy, and scholars of the church. In that regard the statement "The Journey of a Connectional People" added to the *Discipline* in 1988 makes a beginning, but more in the direction of exhortation than as a coherent formulation of the constitutional order of the church.[67]

We believe that an analysis of the three modes of constitutional order described above can help the church sort through some of these major ecclesiological and polity issues. We can only illustrate this with a few examples here.

First, the principles of public assembly within a federal framework require that we look at the local congregation as a participatory assembly rather than as an audience of individuals gathered for evangelistic preaching and mission activism. An ecclesial sense of assembly within a federal framework of covenantal responsibility can enhance the congregation's sense of being the fullness of church in each place through the recognition and exercise of the members' baptismal gifts. The people gathered together can then become the subject rather than the object of ministry and mission. One implication of this would be to base representation of congregations in the annual conference on the number of lay

---

67. *Discipline*, para.112. In subsequent editions this broader effort to explain how connectionalism works has been reduced to a few sentences.

members in a congregation rather than on the number of clergy assigned to a pastoral charge.

Second, a reconstituted understanding of corporatism in ecclesiology would lead us to see the episcopacy not as an executive representative of the church but as a prophetic overseer seeking to bring to bear the mind of Christ in the midst of the people's life and faith together. As the trend continues toward the merger of annual conferences with each bishop presiding at only one conference, it will become increasingly difficult to distinguish between the bishop as executive manager of conference programs and the bishop as general superintendent assigned to a particular area within the whole connection. The integration of corporatist principles with those of assembly and federalism would enable the church to embrace more fully a model of episcopacy centered on proclamation, teaching, sacramental leadership, and spiritual guidance—in short, prophetic oversight of the United Methodist people.

Third, a reformulation of the corporatist aspects of the church within a framework of assembly and federalism can also help us rethink the church's understanding of "service." While service (*diakonia*) is a fundamental practice of Christian faith, many church leaders, agencies, and programs are encouraging the church to adopt marketplace understandings of the "service industries." Thinking about the church as a service provider with clients and customers obscures the baptismal citizenship of believers as participants and actors with responsibility for their own assemblies. It also subjects the church unnecessarily to practices of corporate law whose main purpose is to protect corporations from liability in the marketplace. An ecclesiology rooted more in assembly and federal connection sees the church as an assembly of assemblies within an overarching covenant—a "political" movement anticipating God's ultimate order of governance more than a business corporation providing services within a limited market segment.

Finally, within this perspective the church must take the *Book of Discipline* itself more seriously as its primary constituting covenant. This will require that much greater care be given to the consistency between constitutional order and specific legislation within the *Discipline*. As matters stand now, few United Methodists engage the Constitution in their deliberations. Meanwhile legislative paragraphs are full of inconsistencies and uneven attention to their place within the whole. They are the record of a series of General Conference actions but not the fabric of a coherent covenant of interlocking responsibilities. When no connection is made in the *Discipline* between the chapter on the ministry of all Christians and the paragraphs on church membership; when one has to read more than half of the *Discipline*'s chapter on ordained ministry before finding what the church believes ordination is; or when one looks in vain through the entire chapter on episcopacy for an ecclesiology to support the United Methodist plan of superintendency; it becomes obvious that the church has simply taken for granted some of its most basic elements.

The church needs to achieve clarity about its underlying ecclesiology for its own internal good order, but also for defining itself effectively over against the implicit ecclesiologies present in the way other institutions, especially those of law and business, view the church. Constitutional confusions within the *Discipline* spring from the effervescence of the church's numerous assemblies. We believe, however, that the time has come to attend to these dimensions of ecclesiological order, so that the church can maintain its integrity as it addresses the powerful institutions of American culture.

14

SEALS *and* SPRINGBOKS
Theological Reflections on Constitutionalism and South African Culture

SOUTH AFRICA NOW HAS a living democratic Constitution. Its lengthy period of germination and modification has yielded a young sapling that is already redefining the political landscape of the country. Presidents, parliamentarians, and everyday citizens now seek shelter under its canopy, tended by a Constitutional Court that is also new to these ways. Now its stewards and beneficiaries must ask how this tree will sink deep roots into the cultural soil of the people. How will it be nurtured to full maturity? Certainly the storms of economic pressure will buffet it, sometimes leaving flood, sometimes drought, but what about the deep soil of religious and cultural values? How can the Constitutional tree find deep root in these so that it can weather the turmoil of economic and political conflict?

Political theorists call this rootage "legitimation." How can the Constitution find sources of legitimation in the deep values and religious cultures of the country? Without the fertilization of legitimacy, the Constitutional tree will not become the place where conflicts can be resolved peacefully. It will be unable to organize the comings and goings of people in the public world. If it draws only from one root it will be stunted and weak. If its roots are too spread out and shallow, it will not be able to weather the inevitable storms of political and economic conflict. Legitimation is not merely a question of reinforcing one regime or another, but of providing the authoritative context within which the ordinary conflict and cooperation of human life can be organized for the common good.

Legitimation is mediated by powerful symbols shared by the people. Symbols mediate between the unseen soil of religious intuition, memory, and hope and the visible trees of law, society, and human institutions. Shared images, songs, words, places of memory, and ritual actions sink into people's emotions and minds. These symbols not only galvanize emotional commitment but also organize the way we see the world, each other, and our role in society. They are common reference points for resolving disagreements and organizing collective action. They are compact programs of social behavior.

Religious traditions are usually the guardians of this symbolic world. However, symbols are not merely the property of recognized religious institutions and traditions. They also grow up of their own accord from the soil of human events — traumatic sacrifices, ecstatic celebrations, and recurring acts of interdependence. The old official symbols safeguarded by religion continually shape and are reshaped by these fresh symbols. Some of these symbols are genuinely new, like the new South African flag. Others are turned up in the soil by recent events. They emerge from the depths of memory with fresh power, like the ancient images of *ubuntu* and of "Africa" itself.

In these few pages I would like to strike some old and new symbols together to see what sparks might illuminate the search for roots for the new Constitutional tree. How might Christians in South Africa appropriate and engage emergent Constitutional symbols from within their own forest of symbols? How can the models of right action implied in received biblical and Christian symbols inform the symbols that are emerging to legitimate the nascent common life of the South African people? What reconstructions might these powerful contemporary symbols promote within the traditional symbols, rituals, and models of the churches? How can these symbolic worlds draw together closely enough to shake hands but not so closely that the cultural power of symbols incestuously marries the machinations of power?

To engage in this process of critical reflection I want to explore two important symbolic developments—the re-creation of Robben Island as a heritage center and the struggle for redefining the meaning of rugby. I do so as an American Christian who has explored such symbolic issues not only in America but in Germany and India as well, all of which have struggled with the creation of constitutional orders within a history of warfare, internal oppression, and cultural pluralism.[1] South Africa's recent experience offers all of us a fresh opportunity to try to understand the tangled relationships between religious symbols and constitutional development.

ROBBEN ISLAND

Table Mountain not only shelters the hopes of a new republic at its feet but it also looks out at Robben Island, the island of seals, where prisoners of the old regime were incarcerated, beaten, and nearly worked to death.[2] President Nelson Mandela spent eighteen of his nearly twenty-eight imprisoned years on Robben Island. The seals that gave the island its name were clubbed to death for food and its penguins were butchered for fun by early sailors and settlers. This was the island's first shedding of blood. It would take fifteen generations to begin to end the violence.

Robben Island is now being made into a museum and conservation site. The seals and penguins have returned. Retrieving the island's complex past, however, will require more than recovery of the seals. Not only was it used for prisoners from the seventeenth century onward, but in the nineteenth century many people rejected by the powerful were exiled here—lepers, the insane, and the tubercular. Like the former wildlife of this fragile spit they were forgotten, but now they are being given a new voice, along with the seals, penguins, birds, and vegetation of this rocky remnant of the mainland. Those who were at the periphery of society are now being placed at the center. Visitors are escorted around the prison by former inmates, who can tell of its brutality as well as how it became a virtual university for a government in exile. The reclamation of their years of disciplined survival is mirrored in the reclamation of the island. Their integration into a new society reflects the reclamation of the plants and animals who shared their lives.

At the core of Robben Island's symbolism is the transformation of the outcaste into the participant, the prisoner into the citizen, the rejected

---

1. Everett, *Religion, Federalism*.

2. I am drawing here on several sources: Alexander, *Nation Building*; Deacon, *The Island*; de Villiers, *Robben Island*; Kathrada et al., *Robben Island Exhibition*; Kathrada, Mandela and Mgxashe, *Robben Island*; Mandela, *Long Walk to Freedom*; Naidoo, *Island in Chains*; and Schadeberg, *Voices*. I am also indebted to Dr. Gerard Corsane, Co-ordinator of the Robben Island Training Program, for his helpful comments and information.

into the cornerstone of the new order. It is becoming almost impossible for foreign dignitaries to avoid paying a visit to the cells that held the first president and many of the new leaders of a democratic South Africa. The spiffy new catamaran that shuttles tourists (or are they pilgrims?) across the bay is named *Makana*, after the nineteenth century patriotic Xhosa warrior and prophet, who was imprisoned there by the British. He died when his boat capsized in an attempted escape, a far cry from the easy passage on today's sleek vessel. At its launching it was blessed by representatives from all of South Africa's faith traditions. This reversal of Robben Island's images is now a powerful legitimator of a system of human rights that ascribes to each individual, no matter how rejected, the dignity of presidency.

This bond between past oppression and present election not only resonates with the deepest experience of vast numbers of South Africans but also with the biblical heritage most of them share. The law, the Torah, begins with the words "I am Yahweh, your God, who brought you up out of Egypt, out of the house of slavery." It is this memory of slavery that was supposed to shape, guide, and sustain Israel's life under law. It was the gift of the law that made possible a life of equality and freedom as opposed to one of submission to the tyranny of human beings, no matter how benevolently they might exercise their domination. Law is embedded in a covenant with a God who liberates people from enslavement to false powers. Just as Israel and the Church must frame their understanding of law within a covenant of liberty, so the new South Africa must never forget its miracle of liberation from enslavement. This is the first symbolic function of Robben Island.

Ecological rehabilitation is the second symbolism of the island. Visitors to the island—that is, those who are not gawking at boats, gulls, and the receding lineaments of Table Mountain—view a video in which the island itself recounts its primeval alienation from the mainland. Recent visitors to the Afrikaner monument at Blood River may recall a similar theme in the way the great Trek is framed within the separation of the continents of Gondwanaland hundreds of millions of years ago. The island, once separated from the mainland, then became the repository of all that humans wanted to forget, a rejection of the rich ecology of seals, birds, and plants spawned in its earlier isolation. Rather than being a theater of hospitality for migrants it became a war zone of fearful and aggressive self-defense. The mountains did not clap their hands to witness God's covenant with the people of the earth, nor did the rivers laugh with celebrations of the life that feeds along their banks. The political drive for reconciliation under the new Constitution echoes the ecological struggle to reconcile humanity with a land that it has plundered, exploited, and destroyed. The reconstitution of the people needs to find expression in the reconciliation of the land.

The theme of ecological separation and coexistence is, however, a slippery one. The separate development slogans of racist ideologies find an unfortunate mirror in today's ecological desire to disentangle native plants from "alien invaders," all for the sake of an assumed natural equilibrium. While botanical classification may name the differences among types of beings, the human search is for a way to live in dense integration, for this search for union is as powerful as any thirst for treks and migration. How the new Constitution will integrate the diversities of language, custom, ancestry, and religion with the common voice of citizenship is one of the greatest challenges before it. Perhaps the theme of migration can help the country enter positively into a world of refugees and wanderers. It might be able to cultivate a sense of global citizenship in a world of nations. Memory of its violent past of alienation, imprisonment, and exile must constantly question every separation that betrays our common humanity. The theme of diversity, echoed in the countless protea and fynbos of the Cape, can help us remember and treasure the particular beauties of God's imagination, but it must always remember the common bed which every species shares together.

The land is not only bed but also rocky floor. Visitors must follow the prisoners to the quarry, down into that hole in the earth, where stones must feed our need for building walls and roads. They must get a glimpse of human sacrifice mediating between the earth's indifferent grip and impertinent human construction. The quarry was a "Pool of Blood and Tears." After all those years of blinding, sweating, backbreaking toil only a pile of stones remains—a reunion monument left behind when former prisoners gathered in 1995 with their President to mark their lengthy sacrifice there. It is an altar of their devotion. It is the reminder beside their Jordan, lest their descendants fail to keep their promises. Eight hundred miles away a similar pile of stones stands among the iron wagon monuments of Blood River. Not far away, at Islandwana's windswept hill, the nameless Zulu warriors who fell in their famous victory over British troops find sole expression in those rock piles amidst Greco-Roman monuments chiseled with "For King and Empire," "Queen and Country." The land contains them all among a competition of the gods. Will the piles stand in polytheistic loneliness or will the land provide a common narrative of struggle for the way from guns and spears to the constitution of democracy?

Robben Island is not only about land but also about water. Africa's bonds with the Americas and Antarctica are severed by a boundless South Atlantic. It is the cold Atlantic's arms that made the island not only a refuge of sustenance but also a prison of alienation. The noun that is the island only exists within the sentence that is the crossing of the waters. To get to the island prisoners had to cross the waters in the dark holds of rough ships. They had to live into the middle passage of slave ships to the Americas. They had to enter the belly of the whale. They had to cross the river of death. They had to die into the Jordan of exile to go to the far land where Pharaoh ruled.

To get back to freedom they had to cross the waters again. The life they left had died. They had to get ready for new life, even in the midst of death. The waters were baptism and purification. They were unbridgeable moat of isolation and resplendent frame of the voice of resistance. They were the waters of death as well as of the new creation, for it was here that people lost much of their petty differences and learned the values of shared equality, common humanity, and the precious fragility of freedom. Their random nameless numbers became the alphabet of democratic equality, as prisoners struggled for equal clothing, equal food, and equal treatment. In the darkest days of the struggle for full citizenship the prison, as so many testified, was a water of rebirth amidst terrible pains of labor. The struggle to survive with dignity generated paradigms of responsibility, accountability, care for particulars, and creative ingenuity necessary for a new order. How those values will find sustenance in the new soil of grinding poverty and gravy trains is yet an open question.

Finally, Robben Island symbolizes the struggle between silence and voice. It is not that voices had to overcome the silence. The silence gave people time to think, to cultivate the discourse of the self, even to pray. It was a womb in which memory could gestate hope. To walk quietly on Robben Island is to remember that without the silence of thought there can be no intelligent discourse of liberty. To be a self, dignified within the parliament of our internal deliberations, requires silence. At the same time, the silence exists to make possible a speech that says something. It makes possible a speech that is a reflected response to listening. The remove of Robben Island reminds us that the peril of democracy is first that we would hear too little, second that we would be overwhelmed with the cacophony of too much. It reminds us that the warden's desire for silence already contains the counterpoint of liberation. The abusive father's house of silent children can become a shared household of conversation. The prison could become a university of mutual learning. It could redefine parliament as a realm of mutual education and cooperation rather than a boxing ring of little tyrants. The silent reflection of Robben Island speaks across the water to the Babel of the fractious city.

The Babel of commerce has posed the final challenge to Robben Island as a sacred symbol for a new polity. Some wanted to turn it into a profitable resort, with hotels, marinas, and the bric-a-brac of tourist trade. With that temptation set aside, the island must still negotiate the difficult waters between economic viability and symbolic recollection. It must balance civil sacrality with commercial profit. This, too, is a metaphor of the order to come. The Island's effort to honor the gods of ecology, democracy, and commerce will be a parable for the country. It is a still unanswered question. The Island interrogates even as it remembers. The seals have returned but how will they survive?

THE SPRINGBOKS

Robben Island is not the only symbol of the struggle between commerce and Constitution. The search for symbols of national unity, unrewarded in the tangled undergrowth of religion, language, race, and heritage, seeks resolution on the fields of football and cricket. Though the alien cargo of empire, these sports have long since planted roots in African soil. However, over the past decades different people have gathered under different trees, they played in different fields. Soccer came to be dominated by Blacks, cricket remained the game of the English, and rugby came to be the province of the Afrikaner.[3] During apartheid, the exclusion of South Africans from international sports may have done more to undermine national commitment to the government's policies than economic sanctions. Now, with the lifting of the heavy burden of apartheid and international ostracism, they could take their new flag around the world. The people might find unity in the totems of sport.

---

3. In addition to current newspaper reports I am drawing on Griffiths, *One Team, One Country*; Nauright, "Sustaining Masculine Hegemony"; Nauright and Black, "'Hitting Them Where It Hurts'"; and Grundlingh, Odendaal and Spies, *Beyond the Tryline*.

But there were controversies. In 1992, at the first international rugby game in the country since the lifting of apartheid, resurgent Afrikaner fans waved their old flags and sang the song of Afrikaner nationalism, *Die Stem*, spitting in the face of ANC leaders who asked for silent respect for people killed in the vicious massacre at Boipatong. Even if this old nationalism could be set aside, there were other symbolic hurdles. They involved a graceful animal—the springbok. This dazzling antelope, prancing in enormous bounds through the veld, had come to symbolize the best in any endeavor. It began to grace rugby on the first South African tour to England in 1906. Now there were not only springbok athletes, but springbok wool graders, artisans, businesses, and achievers in every field. Sadly, however, racism had tarnished the sacred springbok, for it had long been reserved for white athletes as the trademark of a government-controlled sporting world dominated by whites.

At first the new national leaders tried to set the springbok aside. The national soccer team became what they were called in Nguni by their adoring fans—Bafana, Bafana ("The Boys, The (Our) Boys!"). The gentlemen of cricket, with their leisurely five-day games punctuated with tea breaks, took the prickly blossom of the patient protea, a flowering bush which grows in unique array around the country. But rugby would not budge. Rugby players don't give up the ball. They hunker down and scrum until a new runner gets the ball. Rugby kept the springbok, though they let it be surrounded by some proteas—the national flower —and a rugby ball, contributed by the old black rugby union. But how was this symbolic compromise to be rooted in the dispositions of the people?

Virtually any South African can recite the story. After his election, President Mandela, recognizing the crucial symbolic importance of sport, showed up at key games to support the team as a unifying force. It was an opportunity to fold old Afrikaners into the new flag of unity. Both he and Archbishop Desmond Tutu accepted jerseys from the players. But the Rugby Union wouldn't move fast enough to shed its white Afrikaner image. So Mandela set up a government commission to investigate the South African Rugby Football Union. Louis Luyt, their burly leader, scrummed down and went to court. Luyt symbolized the strong, self-made man. A man of the soil, he now controlled commercial empires of fertilizer, farming, and sports. Soil, money, and sports formed a seamless garment for his impressive frame. He symbolized the old authorities of paternal strength—the law an extension of his will, the common good his household. This was the commanding Father of Reformed churches and the imperial King of English worshippers. It is not a symbol easily set aside in church, family, sports, or politics.

To show that the new government lives under the law and not above it, President Mandela came to court and testified. The nation was riveted to the drama. Within the struggle for the springbok the nation had to learn the supremacy of Constitution over President and Parliament. Not only was this a symbolism of covenant over would-be kings and chiefs, but of accountability over honor and shame. Both supporters of the President and of SARFU's president, Louis Luyt, knew that shaming was the game. But shaming could only occur within the world of personal honor, a world of personal status. It assumed a world of parental rule and communal belonging. Mandela was trying to change the game. To be ruled by the covenantal bonds of Constitution was to be accountable to the unseen, invisible, unfaced god of the Constitution, not to the person of the chief. It was not even an accountability to law—that the old Parliament could have claimed—for law as the will of the elect had led to the legalized injustice of apartheid. Parliamentary sovereignty had only been a way station on the long pilgrimage from personal rule to democratic self-governance. The English struggle to replace the king with parliament's law had only collectivized the will of the Father. Now the task was to replace the father's family with an assembly of citizens. South Africa had to be born again into a higher covenant of human rights and democratic equality. No fear of shame could break this higher loyalty. And so the teacher of this new covenant came to court, still stern in reprimand but unashamed to bear witness to a higher law. This law was not rooted any longer in a covenant of communal election to racial superiority but in a covenant of accountability for mutual service.

Though the government eventually lost the case, in the new democratic climate it was clear SARFU had gone too far. Louis Luyt, who had brought financial success to rugby, tried to play the hierarchical game of "bowing to no man, except to God," but the country was living into a world, not of submission, but of *ubuntu*. It was a world not of obedience to elders but of mutual cooperation and negotiation. "Submission to almighty God" had already been removed from the Preamble to the Constitution. Now it was removed as a style from sports. Submission to the Father was being replaced by accountability to the covenant of mutuality. The government, through the National Sports Commission, finally brought enough pressure to bear on SARFU to bring rugby back into the fold, enabling them to keep their sacred springbok within the new circle of national unity. A reorganized SARFU tendered a discrete apology to the President. Covering shame had become a private matter, serving the common good the public act.

As the heat of contestation dissipates, the country must confront the question of the degree to which government should control sports. To what degree, moreover, should sports be controlled by commerce? To the degree that sports is the mother of national symbols and unifying rituals, perhaps it too should be separated from profit-making and government control. At the heart of the springbok's sacred aura is the question of the independence of legitimating symbols from the claims of power. It is the new form of the cry for the separation of religion from the state.

Indeed, what institution *is* in charge of the deeper covenants of the people? What priesthood should control the character of its symbols? How do the symbols of a nation emerge? Indeed, what constitutes a nation? Racists had claimed that nations emerge as biological growths. First comes the "race," then the people, then the nation. Some even went on to ask a leader to personify the nation. Others settled with a parliament. But if the citizens must argue freely, is not their public beyond the arm of government's coercion? Is it not beyond the claims of commerce? Are not the people's cults of sport and prayer above the government and parliament? But if they are beyond control, how will the values crystallized in the new Constitution permeate the public's life? How shall the spirit of the Constitution and the deeper covenant of mutuality and democracy be mediated to a public culture that necessarily precedes the law? This was the delicate question underneath the struggle for the springbok emblem, the blessing of the people, and the support of the government. It is a question of how transcendent power creates and governs a people.

The contest over constitution, culture, and underlying covenant rehearses once again the deeply theological question, On what basis can people form relationships of trust beyond the bonds of real and fictive kinship? The biblical answer was to make covenants.[4] Relationships of promise are the alternatives to kinship and coercion. They are the bonds of consent and hope. Such consent is more than the agreements of a Kempton Park or Philadelphia, though they include them. Constitutions are but visible precipitates of promises that seek to cultivate trust and manage mistrust. Constitutions also contain preambles, and in South Africa's case, a "postamble." These "non-enforceable" claims are roots extending

---

4. For these perspectives on covenant see Everett, *God's Federal Republic*; Elazar, *Covenant and Polity*; and McCoy and Baker, *Fountainhead*.

down into the soil of memory, faith, and hope. They seek to embrace the subtle bonds that constitute a unified people out of lesser, limited loyalties. They link the memories of violent oppression to the promises of equality, reason, and persuasion. These are questions for which theology is the suitable language. The conundrums of constitutions seek resolution in the theologies of covenant.

Covenant itself is contested ground. Some people think the theme of covenant election underlies Afrikaner nationalism. Others say it was never used in that way. What is clear is that it was used at most to speak of God's relation to individuals or to a church community.[5] Covenant as a principle of solemn association to create a people across kinship boundaries has not been theological vernacular in a land rent with racial divides. When we start using the language of biblical covenant to talk about the creation of constitutional democracy, we open up a manifold of themes that need new voicing in the present conversation. The struggles over symbols that I have addressed here lead us to two fundamental covenant motifs—sacrifice and new creation.

Authentic covenant-making arises in the juncture between sacrifice and the vision of a new order. In a biblical perspective covenants were "cut." Animals were split open, foreskins were severed, precious beasts were offered up. Only that for which we have given up much can be the enduring object of our loyalty. This is a fact of human existence. For Christians who see in Jesus's death the supreme sacrifice, the new covenantal relationship possible for all of humanity resides in gratitude for a life poured out in non-violent persuasion and utter transparency. The enduring Christian mistake has been to assume that devotion to the cult of Christ should be the explicit basis for political regimes. This was the mistake of Christendom and Christian nationalism. To overcome this myopia we have to see the sacrifice of the past in the context of the new creation yet to emerge fully. It is a recreation of a common humanity, a common earth, a common universe. It is the perfection of our common created goodness. It is not the product of our religion but of God's

---

5. Akenson, *God's Peoples*, sees a deep and powerful covenantal "grid" of election within Afrikaner culture that shapes its sense of peoplehood, land, and rights. De Klerk, *The Puritans in Africa*, and Sparks, *Mind of South Africa*, while not as bold in their claims, hold that some form of this covenantal outlook is part of the baggage of the "Puritan" (DeKlerk) or Calvinist (Sparks) legacy in South Africa. Du Toit, "No Chosen People"; Hexham, *Irony of Apartheid*; and Smit, "Covenant and Ethics?" however, show that the term was never used explicitly in these nationalistic or racist ways and therefore might be available for contemporary associational symbolism.

creative and renewing spirit. Authentic covenants are promises to live into these gracious openings to creation's transformation. Both these accents—of gratitude and of hope—need to be held together in constructing the ongoing covenants of our lives. Our covenants of nationhood are but a fragile earnest of a wider reconciliation.

These are complex and torturous themes. Like all powerful symbols these too can fall victim to our insatiable lust for security. How might they inform the symbols shaping a new South Africa? Here we turn back to Robben Island for themes of sacrifice. It will be necessary to symbolize the sacrifices people made for the sake of a new common order. Sacrifices made by earlier oppressors will have to be pried open to reveal the longing for equality, liberty, and mutual respect they might contain. There will be new saints and icons, blood-remembered places, memorial rooms where screams once echoed futilely. The soil of symbols accumulates but slowly in the seasons of remembrance. Christians need to appreciate the importance of these memories before they rush to cry idolatry and dismiss these earnests of the Cross. Though partial and ambiguous, they are soil for a living re-constitution of a people torn apart by fear, hatred, and greed.

Where then should we look for foretastes of a new creation? If sports were play, the joyful exercise of bodies could indeed be signs of re-creation. If sports were springboks, balls, and proteas, of Bafana Bafana leaping in the fields, then it could be a symbol of new humanity. This is a long way from the competitive nationalism and commerce that dominate it now. To entertain a new humanity, it needs to be freed and given back to people for whom it is just a part of a common life together—in schools, communities, and workplaces. The struggle for a common life in a new covenant should not be reduced to centralized government, for it will fail to be rooted in the people's common sense and their publics of mutual recognition. Constitutions can help expand these publics but the people have to draw on their own roots to act within them. A Constitution is not an order to be obeyed but a tree where people gather to share their life.

So we return to the symbol of the tree. Though I have not seen this in the new symbolism of the country, it is a powerful symbol for traditional village life in Africa. And it frames the biblical life of covenant. There is a tree of creation, for which we are to care and which we are to honor. It is not there simply for our advancement, but for our shelter, nurture, and companionship. There is the tree of the cross—cut, misused, dead, and dishonored. It bears the burden of our sin. And there is the tree of the new creation, "whose leaves are for the healing of the nations," and whose roots sink deep into the river of life (Rev. 22:2). Maybe the constitutional manifestations of our fragile covenants might be trees set between sacrifice and hope, between creation and ultimate redemption. On this basis Christians might enter into the public discourse and the cultivation of civil symbols in a critical and renewing way, not merely with their words but with their worship. Maybe seals and springboks might have something to do with trees, crosses, and a glimpse of a new creation.

15

Reconciliation *as* New Covenant, New Public

The end of the Cold War has let loose a series of heretofore repressed local conflicts whose resolution demands the face-to-face work of reconciliation. Conflicts can now be seen not so much as puppet extensions of the Soviet-American struggle but as the expressions of a variety of long-festering conflicts and injustices among groups and peoples with very different histories. The peculiarly non-violent work of reconciliation has to replace the violence and threats of violence that typified the Cold War. While our world now seems beset with new patterns of local violence it is also open to the local work of reconciliation in a way that seemed impossible earlier. This new fact of global existence challenges both our received practices as well as our traditional ways of approaching conflict. My purpose in this brief essay is to lift up a perspective for viewing this process within the variety of practices that have emerged in the past ten years.

RECONCILIATION IS A RELIGIOUS PROCESS

First of all it is important to realize that reconciliation is not merely a "religious" word, but a human process that requires religious and theological understanding. It is an essentially human process, because humans generally seek peaceful relationships, even when they resort to war to achieve this aim. It is a religious process in the sense that it goes beyond the sheer pragmatism that rests in careful calculation of causes and consequences. It demands that we act beyond the equally typical human impulse toward revenge and retribution. It demands that we approach conflict not in terms of annihilation of the enemy but in terms of reconciliation with him or her. In foregoing the usual instincts of annihilation and retribution, practices of reconciliation assume a basic common humanity with one's protagonist. They assume the possibility of a new life together in a common future. They assume a future that is not totally determined by a response to our past. They assume, in short, the classic religious virtues of hope and faith. Reconciliation assumes a type of love that respects the co-humanity of the other. It is in this sense that reconciliation always remains a religious act, even as it is also a highly human political act.

In order to engage these political practices in a critical way we need to think about them theologically. In doing so we face at least two dangers. The first is that we would simply try to deduce our practices of reconciliation from an inherited theological or even dogmatic formulation rather than to reflect deeply on the practices we have experienced in the past ten years around the world. The problem here is that our received understandings of reconciliation are themselves contextual. They are shaped, for instance, by practices of confession and penance in medieval Christendom, as in the "great churches" of Europe, or by practices of personal conversion in the revivals of America. As processes of interpersonal reconciliation within an existing political order, they may not speak adequately to the kind of reconciliation going on today, in which people seek to mend the relationships among whole peoples, nations, and regions.

The second problem is even more daunting, namely, that we live in a religiously pluralistic world. Reconciliation, simply because of its religious depth, requires the discovery or formation of fundamental commonalities. Religious divisions make this exceedingly difficult. Moreover, religious traditions shape and even cause much of the conflict in our world. This is especially true in the recrudescence of polarization between some elements in Islam and in Christianity. The collision of Christianity, Judaism, and Islam in the Middle East continues to feed some of the most dangerous and intractable conflicts in our world. Therefore, we must take great care to ground our theological perspective in a genuine dialogue among the world religions. This kind of work is just beginning. Not only are Christians searching for analogues in other religions of their received concepts of reconciliation, but thoughtful people in other religious traditions are also seeking points of conversation with Christianity.[1] If we fail to work with this awareness even our own approach to reconciliation will prove to be either fruitless or dangerous. At the very least it will be counter-productive in many of the world's arenas of struggle. Unless theological perspectives are well grounded in the differing Christian traditions, they will be incapable of making any practical difference for the historic protagonists. This is even more important within Christian populations, as in Northern Ireland, as well as preponderantly Christian countries like South Africa.

Thus, we have to pursue our necessary theological work as a kind of "deep anthropology." It has to begin with the common humanity of all protagonists, regardless of their religious traditions. While the assumption of a common humanity is itself a religious conviction, it opens us up to discovering connections between our own conversations and those of others. It does not mean that we must begin with a kind of theological Esperanto of reconciliation but rather that we must form a common conversation with others in which we bring to the table the fruits of our own reflection and practice.[2] That is what I want to attempt to do here.

I want to lift up three elements of Jewish and Christian tradition that may also resonate with perspectives present in Islam and other rich religious traditions. These elements are contained in concepts of *the future*, of *covenant*, and of *public assembly*.[3] More formally, they arise for Christians in perspectives on eschatology, soteriology, and ecclesiology. These concepts can function in our reflections on the practices of reconciliation as frames or even as critical checklists for forming these practices.

RECONCILIATION AS CLAIMING A NEW FUTURE

A great deal of reconciliation work has arisen in the midst of efforts to bring to light and to judge the gross abuses of human rights committed

---

1. For some efforts from the Christian side see Thangaraj, *Relating to People* and Heim, *Grounds for Understanding*.

2. I have in mind here the work of the World Conference on Religion and Peace as well as the recently launched United Religions Initiative.

3. For a fuller elaboration of this approach see Everett, *God's Federal Republic*, and Everett, *Religion, Federalism*. For a German version of *God's Federal Republic* see Everett, *Gottes Bund*.

under systems of oppression maintained during the Cold War. Tribunals, forums, courts, and inquiries in Latin America, Europe, and Africa have tried in various ways to expose the truth of what happened, take legal action against perpetrators, make reparation to victims, grant amnesty, extend pardons, and make recommendations for constitutional, legal, or political reform.[4] Within the inherited Western Christian tradition these can be seen as secular manifestations of the confessional pattern of contrition, repentance, confession, penance, and reparation. We can identify where traditional religious elements are missing—usually those of genuine contrition, remorse, and apology—and where religiously grounded requirements are met, as with public confession and reparation. For many Christians it is thus fairly easy to move back and forth between religious, judicial, and quasi-legal public practices when thinking about how to deal with the wrongs of the past.

Such a pattern is adapted for dealing with the wrongs of the past within a settled moral or legal framework. This usage also limits this perspective. It is not a practice focused on generating or living into a new future. It seeks either to erase the past in some way, as with amnesty procedures, or to repair past wounds so that we can return to some sort of status quo ante. As a religious process it has tended to focus on individuals within a moral or legal system. Reconciliation in this case thus appears either as a pattern for which the law is a partial imitation or as something that must occur beyond the law. It adds personal forgiveness where the law yields only amnesties and punishments.

Reconciliation, however, needs to be seen not only as reconciliation *of* something but reconciliation *for* something. It is not merely the repair of the past but the negotiation of a future. That is, reconciliation arises within the horizon of eschatology. It has to do with how former antagonists construct a new future together.[5] The question is: How are we going to understand and start developing this common future? This work of future-building already resides in our natural life. Flowers and trees emerge from the tortured earth. Women bear babies. Rains and wind wash away the ravages of war. Beyond those natural processes, human

---

4. See for example Harper, *Impunity*; Boraine, Levy and Scheffer, *Dealing with the Past*; Boraine and Levy, *Healing of a Nation?*; and Botman and Petersen, *To Remember and to Heal*.

5. Among contemporary theologians the work of Jürgen Moltmann is crucial in this regard. The perspective opened up in Moltmann, *Theology of Hope*, continues through many of his later works, including Moltmann, *Coming of God*.

efforts at economic survival and acquisition draw together markets, build factories and workshops, and patch together means of transportation. Human hope begins to manifest itself in all the instruments of planning as well as in the speculations of the market. For many people reconciliation arises mainly within these "natural" forces of reproduction, commerce, and economic growth. The stock market becomes the sacrament of the future.

Reconciliation, however, speaks of yet a further hope—a hope in the renewal and recreation of the earth. It is a hope not fueled by human efforts to resist pain and death but by a longing for the ultimate accomplishment of God's creative purposes. Practices of reconciliation, therefore, have to be framed by this anticipation of creation's perfection rather than by the mere extension of our own struggle for survival.

In this sense, the manifestation of such a future is a miracle. The process of reconciliation must find an anchor in some miraculous interruption in the patterns of oppression, betrayal, mistrust, and despair that fuel conflict and violence.[6] It may be the miracle of Leipzig's marches or the fall of the Berlin Wall. It may be the miracle of Nelson Mandela's release. It may be the miracle of a charismatic figure like Mahatma Gandhi or Hector Romero. It may be the slow and patient miracle in which persistent and patient non-violence finally finds a public voice, as in Argentina. In some way, the work of reconciliation begins as response to a gracious and miraculous intervention in the vicious cycle of violence and revenge. The incubus of the past is lifted from our shoulders by a vision of future possibility. Theologically speaking, this is the divine redemptive moment in the process of reconciliation. Such a miraculous intervention concretizes the impulses of hope in a public image that becomes a vision of the future.

Such a vision, it seems to me, is generated by two classic biblical themes—Sabbath covenant and Pentecostal assembly. Each theme lifts up distinctive elements that need to be present in the process of reconciliation. They connect deep religious symbols to the general publics in which reconciliation must proceed in a world of religious, economic, and ethnic divisions. They also are intertwined concepts. They need each other even as they correct limits in each other.

---

6. See the importance of "transforming initiatives" in Stassen, *Just Peacemaking*.

RECONCILIATION AS COVENANT-MAKING

The Israelite concept of covenant is rooted in the Abrahamic and Exodus stories, the Deuteronomic code, and the prophetic literature of renewal, as in Jeremiah and Isaiah. While it is a rich symbol and tradition I only want to lift up some selected aspects relevant to our purposes here. First, it is important to see that the covenant stories in Genesis and Exodus deal with the construction of a new order for life in a new land.[7] In the assembly at Sinai this covenant teaching becomes the means for reconciling the diverse interests of the tribes. Their reception of the covenant law becomes the basis for arbitrating their future disputes as well as forming them as a single people. The demand to form a new future that is not simply a return to a past ideal makes the covenant essentially voluntary. It is not merely rooted in the people's biological past. It is not simply rooted in their escape from bondage in Egypt. It is rooted in the gracious miracle of YHWH's intervention in their lives to promise them a new future.

As the means for binding people together in a common life, covenant is not merely a matter of individual existence, even of individual existence before God. Covenant always involves people's lives in groups. It involves peoples and nations. Thus, a covenantal approach to reconciliation resists the reduction of reconciliation to interpersonal forgiveness. It sees in reconciliation the construction of a common life together based on promises about the future.

Living according to promises demands that we live by reference to expectations about a common future more than by appeal to characteristics inherited from a fractious past. For humans this is extraordinarily difficult, since our expectations generally flow from our past experience rather than from faith in a God of a new future. We trust our own kin and kind rather than strangers. We trust our memory more than our hope.

This characteristic of covenant-making has immense political ramifications. Constitutionally, this tendency to live by inherited characteristics rather than by promises expresses itself in the dilemma of federal structures rooted in ethnic and tribal divisions rather than in common agreements about how to live together in a pluralistic world of many tribes, ethnies, and nations. Most of our political conflicts seek reconciliation through the creation of federal constitutions that provide safe spaces for self-identified "peoples" within an embracing common law for

---

7. For a thorough review of this development from the standpoint of political theory see Elazar, *Covenant and Polity*.

adjudicating disputes and serving common purposes. This federalism is itself rooted in the traditions of covenant. It is a political structure based on promises rather than biology and ancestry. These constitutional orders for reconciling differences break down, however, when the federal order is based simply on ethnic, racial, or biological claims. Federal solutions to conflict, like their covenantal roots, can only function when they are rooted in free promises oriented to a new common future. They also rest, as I shall claim subsequently, on open publics rather than biologically closed communities.

In this political understanding, covenant, as a basis for understanding reconciliation, creates a "higher law" as the basis for human resolution of disputes. This concept of a "higher law" has been indispensable for the reconciliation processes of the past ten years. It appears first of all in the higher law of human rights, concretized in the UN Declarations and given legal form through various international covenants. Without them we would have little basis for moral and legal judgments of atrocious acts carried out according to the "positive law" of oppressive and corrupt governments. There would be little basis for publics that can make judgments about truth and justice independent of governments and other agencies of coercion. This higher law also finds concrete expression in constitutions and constitutional courts that can regulate the statutes and policies of legislatures and governments. Germany and South Africa present us with clear examples of these developments. These products of negotiation and profound collective promise are also works of reconciliation.

Through both the doctrine of human rights and the creation of constitutions based in them reconciliation appears as more than a process to deal with the limits of judicial process. It appears as the very basis for law itself. Reconciliation does not begin where law leaves off but is itself the beginning for the creation of genuine law. As the work of covenant-making, reconciliation emerges as the root of constitution-making. Under constitutional government law flows not merely from the will of the ruler or the legislature but from the settled agreement of the whole people. Even beyond this, it flows from the settled convictions and promises of people who were wounded by the oppression that sought to eliminate them not only from public life but from the earth itself. In the exodus from these afflictions people come to agree on the principles of a common life which has room for all as fellow human beings. It is at this point that we begin to see the connection between the reconciling work of covenant and the reconciliation manifested in the Pentecostal assembly. Before exploring that connection, however, we need to speak about the common earth in which this covenantal reconciliation must occur.

Covenant and sabbath are closely connected in the biblical vision. One version of the institution of the sabbath sees it as a remembrance of YHWH's covenant with Israel. The purpose of the sabbath is to remember and meditate on God's redemption of Israel from bondage in Egypt (Deut 5:15). The purpose of the covenant is to lead Israel to rest from its ordinary labors and remember God's redemption. In another (probably earlier?) account, the sabbath was instituted to commemorate creation itself (Exod 20:11, 31:17. Cf. Gen 2:2-3). It is the fulfillment of creation's purpose, which is to enjoy God's creative goodness. It is itself part of the covenantal bond of creation and redemption central to Israel's faith.

Sabbath thus poses limits to the infinite expansion of the market and economic growth as the primary basis for effective reconciliation of warring people. It calls into question their claim to be the doors of the future. It presses us back to acts of gratitude for the miracle of a common earth. It presses us to cultivate a sense of limits on our greed and on our search for absolute security. In whatever form this sabbath rest may take in various religious traditions, it helps shape an awareness of the limits within which we have to find our shalom.

This sabbath limitation also forms the biblical image of Jubilee—the sabbath of sabbaths. The Jubilee mandate (Lev 25) requires that we restore the basis of life for the primary communities in which people live. It is a return to some envisioned "original position" to overturn the oppressive indebtedness that inevitably seems to arise in human relations. Jubilee is a form of collective reparative or restorative justice. It takes into account not merely the need for individual reparation but for an equitable sharing of the earth's resources by the whole society. Jubilee itself is a model for the reparation that reconciliation demands. However, jubilee is limited if it rests only in the search for an absolute original position. Jubilee has to become, as it does in Jesus' preaching, an image of the future renewal of all creation as the basis for a common life for humanity.

Covenant, creation, redemption, and sabbath form an interlocking set of symbols for guiding life and pursuing reconciliation. The point of these connections for the process of reconciliation is manifold. Most important for our concerns here is that the covenant-making which constitutes reconciliation involves connection to the wider creation. The renewal and common stewardship of the earth is intrinsic to the process of reconciliation between human beings. No matter how conflicted on other matters, human beings still inhabit a single, common earth. If they are to be reconciled they must begin with their common affirmation of their dependence on the integrity and fruitfulness of the earth—its waters, air, and land. The drawing of boundaries to curb aggression is not enough for reconciliation. To live together in peace humans must find a way to share in the stewardship of the earth. To see themselves as earthlings is the first step toward an affirmation of the co-humanity underlying reconciliation.

This restoration of the earth appears clearly in South Africa's efforts to repair the damage of apartheid, which tried to force the majority of the population onto patches of marginal soil, creating enormous degradation of the country's ecology. It is also recognized in the way the Vietnam Veterans Memorial in Washington seeks a harmonious relationship to the earth, in distinction to the ecological destruction that characterized American military policy in Vietnam. Reconciliation is always bound to the sabbath covenant with creation.

Grounding reconciliation in understandings of the future and of covenant emphasizes the way reconciliation is a life of promissory trust beyond retribution and reparation. It involves not only the common life of people but also their bond with creation and with a purpose of life that goes beyond mere material acquisition. It frames an understanding of reconciliation as a process creating law rather than merely going beyond it.

But covenant as the only concept for approaching reconciliation also has its limits. Even ancient Israel had the problem of relating its covenant to the question of reconciliation with the Canaanites—a tortured question that besets us to this day, whether in Israel or in the "new Israels" of America, Northern Ireland, or South Africa.[8] The pattern of biblical covenant-making raises the question of who covenants with whom. What kinds of groups can engage in reconciling covenants? Here we require additional perspectives. These perspectives emerge from a vision of the new assembly that reconciliation requires and creates. Reconciliation requires publics.

RECONCILIATION AS PUBLIC ASSEMBLY

Covenant-making and commitment to a common future require open publics. In order to go beyond inherited and ascribed identities

---

8. For these negative effects of the "covenantal grid" see Akenson, *God's Peoples*.

covenants have to occur within a public where persuasion, negotiation, and a search for common truth can occur. Even the covenant in Exodus exhibits an elaborate negotiation between Moses and YHWH! Covenant-making cannot be restricted to the models of parent and child, or judge and defendant. It demands mature participation from those who are to be bound in covenant to one another. Moreover, it demands movement beyond the "ascriptive solidarities" of race, caste, ethnic community, and gender to participate in the voluntary association that arises in promise-making itself. If reconciliation is to occur it has to bring together in a new way people with very different inherited characteristics. The negotiation of the new South African Constitution, for instance, with its covenant-like basic principles and its effort to involve all parties in the process, is the latest exemplification of the work of reconciliation as a public process of negotiation and agreement.

Within this process of publicity the search for truth, which has been so central to many reconciliation processes, takes on a new light. It is not enough to find some objective forensic "truth" if this discovery does not occur as a public act. When rebuilding the common trust people need to agree on the validity of claims before them through open examination and debate. The word of discredited authorities is not enough. Moreover, the journey from oppression to reconciliation usually means a journey from silence and secrecy to voice and mutual recognition. The "truth" must be part of the formation of an ongoing public. Thus, it was very important in South Africa that the victims of apartheid atrocities be given a forum in which to articulate their experience and gain admission into a common public. It is the absence of this public voice for Vietnam War veterans in the US as well as for their Vietnamese counterparts that has made the work of reconciliation so difficult in the wake of that war. The failure of reconciliation in America has damaged subsequent efforts to reshape America's relation to the rest of the world.

Within such open publics people can engage in the work of confession, not merely as the request for forgiveness but as a quest for genuine conversation about the basis for a new beginning. That is to say, the confession that so often figures in standard models of reconciliation is not merely for the sake of forgiveness but for entry into a new public realm. In this sense, it is connected, for Christians, to a confirmation of baptism. The forgiveness sought in reconciliation needs to be seen as initiation into a shared public. The forgiveness of sins flows from the possibility of a new life in the Christian assembly. For Christians, the symbol of

Reconciliation as New Covenant, New Public 265

this baptism into a transcendent public is the Pentecost experience recounted in Acts 2. In this peculiar public infused with God's re-creative spirit people speak out of their own experience and can understand one another, transcending their normal divisions of language, nationality, gender, and race. It is in this new linguistic beginning that the ecclesia of Christ's spirit is formed.

In this sense confession and truth are not enough for reconciliation. Neither does reconciliation necessarily result from the discovery of truth about the past. Reconciliation occurs in the creation of a common public in which to work out trustworthy promises about the future. In this sense the work of reconciliation is not simply the end point of harmony, in which all our competing interests meld into one. Reconciliation is the very act of entering into respectful argument in a common public. This is a very "political" understanding of reconciliation rooted in notions of covenant and public assembly. The work of reconciliation involves the continual reconstituting of these publics through the covenantal dynamics of negotiation and promise-making. It involves the promise not to use violence or secret manipulation and coercion to reach agreement. It requires keeping the underlying covenant by which humans bind themselves to each other in co-humanity and with the earth as co-creatures.

Thus, the creation of new publics of mutual engagement rests in deep covenants binding humans and the creation to a new future. Within this constellation of religious images reconciliation emerges not as the overcoming of conflict but as the agreement to constitute non-violent means for adjudicating disputes. Reconciliation also emerges as the very human and earthy effort to achieve a common life with the creation, whose injuries are also our own.

I have tried to sketch here the basic elements of a framework for thinking about the practices of reconciliation in our present situation. While mindful of the religious plurality of our globe I have drawn on some key elements in Christian tradition to illuminate this human process. In lifting up the importance of the future I have tried to provide a critical perspective for reconciliation that focuses on more than discovery of and retribution for the past. In focusing on covenant, I have tried to emphasize the importance of constructing new constitutional frameworks in which to adjudicate disputes peacefully and to care for the earth that sustains us. In drawing in the idea of pentecostal assembly I have tried to indicate the crucial importance of moving from ascriptive solidarities based in biological appeals to open publics of civil debate and negotiation. Without these publics neither adequate covenant nor its intended reconciliation is possible, for reconciliation is finally a participation in the cosmic process of the realization of God's creative purpose.

16

SERVING *the* CHURCH *and* FACING *the* LAW
Virtues for Committee Members Evaluating a Pastor

You are a member of a denomination's regional Committee on Ministerial Certification and Appointment, which oversees the movement of candidates toward ordination as well as their general progress through their career. It is in charge of the certification of clergy and oversees their placement in local churches and church organizations. Its members include clergy as well as lay people. In problematic matters your Committee can turn to the regional legal counsel to the denomination, who relies both on relevant civil law as well as on the church's internal law, which has developed over many years.

In the course of its work, the Committee is asked to deal with a conflict between a congregation and its pastor. The claims are complex and manifold and evidently have some history to them. Following ordinary procedures of mediation according to church law, the Committee helps resolve the conflict by enabling the congregation and pastor to reach an agreeable separation that honors the needs of all parties.

Subsequently, the pastor requests approval for seeking appointment to another congregation. In the course of its routine inquiry, disturbing news begins to filter back to Committee members informally from members of the previous congregation. Underlying the more public claims of the congregation made earlier against the pastor there seem to have been various sexual indiscretions and behavior by him that had led over the years to an elaborate pattern of secrecy and manipulation. The members in question could not raise any of this at the time because of their personal embarrassment and their fear of retaliation from their former pastor. In the earlier proceedings their personal objectives for separation from the pastor could be met through the formal procedures of the church. Now, however, they feel an obligation to find a way to prevent this pastor from visiting the same behavior on others.

At first the Committee, although it has received these reports only as informal information, feels it can move ahead to a formal investigation according to church law. The sources of the information seem credible enough that the matter should be taken up, though discreetly, as a personnel matter under the cloak of ordinary procedures for re-appointment. The members sense an obligation to the congregation's members as well as to the pastor and think it is important for the sake of the denomination's credibility to find the truth.

At this point the Committee hits a snag. All of the alleged actions occurred several years earlier. According to church law, a statute of limitations disallows actions regarding such claims. To take formal action would violate this rule.

The Committee members are convinced that the parishioners were not far off the mark in their allegations, even if they had initially accepted some of the pastor's advances. Other circumstances of the pastor's life and separation from the congregation seem to confirm these claims. Moreover, the Committee members feel they ought to honor the parishioners' deeply felt desire for anonymity in the face of possible retaliation or humiliation. At the same time they want to protect other congregations from possible misbehavior. Indeed, in some sense they feel a civil obligation to avoid lawsuits if they do not take action, regardless of the church's statute of limitations. Finally, they believe that they have duties toward the pastor to abide strictly by church law, which exists not only to protect congregations but also pastors. Here too, their actions might lead to civil suit if they violate well-established rights of citizens in areas of employment, reputation, and livelihood. Theologically and ecclesiologically, they also sense the need to uphold cherished parts of their heritage, which may well be at stake in proceeding precipitously or clandestinely toward the pastor.

The Committee begins to work through a series of considerations that bear on how it ought to weigh claims about the conduct of the pastor over against the claims of proper procedure. The procedural limit seems to pose an absolute barrier to formal investigation, even if in this case it stands in the way of ascertaining a matter of substantive justice. Its claim is clear, public, written, and observed in fact. Its claims carry with them the force of justice. The body of church law that contains these procedures is a formal declaration of its commitment to all members of the church, including those suspected of wrongful behavior.

Moreover, failure to comply with its own legal requirements could subject the Committee or denomination to civil suit, since American courts have increasingly been willing to hear civil cases for breach of contract and violations of by-laws, even in ecclesiastical cases. Over the years the distinction between civil justice and ecclesial justice has become blurred in the popular mind as well as in the courts. The room for discretion in church decisions has been narrowing. The Committee does not want to expose itself on this flank, an exposure that would in fact exacerbate the situation even more through the publicity it would arouse.

Even short of a civil suit, moreover, the Committee is bound by church law to make a public report of any formal investigations. Centuries of suspicion about secret decisions and tribunals have led the church to make public disclosure for all its decisions and official actions a top priority. In some sense, everything the Committee does falls under the requirements of publicity. Its discretionary room is constricted. Its capacity for secret negotiations is quite limited.

However, once having decided that it cannot launch a formal investigation, the Committee still feels bound by its own judgment about the specifics of the case and its concern for the good of the church. It can not simply pass the pastor on to another congregation. The weight of the evidence, even though resting on the character and credibility of the parishioners, is too great to be set aside. However, if the Committee does act in some way outside the formal rules it also has to keep in mind the parishioners' needs for anonymity, the needs, claims, and rights of the pastor, as well as its own fiduciary obligations to the wider denominational structures.

How then, should the Committee proceed? How would you act in such a situation? What kinds of values, perspectives, assumptions, and loyalties would inform your decision?

SPECIFICALLY, THIS CASE PRESENTS us with two main problems. First, to whom or to what institutions are the committee members responsible? How do they balance these various responsibilities with each other? Secondly, what virtues should or can they rely on in moving through their decision? What are the qualities of character and disposition that will best enable them to do the right thing? With these two concerns in mind we turn first to the question of how these virtues relate to rule-oriented conduct, which in this case has brought the members to some difficult dilemmas. I shall then frame the Committee's questions of responsibility and virtue in terms of accountability within the various publics shaping their lives.

Much of our ethical life is governed by rules. They are the webs of expectation that make institutional life possible. Without them we return to the arbitrary decisions of parents and monarchs or, more chillingly, of dictators. Religious movements, especially because of their appeal to personal charisma or to parental nurture, have often developed rules, laws, canons, and judicial procedure in the face of moral turpitude by their leaders. These rules arise because appeals to the leaders' virtue has usually fallen short of securing the accountability required by a wider public. The norms of justice, it seems, require more than virtuous leaders. They require public, rule-governed behavior.

No matter how ecstatic their beginnings, religious movements of innovation, renewal, or reform inevitably develop the routines of law in order to preserve norms of justice within the religious group. At the same time, these rules have also sought to preserve the peculiar principles that generated the religious movement. Catholics seek to preserve the integrity of sacramental life, Methodists the life of conference and itinerancy, Presbyterians the pure preaching of the Word, Lutherans the free operation of grace, and Pentecostalists the free movement of the Spirit.

But are laws enough? And are they enough for the church? What ethical thought should guide us at the limits of the law? In each denominational tradition, as in every polity, the rules of the church inevitably find their limits as well. While virtue may not be enough for organizations it is necessary for their actual daily survival, especially where their laws and rules fall short. Virtues speak to the qualities necessary in leaders, laws to the structures of accountability within which they must act. In this brief essay I want to explore the kinds of virtues necessary for leadership where the law's workings are ineffective or violate key goods of ecclesial life. These are the kinds of issues raised in the hypothetical case before us.

The case confronting this denominational committee reveals the limits of an ethics of rules and law in several ways. First, we see a conflict among rules, such as "respect confidentiality" and "respect full disclosure." We also see conflicts between ethical rules, like "seek the truth," and legal rules like "observe statutes of limitations."

Secondly, we see at the institutional level possible conflicts between legal systems designed for the church and those designed for civil society. What are the claims of church autonomy and what are the claims of civil law?

Thirdly, we confront ambiguities about the application of rules within institutions. How far does a committee's mandate extend? When is it exceeding its authority? How narrowly should the rule on limitations be applied?

Fourthly, we find the distinction between law as "the protection of rights" and law as "a means for seeking the common good." Much of our civil law has arisen to protect the rights of citizens over against governments as well as other persons and groups. Thus, the statute of limitations exists to protect rights to a fair trial. However, much law also exists to seek certain goods, whether of health, survival, beautification, environmental integrity, or economic well-being. We are constantly judging these laws in light of the goods they purportedly seek. The Committee in this case is seeking "the good of the wider church" in the midst of laws that protect "ecclesial citizens" as well as seek that ecclesial good.

Finally, these limits of the law and of rule-guided behavior bring us to the boundary question of obligation. Why are we obligated to certain rules or laws? How do we rank the publics to which we are obligated? How do our obligations to God, to Christ, the church or to persons shape our ethical decision-making at the limits of the law? All of these limit questions raise issues addressed by ethical concepts of virtue.

One way of framing the question of how to proceed in this case is in terms of the virtues of the persons involved in the question as well as the virtues of the institution in which they are acting. Let us begin with the simple definition of a virtue as a strength of character that creates the capacity for moral existence. This strength in persons can be seen in terms of the character of their will, disposition, imagination, and purpose. All of these psychological attributes are formed over time into habits. Similarly, institutions can be seen as exhibiting virtues. These are institutional strengths that make moral existence possible for their members and for other institutions as well. Institutions may seek to cultivate virtues of justice, fairness, or openness. Both of these kinds of virtue are at work in this case.

Institutionally, this ecclesial institution has the habit of treating its members as citizens with a wide panoply of rights. Thus, the institution is restrained in disciplining and guiding its members because it assumes, let us say, that they have an independent theological existence rooted in their baptism. This is quite different, for instance, from an institution that treats its members like children of stern but loving parents. There, the habits of nurture, direct chastisement, guidance, tutelage, and protection are much more evident in the policies and practices of the institution. Both institutional habits have been important in the historic life of Christianity and also appear in this case, though it is clear that the citizen habits of this institution dominate significantly the parental habits. Not only do the parishioners have citizen-like rights but also the pastor, who is not merely an employee to be discharged at will but a kind of citizen in an elected office in the church.

This mixture of citizenship and parentalist virtues in the institution confronts the Committee with personal as well as group challenges. Clearly, the Committee, as a group with its own character, has to chart its way into the gray area beyond the law's limits. Essentially, it must navigate according to the lodestar of ecclesial good. But the night is dark and cloudy. Each individual in the group must also draw on certain virtues to participate in this voyage. How might we speak about these virtues?

In order to get at these virtues we need to frame their institutional setting more precisely and draw to some degree on a theory about institutional life. We can envision institutions as a complex web of mutual expectations and obligations. These expectations crystallize around particular structures, like committees, boards, councils, senates, conferences, and the like, as well as around roles, such as president, secretary, treasurer, pastor, deacon, and trustee. The function of law in this network is to facilitate trust among the members so they can work together for the common good of the institution. Without this primordial trust and its daily ratification through law-abiding behavior in the institution, the institution itself breaks down and it fails to fulfill its purposes.

Theologically, we can see this centrality of trust in our concepts of faith, of God as covenant-maker, and of us as covenant-creatures. We can see it in our concepts of grace as the capacity to entrust ourselves to God in response to God's entrustment of the divine mystery to us. While this theological ground can be nuanced in various ways, it is a central way we can get at the connection between our life in faith and our life in institutions. In either case we live in a web of trust that seeks durability in law.

At the same time, our life in trust lives with constant mistrust. Our life experience and our knowledge of history presents us with trust that is broken, with trust placed in false places and people, with trust that disables us. One expression of this mistrust in our institutional life can be seen in how we decide which groups or officials we can trust with what knowledge. How public can we be about a matter? Some statements can appear with little fear in the widest, most expansive publics. Scientific claims are like this. Other claims, as in a criminal trial, require the more limited public of a courtroom, lest an inflamed mob or a callous media horde carry out its snap judgments. Still others, like a shameful violation of other people's trust, need the shelter of private confession, which can only be a "public" in a representative way. In extremis we finally stand alone in "God's public" through prayer and contemplation.

How we negotiate the tensions and differences among these publics requires great discretion. The Committee faces the tension between publicity and secrecy in its work. Its wider institution strongly evidences the virtues of publicity, such as open disclosure, full debate, and collective judgment. These institutional values generally operate to enhance trust by giving each person some control over how they are treated in the institution, even if very indirectly. However, there are times, as in this case, when too much publicity threatens the fragile network of trust, either between pastor and parishioners or among the Committee members themselves. The Committee's search for the good of the institution, balanced with the good of the members in their various roles, demands a strictly limited publicity for their activity.

This kind of situation highlights the importance of the classic virtue of prudence. I would like to specify it in this situation as discretionary judgment. The virtues of discretionary judgment require a capacity to hold opposing values in tension as well as to select the appropriate level of publicity in which to seek the overall good of the institution. This imaginative work of discretion also requires the fundamental classical virtue of courage. Courage itself, aside from the sheer disciplining of our natural fears, can be seen as a choice of which public we wish to be remembered by. It is not merely a matter of avoiding potential shame and disapproval, but of which public we want our deeds remembered in. To whom, finally, are we accountable?

Each of these publics, no matter how small, requires this exercise of courage. Each public requires in us the capacity to give intelligible voice to our convictions in the face of possible disconfirmation by the claims of others. Legal publics exist to manage the scope and severity of this disconfirmation, even as they shape how we express our claim. Ecclesial publics, because they expect even deeper self-disclosure, must provide a greater array of small and large publics in which people can gain the virtues necessary for faithful existence. It was the violation of expectations within some of these limited publics that generated the Committee's quandary in the first place. Now it must draw on the virtues of its members as well as of itself as a group to negotiate the passage out of these violations into a possible regeneration of trust in the wider institution.

How, then, might we speak of the appropriate virtues for this discretionary work among the little and large publics of the church and civil society? We could imagine the members of the Committee proceeding in this way. They might first try to establish some common assumptions in the group about the nature of the challenge they face. I have cast the problem as one of public trust. Both the alleged wrongdoings and the process by which the Committee deals with these claims involve the violation and reconstruction of public trust in the institution and between it and the environing society. This is the ecclesial good that should guide their discretionary judgments. The ethical question poses itself in terms of what personal and group virtues are necessary to the achievement of this higher level of public trust.

In a situation at the boundaries of clearly defined law the members of the group need to heighten their capacity to trust each other. This begins with an agreement about the situation they face and the primary good they seek to advance. In order to heighten their internal trust they need the virtues of honesty, clarity of expression, sincerity, ability to listen, and the ability to persuade and be persuaded. The latter two strengths are ways of understanding the virtue of humility within a network of communicative trust. Sometimes, to achieve this trust, groups will ask everyone to recite their life story and reveal any possible conflicts of interest relevant to the case at hand. Sometimes this self-revelation is helpful but other times it avoids the question of whether the members have the capacities to help the group move ahead with its ethical task and aim. In any event, this formation of what the German philosopher Jürgen

Habermas calls "communicative competence" and of simple trustworthiness cannot occur without these communicative virtues.

Notice here how this task draws on classic virtues like humility or honesty but recasts them as the task of building trust through communication rather than through obedience to external orders. The disciplines by which virtues crystallize as habits is not modeled in this case on the image of making our passions instruments of our reason and will. That, it seems to me, is the received, somewhat unconscious way we perceive them. Rather these virtues arise in the discipline of communicative interaction. They arise in the process of self-disclosure, listening (with more than our ears), and struggling together for a common world of understanding. This decisive shift in the way we understand these virtues occurs because the wider framework of understanding rests on a conception of public life rooted in ultimate convictions about God and God's ways with the creation. The ecclesial institution is seen here as a testing ground and rehearsal space of our capacity for this public life as well as of our capacity to deal with our sin and needs for secrecy.

Beyond its internal work of trust-building the Committee needs to keep remembering that it is in some sense a representative assembly. Its members and the group as a whole are the means by which the larger institution has assembled to deal with its common life. The imaginative work of representation is crucial to the moral life, for it establishes in our characters the publics to which we are loyal and accountable. It shapes the kinds of concerns that are uppermost in our minds when we speak, listen, argue, and negotiate in our immediate assemblies and meetings. In this situation the members imagine themselves as representing the church's wider assembly, the relevant congregations, and also the civil public. Beyond that they also need to imagine themselves as representing the universal church, past, present, and future. Finally, each of them remembers himself or herself as representing the Christ into whom they have been baptized. That is, each of them is to represent the spirit of Christ in that meeting. Each of them is a representative of Christ's presiding power in that assembly. In this sense, each member understands himself or herself, to extend this conceptual scheme, as participating in the very public where God's Spirit presides. This is the kind of imaginative representation that shapes our understanding of conscience, which is, to reframe Pope Paul II's phrase in *Veritatis Splendor*, an imaginative reconstruction of our internal conversations with and before God.

This indeed is a tall order. This virtue of representation rests on our imaginative capacities, our capacities for empathic understanding, and our capacity to be transparent to our represented publics without yielding our own unique creative capacities. Both the Committee as a whole and the persons within it need to manifest this virtue of representation. The capacity for representation is not unlike that of acting, in which we take on a role and try to re-present the persona of the character in a play. It requires the work of rehearsal and the acquisition of an imaginative repertoire that is often lost in a world focusing simply on our own expression of our unique selves.

In seeking to represent the conversation and argument of these wider publics in its own deliberations, the Committee goes not merely beyond the law but beneath it, back to its own grounds and purposes. The law is seen as existing to preserve the basic web of trust that makes social and institutional life possible. By seeking to re-present the basic publics of the church's life the Committee, though operating discretely and in that sense in some secrecy, can also anticipate and guide whatever degree of public argument might arise out of its decisions and actions. Its secrecy is not a secrecy of hiding from the law but of working beyond its limits. It is a return to the embryonic assembly, with its principles of public trust, that originated the church's law in the first place. The two forms of secrecy may look the same to the outside publics, but the latter can stand the light of publicity and the former cannot. It is a crucial difference and one that we can often collapse when we forget that it is our task to represent these wider publics rather than our own narrow interests.

The case with which we began poses some classic dilemmas of institutional life, whether in the church, in secular corporations, or civil life. In framing these dilemmas as dilemmas of publicity and secrecy, of discretion and simple adherence to law, and of procedure and institutional goods, we claim some new angles of vision on classic virtues of humility, courage, prudence, and honesty. The virtues of public trust and of representation, though not spelled out at this point in detail, point us to the way we need to rethink our understanding of virtue in terms of capacity for trustworthy communication and imaginative participation in the ultimately open public to which God calls us. While we cannot determine a successful outcome to the Committee's deliberations, we can hope that reflections like these might helpfully inform their search for the good of the church within the complexities of human aspiration and human failure.

17

Public Works
Bridging the Gap between Theology and Public Ethics

THEOLOGICALLY TRAINED ETHICISTS GROAN frequently during the evening news. Announcers tell of "Muslimfundamentalists" and "Rightwingchristians" terrorizing civilians in the streets and parliamentary chambers. "Catholicantiabortionists" struggle with "prochoicehumanists" over the status of life in wombs and prisons. Billy Graham and white robed Popes float like icons through a fog of detached personal piety or collective enthusiasm. The discomfort we feel is not merely due to the liberal bias of media purveyors but to their ignorance and virtual illiteracy about matters religious. They evince little awareness or understanding of the long history out of which religious movements and institutions develop, nor do they have any familiarity with the way theology, ethics, and religious organization work together to shape persons, institutions, and culture. When this foggy consciousness descends upon much of American religion it is merely patronizing or nostalgic. All nuns, like Whoopi Goldberg, must wear archaic habits; all preachers must have white shoes. At its worst it is unable to understand, for instance, the religious culture out of which impeachment was brought against President Clinton. But when it shapes the way American foreign policy deals with Muslim cultures and their governments it becomes deadly. When it doesn't grasp the religious depth of African cultures it cannot penetrate what it relegates to ethnic and tribal warfare. This illiteracy is not merely annoying, it is dangerous. How can we think theologically about public affairs in a language comprehensible to a more general public?

This split between theological memory and contemporary public discourse is a fracture in a complex history. Therefore, our first response must be to heal this amnesia and try to put together again how we got here. The second step is conceptual. It requires that we think more clearly about the connections among religious and political ideas as well as between religious and civil institutions. First, let us turn to the problem of memory.

RECOVERING OUR THEOLOGICAL SENSES

The problem of finding theological literacy in civil discourse is rooted in the successful effort to distinguish these two spheres. For fifteen hundred years Europeans carried on their political, cultural, and ethical disputes in what we call today religious and theological language. Questions of authority were traced back to theories of God, Trinity, and incarnation. Struggles over the nature of law and the limits on governing power were discussed in terms of ecclesiology, scripture, sin, and the meaning of creation. It is impossible to understand how we received key political ideas and values without awareness of this theological matrix.

Americans inherited a pattern of relationships between religion, government, and civil society that resulted from the 1500-year European experience of Christendom. That experience, though shaped profoundly by a distinction between the Roman Church and governing monarchs, finally produced the collective decision to separate them more clearly and cleanly. What began in the seventeenth century as an effort to end the "wars of religion" between Catholic and Protestant monarchs led eventually to the institutional "separation" of the American experiment. What we often overlook are the stages in this development. At the treaty of Westphalia (1648), which ended the Thirty Year's War (think Cambodia or Vietnam for brutality), each monarch was allowed to choose his own religious affiliation and, with that choice, the established or state religion of his people. With the end of the English Civil War of the same century, the political theorist John Locke laid out principles of religious toleration that would make possible a governmental order based on rational consent, contract, and public debate.[1] While there would be an established

---

1. Locke, *Letter Concerning Toleration*.

church to provide the cultural glue and ground for proper governance, competing religious and cultural visions would be allowed to carry on their work and witness so long as they did not attack this order itself.

A century later the Americans, influenced as much by French anticlericalism as by English toleration, took the next step and severed religious institutions from state sponsorship or control in their new federal republic, though many states maintained religious institutions until well into the nineteenth century. The institutional differentiation of religion from government was necessary if government was to be based on the arguments of the people rather than on the authority of traditional leaders and customs. Personal autonomy, though secured earlier on the basis of appeals to conscience, baptism, the work of the Holy Spirit, and entry into personal covenant with God, had to be freed not only from church control but also from its theological basis if politics was to be pliant, argumentative, and pluralistic.

In this important process, "disestablishment," which originally meant only withdrawal of state support for a church, increasingly took on the secularist agenda of hostility to religion in public life in the twentieth century. Public life was to be cleansed of religion, whether as crèches on courthouse lawns or prayers in government run schools. Toleration became an agenda of ignorance and amnesia as history books left out the crucial religious motivations, institutions, and intellectual currents that actually shaped western civilization. Religion was allowed to stay public, as in the military or on dollar bills, only if it clearly supported the existing political or economic order. The political separation of institutions created a cultural amnesia about public religious discourse.

We are now experiencing a reaction against this hostility and amnesia, not just from those who would reimpose some sort of governmental orthodoxy, but from those who see in this amnesia an inability to deal with the richness of human life and the organization of other cultures in a pluralistic world. The public spaces of education, media, and public discourse are filling up again with references to religion, theology, spirituality, and ethics. Religion is back, not only in the sensationalist packages of the mass media but in serious efforts at religiously grounded public discourse. The "naked public square"[2] is full of religious costume, gaudy and plain. The question we now face is how to speak within this public

---

2. Neuhaus, *Naked Public Square*.

discourse? What is the kind of "literacy" that can guide us in recovering from the amnesia effected by the well-intentioned effort at separation?

FINDING THE BRIDGES

This question leads us to the problem of concepts for speaking theologically and ethically in the general publics we inhabit. I would like to frame our problem in terms of the construction of bridge concepts and symbols between traditional religious discourse and that of our governing publics. From this perspective we can then see three important dimensions of theological literacy in public life. First, we can see how these bridges function in the creation and shaping of civil society. That is, we can see them as institutional vocabularies for shaping how people create and maintain social structures in accord with an ethical vision. Second, we can explore the role of theological concepts and symbols in the vitalization and transformation of political culture. Bridge concepts work at the deep level of symbols, collective psychology, and fundamental orientations. They connect individual psychology to groups and institutions. They create loyalties, habits, and customs. Third, we can inquire about the language with which we can open up these societies and cultures to the mysterious transcendence of the divine purpose and power. Bridge concepts can open up the pragmatic, utilitarian, and limited concerns of ordinary public life to the wider questions of existence. Laying this out in a brief article is a tall order, but at least these four points can form a framework for struggling with the problem before us.

CROSSING THE BRIDGES

In speaking of bridges between theological and civil discourse we first of all are acknowledging the relative integrity of these two vocabularies, each with its own distinct grammar. Yet we also acknowledge they inhabit a common geography, a common world; hence the need for bridges for those of us who live in both countries. This is not an unfamiliar idea, going back at least to Augustine's *City of God* and coursing through the whole distinction between "nature" and "super-nature," between natural law and the law of Christ. However, the disparateness of the languages, their grammars, and the intellectual methods they entail requires an even more careful attention to the bridges between them.

Frequently what the journalists call "fundamentalists" are people who seek to replace civil discourse with a religious language, along with its attendant practices. However, instead of a bridge, they create a landfill or isthmus merging the two. The Constitution is overlaid by the Bible or the Qur'an, and customary marriage and contracts are replaced by religiously authoritative prescriptions. They may do so for a variety of reasons, but none of them contain a concern for maintaining the fragile pluralism, consensus, and shared constructions of the civil arena. The search for a bridge concept, however, already rests in a theological decision to treat the life of civil government as a valid part of creation and of human reason and creativity. From the civil side, of course, this implies that "secular" society itself rests on a religious conviction about the goodness and necessity of these civil constructions.

Within that perspective we can identify several clusters of religious concepts or concerns that have served as bridges in European and American civilizations. (How these might then be bridges to the other historic cultures of Africa, Asia, aboriginal America, and the Pacific Islanders is yet a further question that needs development elsewhere.) I will only cite four concepts here to illustrate this dynamic: covenant, assembly, household, and nature.

The Concept of Covenant

In English and American public discourse covenant has played an exceedingly important role. Indeed, it has been so pervasive in both religious and civil contexts that we now must explore its meaning very carefully if we are to understand its role as a bridge between the two realms. Its connection with civil life can best be understood by a brief rehearsal of its history.[3]

Since the early Hebrews and Greeks, governing publics have turned to federal forms such as leagues, confederations, and alliances to maintain themselves internally and externally. Federalism itself reaches back to the ancient Hebrew concept of covenant (*b'rith*). Covenantalism was derived from ancient treaty formulas and then given its theological meanings in Hebrew culture—its first function as a bridge. From Hebrew covenant and Greek and Roman leagues came the evolving modern

3. See the masterful presentation of this history in Elazar, *Covenant Tradition in Politics*.

forms of federalism. The Latin word for covenant is *foedus*, from which we have the English word "federal." Federalism is thus not only a system for sustaining and interrelating publics but a framework for religious thought and practice as well. It is one in which the very understanding of God in Israel was shaped by the federal orders of the ancient world. It is thus a key metaphor for bridging faith and politics.

A covenant is a set of promises among independent parties to secure the future. Like publics, covenants were historically the first step beyond kinship for establishing durable and trustworthy human relationships. That is, they began to replace the "essentialist" categories of race and tribe with those of promise and negotiated agreement—a fundamental basis for civil life in a multi-tribal world. In their ancient biblical form covenants were ways of relating people to the center of their deepest commitments and of relating themselves to each other, to their governors, and to the land on which they live. Covenantal thinking is thus a complex way of approaching a variety of relationships in terms of mutual promises. We can think of it in terms of the parties to the agreement.

Covenantal relationships assume that there are at least two relatively independent parties to the set of promises. In Israel we find that this mysterious God YHWH is one who seeks to enter into covenant. This is not merely a mode of God's activity. It is essential to who God is. God is a covenanting God. Human beings do not merely flow from the divine essence but exist with freedom before the Divine. The relationship of divinity and humanity comes to focus in the ever-surprising story of their promises. This means that the eyes of faith focus on the expectations they have of each other, the hopes that sustain them, the way promises are broken and the way they are taken up again.

Thus, there are two parties—God and the people with whom God covenants. This "people" itself is not an automatic unity. It too is a result of a covenantal process that is known to them through their story—not the timeless round of seasonal necessities but the unique and surprising ways they have been sustained over time. Their history is the outcome of promises made and kept, broken and renewed. Thus, the people are also intrinsically covenantal inasmuch as they take their historical existence seriously. Thus, covenant-making in the Bible usually begins with a recitation of the story of grace, sin, and liberation that brought Israel to that point. That is how they know who they are as a party to covenant.

Finally, there is another party, often overlooked—the land. In one sense it is the "property" of the trust between YHWH and Israel. It is the gift or trust that goes with the promises and their honoring. But in another sense it too is a party to the covenant. It cries out, it dies, it flourishes. It witnesses. It participates in the covenant according to its own dynamics and integrity. It is not simply at human disposal but precedes humanity as God's first covenantal partner in creation. From the covenantal perspective, it has legitimate claims and standing in the web of covenant upholding life.

God, people, land—these are the parties constituting a "full covenant" in Israel. It is among these three parties that the web of covenantal promises is spun out. It is this intricate set of obligations that supports the system of law obligating all the parties to common action. They are to be faithful to each other through this law, this Torah. In the biblical concept of covenant we find the peculiar sense of law as binding, voluntary agreement. It is law as "constitutional," which means that the people are bound together not simply by family, tribe, race, inheritance or nature, but by a legal framework which "constitutes" them as a people. From this covenantal understanding of law we have gradually come to a view of a higher law that stands in judgment over the laws made by an individual nation. This has been a powerful religious factor in the generation of the modern conception of human rights, which the United Nations has then spelled out in international "covenants." Since World War II, national constitutions have increasingly contained extensive Bills of Rights incorporating these claims, the most recent and extensive being the new South African Constitution.

In the interaction of these three parties, we recognize that at some points Yahweh's claims and initiatives are stressed and at other times the human parties emerge with greater equality. Sometimes even the land's claims are central, as in the prophecies of Isaiah (Isa 49, 55). The welfare of each of the parties depends on the faithfulness of the other two. Covenant always implies some kind of mutual interdependency maintained through faithfulness to promises that secure the future. This is the heart of covenantal, or federal, theology.

Covenant-making itself usually occurred within a sequence of actions that begins with a narrative of the gracious history by which the people of Israel formulated their image of the God of their covenant. Covenants are historical expressions of a tested trustworthy relationship. The second step, also incorporated in this narrative, is the concept of the "people" that emerges. A people is not strictly a category of race, ancestry, or biology, as many intoned in the nineteenth and twentieth century—with disastrous outcomes – but a binding association that emerges in historical struggle. It is in this history that an identity is forged—the second party to covenant emerges.

The third step is the formulation of mutual promises about their common future. These are the "commandments" of Exodus but also the promises of Deuteronomy and Isaiah. For Biblical life, the land is both trust and partner in this covenant making.

The fourth step is the calling of witnesses. Here the land also plays a part, either as the witness of which Isaiah and the Psalms speak, or the silent proof stone of Israel's fidelity. The breaking of covenant causes the land to wither and die.

The fifth step, of blessings and curses, is the part we would rather not mention. Covenant-making is serious. It is the act of entrusting our life to another or to others. Thus, its failure is also a diminishment of our own. Covenants, because they are extended entrustments, also contain obligations to future generations. Their voices and claims, yet unknown, are in some sense anticipated in the covenants of the present.

Finally, there are usually provisions for remembering the covenant and continually recalling it. This also means that it is continually re-worked in light of our historic experiences and emerging understandings about who we are, who God is, and what the land speaks to us.

Each of these moments in covenant-making can lead us to ask questions about this process in any polity. They press us to ask deeper questions about the basis for a public identity and political commitment. They ask us to probe the deep conditions that make constitutional democracies possible and viable. They push us to basic questions of political loyalty and personal sacrifice.

We see this in the fact that biblical covenants are "cut." They are sacrificial acts that end an old life and begin anew. Even more so, the establishing of new covenantal orders has always entailed great sacrifice. This work of sacrifice is not merely the brutal murders and exterminations that have dominated our century. They have produced victims and perpetrators but not necessarily new orders. The work of building up new relations of covenantal trust has required that we become vulnerable through the extension of forgiveness, of truthfulness, of apology, and repentance. It demands the work of reconciliation.

Covenant is a bridge term because it relates this rich religious history with widespread political experience. It helps us talk about the basis for law, civil relationships, free association, limited government, contracts, and social breakdown in a comprehensible way that leads us back to the religious ground for these human civil endeavors and experiences.

The vocabulary and grammar of covenantal thought has both institutional and cultural dimensions. Covenantal trust is a particular way of approaching the task of constructing institutions in a world that has usually depended on kinship, race, gender, and other biological markers to create trustworthy relationships. Thus, it has an affinity with the voluntary associations, professions, corporations, and congregations that constitute civil society. Covenant-making is not a dispensation from "the powers that be" but the practical expression of the life of a people who see covenantal association in their fundamental relationship to God. Their history is a lifetime of experiments in covenant-making, of covenantal failure, and of the struggle for reconciliation and renewed association. Thus, the covenantal work of civil association precedes in an ethical sense the covenantal construction of government, a key concept in the republican heritage flowing from that child of Puritan covenantalists, John Locke.

The formation of these covenanted associations has always carried with it an impulse toward federation. The little assemblies, associations, and societies of the people, if they are to live together in any semblance of wider peace and security, have sought solemn league and mutually consensual relationships with each other. This is what federalism is all about. Our world is now engaged in a massive federal struggle in order to hold together people's legitimate desire for participation and public life with their need for common action to deal with shared problems of ecology, population, and global markets. How to maintain vital publics within a federal framework lies at the heart of the human struggle today. It is also for many people the earnest of their ultimate hope. Covenant-making and federalism are not only religiously grounded practices; they are vehicles of a more transcendent vision.

These associational and federal structures thus rest in a deeper cultural ground of commitments and orientations. The cultural paradigm of the three parties to covenant always presses us to questions about what constitutes a people, their center of loyalty and trust, and the nature of the land with which they are entrusted. The cult of covenant-making, covenantal memory and acts of repentance and reconciliation shape deep paradigms for the maintenance of constitutional orders. This mirroring of covenantal practice in the creation or repair of constitutional orders is vividly evidenced in the contemporary widespread religious and civil interest in reconciliation. Knowledge of the covenantal roots of these efforts can help us sort out what to do in these circumstances.

Finally, this deep work of covenant-making opens us up to claims and values that transcend our everyday covenants. It provides a critical dimension for all our efforts to construct constitutional orders that respect human rights, the claims of the land, and of future generations. As a bridge concept covenant illuminates the deep presuppositions of a constitutional civil order. It can both undergird this constitutionalism and also, when connected to claims of human rights and ecology, can critique our present constitutional orders. Indeed, it provides the context for relating the legal structures of our time, with their ascending hierarchy of positive statutes, constitutions, and human rights covenants, with the even more expansive systems of interdependence in the natural universe and history.

Covenant is a rich bridge between religious and civil discourse at all three points of institutional structure, civic culture, and openness to transcendence, but it also has its limits. It can fall prey to legalism in spite of its sense of God's mysterious workings. It can even more frequently cultivate a narrow chauvinism that elevates the covenanted people above their neighbors, thus creating a deep incivility and hostility between them. This limit opens us up to a second bridge concept, which, from the religious side is contained in the idea of assembly.

Assembly

From its earliest times the Church called itself by the Greek term ecclesia (*ekklesia*), public assembly. Once again, a crucial religious term was borrowed from the life of civil governance. Ancient Israel spoke of this assembly as the *kahel* of the people. In Jesus' time, with the removal of civil autonomy, it yielded up the idea of the synagogue (those gathered together), which formed the root pattern of Christian assembly. The Greek *ekklesia* was the assembly of those "called out" from their households and from ordinary economic life in order to appear as equals engaged in argument rather than as kinfolk obeying the elders. In the traditional household people related to each other in a hierarchical pattern of subordination—elder over younger, men over women, parents over children and slaves—but in the ecclesia at least the male citizens had the possibility to be equal, not because they were equal in physical ways but because they were equals before the law. It took many centuries before this principle could be extended to all adult humans, a process hastened in good part by the Church's adoption of this institutional model in its early self-understanding. Thus, in the very concept of a separate institutional sphere from family and monarch for the expression of fundamental loyalties we see a critical ingredient for the institutional differentiation necessary for public life in a pluralistic world.[4] The ecclesia, like covenant, created a historic departure from family relations as the means to order the way we govern ourselves. The Christian ecclesia carried within it the possibility of being a kind of "proto-public," a peculiar public (in Roman terms) gathered in the Spirit of Christ.[5] Ecclesiology is simply the study of the purpose and structure of this public assembly called the church.

This dynamic of assembly, or "publicity," was intensified in the church's history through the experience of Pentecost, in which people from all over the known earth could speak out of their own culture and language and yet also be heard and understood by others—surely a paradigm of public life for today. Thus, the history of the church as assembly was inextricably tied up with the origins of our modern concept of a public. The early church put proclamation—public speech—at the heart of its life. With it came appeal to individual conviction as crucial to conversion and faith. It was this activity of publicity that upset the imperial order and brought persecution upon the Church.

One of the basic claims that made its way across this bridge of assembly, with its companion concept of public, was the Church's often contested claim that the Holy Spirit is fully the presence of God. The Holy Spirit infuses each person and gives him or her voice in the ecclesia. It almost forces them to speak, sometimes in unknown tongues going beyond their linguistic barriers. The Holy Spirit continually opens up the assembly to new voices and widens the bounds of participation. Here we see a religious ground for a life of democratic participation in public assemblies. The Church's often halting struggle to manifest free assembly in Christ's Spirit resonates in complex ways with the human struggle for free assembly in civil life.

---

4. This is a favorite theme of Talcott Parsons. See Parsons, "Christianity and Modern Industrial Society." See also, from a theological perspective, the work of James Luther Adams in Engel, *Voluntary Associations*.

5. Smith, *Where Two or Three Are Gathered*, presents a whole ethical outlook rooted in this act of public assembly.

We cannot go into the rich and often highly conflicted way that disputes over the nature and form of this ecclesial life shaped political theory. One would have to go back to the writings of Augustine, Aquinas, the fifteenth-century ecclesiastical councils, John Calvin, Johannes Althusius, and then in the modern era Thomas Hobbes, John Locke, and Roger Williams, to name only a few. At almost every point theological ideas about ecclesiology struggled to stake out a sphere for the cultic and ethical enactment of ultimate claims distinct from those of family and governance, whether they be princely, statist, or corporate. Without this fundamental differentiation, as many have pointed out, there would have been little social and cultural space for a civil society independent of the state or a civil society that could transcend national boundaries. For instance, the church's provision of experimental religious groupings around a particular vocation was nurtured in the monastic life for centuries before spawning the modern university and professions. Recall of the covenants and ecclesiologies by which they developed that life can both revitalize and critique our professions and universities in our own time.[6] In Germany and South Africa, not to mention the Civil Rights movement in the United States, many churches staked out spaces of free assembly that could nurture the sparks of civil freedom.

Ecclesiological debates also contained within them critical differences over the proper ordering of not only the religious life but of civil life in general.[7] The synagogue assemblies of the early church soon yielded to the rise to patriarchal and monarchical forms of church governance.[8] Organic notions of the church as the Body of Christ competed sharply with those of charismatic societies of the saints. St. Paul's egalitarian notions of church membership (Gal 3:25-28) competed even in his own writings with hierarchical images of obedience and service (1 Cor 11:3-12). All of these disputed ecclesiological conceptions have informed and contested their secular counterparts in European (and now African or Asian) monarchies, republics, and confederations. Without knowing the language of ecclesiological debate, we cannot understand the depth of contest today between those who want to govern by personal, usually parental, rule and those who seek a government tightly governed by codes, laws, and covenants. How general councils, federated units, and popular

---

6. See the covenantal treatment of professions in Mount, Jr., *Professional Ethics*.
7. See for example Stackhouse, *Ethics and the Urban Ethos*, and Stackhouse, *Creeds*.
8. Schüssler-Fiorenza, *In Memory of Her*.

participation should be mixed together in civil life has rich ancestry in debates over the role of the Holy Spirit in church assemblies, the relations of councils and popes, and the role of canon law in regulating church life and worship. This radical congregationalism, with its "bottom up" federalism, continues to contest with hierarchies of national regulation, control, and corporate administration.[9] Each brings with it deep reservoirs of historical experience in ecclesiological controversies and their associated penumbra of theological claims.

Ecclesiology has shaped both people's practical forms of association and the deep cultural orientations toward the form, relative value, and function of civil institutions. It also functions to introduce a transcendental element into these institutional and cultural debates, namely, because it is always connected to the symbolic rehearsal, in worship, of the connection between human association and ultimately divine purposes.[10] There is a necessary connection between these symbolically envisioned ultimate relationships and how we relate to one another. A worship focused on a relationship with Jesus as our friend sustains commitment to the priority of personal relationships, often among equals. A worship permeated with themes of parental care can then undergird commitment to a government that cares, even parentally, for the welfare of the people. A worship that rehearses apocalyptic destruction can fuel incentives to the warfare of Armageddon or the delegitimation of civil life in whatever form—a pressing issue for Middle Eastern politics.

On the religious side, this connection of liturgy and governance has permeated arguments over the relation of priests, ministers, bishops, laity, and councils. On the civil side it gives rise to the development of civil religion as a means to legitimate and critique the forms of civic and governmental life. Ecclesiology, by drawing the connection between worship and organization, awakens us to the rich relationships between civic institutions and their ultimate forms of devotion.[11]

In this discussion of ecclesiology I have, for reasons of personal (though not idiosyncratic) conviction, emphasized the connection between religious assembly and "publicity," the life of civil society and association. However, I have also indicated the contrast between those who have seen the church as a public and those who have understood it as a

9. See Everett and Frank, "Constitutional Order," also appearing in this volume.

10. See Everett, *Politics of Worship*.

11. See the essays by Robert Bellah and others in Richey and Jones, *American Civil Religion*.

kind of household. The concept of household, our third bridge concept, both complements and contradicts that of public assembly, which is most closely associated with governance.

Household

Numerous recent theologians have drawn on the concept of household and its Greek root word, *oikos*, to form a bridge language between ecclesiology and ethics on the one hand and general discourse on ecology, feminist theories of society, and governance.[12] In the ancient Greek world the word *oikos* enabled people to move easily from household management (*oikonomia*) to questions of care for the civilized world (*oikoumene*). Christians saw their own assemblies as gathering in the "*oikos ton theon*" (household of God) and later spoke of God's saving word in terms of the divine *oikonomia*. In a differentiated world we can draw on this integrating term to make connections between what have become disparate languages of faith, politics, and ecology.

In many respects the household language contrasts sharply with that of ecclesia and public, with its distinctions between private and public, autonomous citizenship under the law and caring parenthood within the family. While it resonates especially with women's experience in the household it also can undermine the public in which they seek to exercise their public citizenship. In other respects, however, it draws attention to much of the substance of what has to be argued out in the public order—how to educate the young, how to care for the earth, and how to protect the weak. It is strong on the purposes of public policy even as it may introduce paradigms of inequality into our public discourse. Household concepts enable traditional religious concerns for care, limits to life, dependency, and the need for wisdom's guidance to inform public discourse. They raise questions about our ultimate needs and interests in the middle of the narrow, market-driven interests of much of contemporary political life. They raise questions of enduring relationships in the midst of competing claims made by supposedly autonomous, indeed isolated, individuals. Household, with its emphasis on caring relationships, whether egalitarian or dependent, contrasts sharply with legal

---

12. For example, Mudge, *Church as Moral Community*; Russell, *Household of Freedom*; Raiser, *Ecumenism in Transition*; and Everett, "OIKOS: Convergence," also in this volume.

arrangements of adversarial truth-finding, retributive punishment, and monetary compensation. The symbolism of the household points to a justice of repair, restoration, and relationship that both transcends and critiques the ordinary justice of the law. In these ways we can see institutional, cultural, and transcendent dimensions of public life that are illuminated by the household concept. It is especially important to see how household and family metaphors have shaped so much of Christian discourse and how this discourse has implications for these dimensions in public life. Theologians and ecclesiologists draw on family metaphors to speak of God, of Trinity, of ethics, and of church authority and membership (Popes, abbots, fathers and mothers, not to mention the "family of God"). But these were also contested symbolisms. Let me give one example.

The use of family symbolism to talk about authority and membership in God's assembly was already contested by Jesus in his rejection of family ties or titles (Matt 10:34–37, 12:46–50 and parallels; Matt 23:9). In subsequent theology the relation of the Holy Spirit, a potent "democratic" element, to Christ and to "the Father" fostered the critical break between the Eastern and Western churches. The lack of fit between the Spirit and the two male figures in the Trinity became the symbolic matrix for arguing out the relation of paternal authority to democratic participation. In this regard we need think only of the immense controversy between the seventeenth-century Society of Friends and the Church establishment of England or New England.[13]

Today we face in our public life the equally vexatious problem of how to relate the democratic processes of government not only to the problems of caring for the earth and the poor but of relating "private" corporations, with their frequently parental or authoritarian administration, to the claims of the common weal.[14] The Trinitarian struggles to hold these together with respect for their differentiation can be a helpful contribution to public debate. The household bridge concept can mediate between these two discourses.

---

13. John Locke's attack on patriarchy in Locke, *Two Treatises*, still left household authority proper intact. It was the Quakers who took the theology of the Spirit to its democratic conclusion and were persecuted mightily for it.

14. Goudzwaard and de Lang, *Beyond Poverty and Affluence*.

Nature

Finally, we turn to a bridge concept with a rich and lengthy history, but one which is bound directly to the problems of parental care for the earth, the weak, and those deprived of a household. The concept of nature is itself contested mightily. From the religious side it is rooted in the doctrine of creation and the unity of the Creator's purpose. Questions of "fact" and "value" are inseparable in nature, for we ourselves are valuing, purposive agents. In some sense the rest of creation shares this with us. All beings live in relationship with one another and with their source—God. So-called "creationists" may have erred in translating these questions of unity, value, and purpose into a temporal framework concerning our origins, but they are correct in maintaining that our nature involves questions of wholeness, value, and purpose that go far beyond the mechanical and segmentary ways of understanding us that have shaped modern medicine, technology, and education.

On the civil side, nature has often been appealed to as a stand-in for God. Our rights, our legitimate claims and interests, our proper needs—all can be traced to a publicly discernible "nature" that doesn't need a Creator behind it. Our ethics can be drawn from a "natural law" without appeal to special revelation or claims of particular church authorities. Appeals to some kind of natural law are thus indispensable in the drive to sustain a conception of human rights that transcend the positive law of the nations. Thus, natural law as a body of philosophical and ethical reflection has been one of the main bridge concepts between religion and civil life since the early years of the church's existence.[15]

This in no way implies that the civil order has a settled conviction about our nature, about the way our nature fits into nature in general, or for the way this should be translated into institutional, legal, or political forms. Merely a glimpse at the heated debates over abortion and homosexuality, not to mention genetic manipulation, shows that these questions are far from resolved. However, the point here is that the concept of nature itself is a way of introducing classic theological and ethical claims into public discourse in a way that makes them publicly arguable.

The outcomes of these debates can have important cultural and institutional consequences. Family policy based on our nature to live in families, indeed families of a certain form, can be reasoned out of theories

---

15. For the classic exposition see Troeltsch, *Social Teachings*. For a current use see Mudge, *Church as Moral Community*.

of human nature. Ecological policy is deeply shaped by theories about the way "nature" evolves, seeks equilibriums, deals with instabilities, or manages complex interdependent relationships. Concepts of nature can reshape our basic cultural orientations in ways that put them in conversation with classic theological concerns.

Arguments from nature also provide an entrée into classic concerns of transcendence, at least in an important limited form. In placing "the human" or "nature itself" over against our settled statutes, practices, habits of thought, and parochial customs we gain a certain transcendence over our limited place in history and the world's cultures, not to mention curbing the arrogant pride that ignores our tenuous ecological perch in the cosmos. The nature-related concept of Gaia, for instance, plays an important role in our imagination today, exercising something like God's judgment over our careless destruction of our habitat.[16] Questions about the purposes of human nature can enable us to entertain alternative notions of the good society, of possible utopias, and ecological scenarios that might otherwise be dismissed by appeal to our intractable sin.

These four bridge concepts—covenant, assembly, household, and nature—offer a way of thinking about the problem of theological literacy in public life. Not only do we need to understand how these concepts have functioned on both sides of our present cultural divide, we need to see the ethical issues they have engaged in history and in our present debates. There is, unfortunately, no quick fix or twelve steps to recover from our amnesia about these debates and the language of the protagonists. It will take practical efforts in education and in religious communities that seek to enable us to find our many voices and take up the perennial task of living together on a beautiful but wounded globe.

---

16. Reuther, *Gaia and God*.

18

RECONCILIATION *between* HOMECOMING *and* FUTURE

A Case Study from America's Struggle with the Vietnam War

THE PRACTICE OF RECONCILIATION is rooted not only in theological concepts but also in deep symbols. These symbols give a strength and broad orientation to the practices of reconciliation within a distinct cultural context. The work of reconciliation needs strong symbols so that it always contains an eschatological dimension as well. Reconciliation seeks to bridge the chasm between our present broken circumstances and the ultimate purposes of God. This bridging is possible only through symbolic rituals and events, because we cannot bring about complete justice in our world as it is.

Working with these theological understandings, I have selected a case study on the establishment of a memorial as a way to inquire into the possible process and dynamic of reconciliation. With memorials and their associated symbolic actions, we experience the deep symbolic dimension of the practice of reconciliation. These symbolic aspects of the human search for reconciliation bring Christian and other religious and cultural symbols together to anchor the deep religious meaning of reconciliation in particular human contexts. The case study we have before us brings these elements together in striking relationship.[1]

On November 13, 1982, thousands of women and men assembled on the grassy mall in Washington, DC. Before them lay two black walls bound together at one end like wings or plowshares sunk deep in the earth. The wings of the wall were divided like pages of a book. Only names broke the highly reflective surface of the stone. It was the Vietnam Veterans Memorial that was being dedicated on this day.

Three years earlier it was only an idea in the mind of Jan Scruggs, one of the Vietnam veterans who had decided to find a public means for remembering the Americans who had died and those who were still missing in Vietnam. With veterans John Wheeler and Robert Doubek he established a foundation that collected $8,000,000 from over 650,000 citizens. After three years of disagreements and conflicts about the plan and its location, the repressed feelings about the war came to expression in this symbolic construction.

In describing the purpose of the memorial, Jan Scruggs said "We do not seek to make any statement about the correctness of the war." "Rather, by honoring those who sacrificed, we hope to provide a symbol of national unity and reconciliation."[2] Robert Doubek spoke in a similar tone: "The Vietnam Veterans Memorial is conceived as a means to promote the healing and reconciliation of the country after the divisions caused by the war . . . ." "The Memorial will make no political statement regarding the war or its conduct. It will transcend those issues. The hope is that the creation of the Memorial will begin a healing process."[3]

In addition, there was also the hope, as expressed by President Jimmy Carter, that "[The memorial would stand as] a reminder of the past, what was lost, and a reminder of what we learned." However, *what* there was to learn differed among the participants. Some remembered the error of the war. Many, like Senator John Warner, stressed that "We learned that we should never again ask our men and women to serve in a war which we do not intend to win. We learned that we should not enter a war unless it is necessary for our national survival. We learned that, if we

---

1. This account draws on Bellah, *Broken Covenant*; Capps, *Unfinished War*; Hellman, *American Myth*; Scruggs and Swerdlow, *To Heal a Nation*; Wheeler, *Touched with Fire*; and Wheeler and Horne, *Wounded Generation*.
2. Scruggs and Swerdlow, *To Heal a Nation*, 43.
3. Scruggs and Swerdlow, *To Heal a Nation*, 16, 53.

do enter such a war, we must support our men and women to the fullest extent of our powers."

The Memorial itself soon became a recapitulation of the war and the questions the war itself had raised only ten years earlier. In the wake of the conflict over its design, the foundation and its critics decided to erect a flag pole and figures of three soldiers at the entrance to the memorial. In the words of Jan Scruggs "[The anger] shows we need to do a lot more healing."[4] John Wheeler saw further, saying "Grief means looking at the truth and brings at time anger at the loss, and the anger takes different forms in each person. The hardest part of the job of building the Vietnam Veterans Memorial is to face the anger involved as our country does this work."[5]

Several saw the black wall as a fallen bomber, a "black gash of shame," or simply as a grave—the downfall of the American identity. The architect Maya Lin saw it otherwise. Scruggs wrote later: "She decided that the way to build a memorial would be to cut open the earth and to have stone rise up as part of the healing—something that would be like two hands opening to embrace people. She saw the Mall as a living thing that should not be disrupted or destroyed."[6] For her the winged wall was more seed than grave—or perhaps a grave that is also a seedbed. Whether it would be that for America was a very open question.

In its garden-like setting the wall would create reconciliation also for the invasion and destruction of the land of Vietnam. The earth itself was respected. The wall is a part of the land around it. It arises out of the earth and brings the lost and fallen back to the earth. Humans and the earth find reconciliation. Similarly, the tall, masculine aspirations of the Washington Monument found a connection to the calm, meditative words and figure of Abraham Lincoln at the other end of the reflection pool.

The placement of the memorial also evidenced the deep anger, grief, and loss of meaning fostered by the war. Senator Charles Mathias, of Maryland, an early supporter of the memorial, observed: "[The site near] the Lincoln Memorial is also fitting, for not since the Civil War has this Nation suffered wounds and divisions as grievous as those endured

---

4. Scruggs and Swerdlow, *To Heal a Nation*, 87.
5. Scruggs and Swerdlow, *To Heal a Nation*, 132.
6. Scruggs and Swerdlow, *To Heal a Nation*, 54.

over Vietnam."[7] Moreover, the Mall was a place of protest and struggle, of Martin Luther King's 1963 "I Have a Dream" speech and demonstrations. The feelings of a people divided over the war had made it holy. No other place could serve for this purpose. The Mall was more a place of the citizens than a place of official and governmental organizations.

Here we would only find the names and the quiet reflections of our faces in the names of the dead, the victims of a contested war. According to a Gold Star Mother cited in a lead article in The New York Times: "[The Memorial design seems to be] a lasting and appropriate image of dignity and sadness. It conveys the only point about the war on which people may agree: that those who died should be remembered."[8]

The soldiers and veterans were not only to be remembered, but also welcomed home. There had been no real homecoming in the sixties and seventies. The soldiers had simply dived back into civilian life. Shortly before the dedication of the Memorial the foundation held a Vietnam Veterans Reunion with a parade, a convocation of friends bonded in fire, and seminars and meetings over the laments, griefs, and lessons of the war. In this way the invisible veterans experienced a homecoming, a welcome, and thanks. Veterans of the world wars and Korea also bridged over the chasm that had separated them from the soldiers from Vietnam. It was a time of reconciliation among soldiers before the eyes of the whole nation.

From the beginning the directors of the foundation sought simply to honor the fallen soldiers. No political statement should be evoked, no interpretation. Only the names. Therefore, they thought, the memorial should be more abstract than figurative. Nevertheless, they announced an open and blind competition that was eventually won by Maya Ying Lin, a twenty-year-old architecture student from Yale University. Maya Lin was a first-generation Chinese-American who was, according to her own declaration "a-political." Her face displayed the complex reality of a country that had sought to cover over and forget the nightmares of Vietnam.

As soon as the committee had agreed on the simple conception of the names, other questions emerged. In what order? Order means interpretation. It was decided to follow a chronological order—from the center left, then in an invisible circle that enfolded the onlooker, and then moving from the right end along the wall back to the center. The list

---

7. Scruggs and Swerdlow, To Heal a Nation, 18.
8. Scruggs and Swerdlow, To Heal a Nation, 71.

of names began in 1959 and ended in 1973—America's longest war at that time. And title? Rank? Should the names be engraved with title and rank? In spite of protests from politicians the committee decided to forgo the naming of title, rank, and position. They belonged to individuals. So were they born, so they died. The wall should display a democracy of the fallen. Not one of them was unknown. They would not be forgotten. There are no unknown soldiers from Vietnam. The gravestone for him was removed from Arlington military cemetery in 1980. Each one should be remembered. In the fifty hours before the dedication of the memorial every name was read at a night vigil in Washington's National Cathedral.

In spite of agreements about the names, the foundation decided to inscribe a simple explanation at the beginning and end of the over 58,000 names. The prologue simply said: "In honor of the men and women of the Armed Forces of the United States who served in the Vietnam War. The names of those who gave their lives and of those who remain missing are inscribed in the order they were taken from us." The epilogue said: "Our nation remembers the courage, sacrifice, and devotion to duty and country of its Vietnam veterans. This memorial was built with private donations from the American people." All other interpretations are left to the feelings and thoughts of visitors, whose fingers softly touch the names and place flowers, memorial tokens, and pictures in front of the wall.

After construction of the memorial an additional sculpture was installed for the women of the war. In 1998 a memorial was dedicated to the African-American soldiers killed in the Civil War. Thus, the aspiration to bring the forgotten and minorities into public awareness reached further. All victims—as individuals—should be kept in our memories.

THE CONTEXT

The Memorial Project had to be driven by a voluntary non-governmental initiative, since the war itself was so disputed and fruitless that few politicians wanted to touch such an idea. The experiences of the war were also the or one of the important catalysts for a splintered generation—on college campuses, in the hippy culture, in the civil rights movement and in the reawakened women's movement. All of this separated this generation from the culture of their parents, who in their own time had been victors over economic depression and the Axis powers of World War II. From the standpoint of a veteran this was a chasm between those who lived out their duty and their promises and those who simply exercised themselves in opposition. In fact, this was a deeply rooted opposition between two fundamental loyalties. That this opposition stood in a long line of conflicts in American life was seldom mentioned. Revolutionary "Tories" against "Patriots," Civil War Unionists against Southern-sympathizing "Copperheads," conscientious objectors against war in the twentieth century, not to mention Henry David Thoreau.

For the majority of the population, especially the politicians, the Vietnam War was America's first lost war. It was simply a disgrace. Because of the depth of domestic political conflict, a hunt for traitors or internal enemies was impossible. On the one side stood those who saw the loss of the war as the consequence of the lack of domestic support for the war. This perspective was often emphasized by military or political leaders as the "lesson" of the war. On the other side stood those who considered the war as a mistake, an illegitimate action. This group included military opponents of the war and almost all the resistance groups. Behind this split lay very different concepts of America and its role in the world. Because of that, victory was impossible.

The struggle to come to terms with the war was also very difficult because it was really not a genuine "war." A "war" as such was never declared by Congress. Here many saw a failure in the American political system itself. An "action" became an actual war through the "Tonkin Gulf Resolution" of 1964 and the expansive Cold-War system of secret state operations without any formal approval. The system itself had failed "the People." It was fundamentally a revolutionary situation. Either one searched out the implications of the system failure or one had to remain silent and live a lie. Most chose the latter. The Memorial confronted this problem. It called it a "war." In this way the soldiers could be considered the same as other veterans of war.

These factors also made the question of responsibility unanswerable. Usually, the search for evildoers or traitors plays out within a system of accountability. The sheer examination of the facts about Vietnam revealed the honored and murdered President Kennedy as the one who took the first decision for American involvement in Vietnam. He could not be found guilty. In a sense the political system itself executed its judgment: Lyndon Johnson had to give up his candidacy, an end to the war was resolved, and in the end Richard Nixon had to withdraw from Vietnam. The protest and opposition against the war was a success of the media—the watchers and simultaneously the manipulators of the public.

Access to the "truth" of the war could not be won through the question of accountability. The problem was not a problem of the truth itself—especially of true responsibility—but of meaning and re-orientation. It was a problem of fundamental values in American culture. This is why it was so difficult to foster an open discussion about the war or publicly to recognize those who had fallen in it.

For the soldiers there was no homecoming, no gratitude, no recognition of their wounds, such as the effects of Agent-Orange. Consequently, Americans could not achieve any reconciliation, not only with respect to the wounds that the war had inflicted, but also with respect to the deep conflicts of values and perspectives that the war had unloosed. That people drew back from taking in the situation of the veterans reinforced the repressed memories of the Civil War, the genocide of Native Americans, of slavery and racism. Their effects were still manifest—in the color of the faces of the soldiers, in the financial destruction of the "War on Poverty," and in the engrained American attitude toward the population of Vietnam, whose lives, forests, and cities were annihilated daily. None of this could be acknowledged. All these repressed memories made it impossible to understand the war and work through it appropriately.

Out of the Civil War and earlier Southern military traditions came an intensive military culture of personal honor, virtue, duty, courage, and obedience. Mixed in with this strong tradition was a kind of Christian popular religion rooted in local customs tied to a strong defense of the American Way of Life. Thus, American flags are displayed in churches, civil holidays evolve into religious holy days, and a religion of interpersonal love and attachments becomes the motivator of a national sense of belonging. This Super-Patriotism, especially in the Sunbelt, has served to a certain degree as a substitute reconciliation in place of the shame evoked by the Civil War. In the framework of this form of popular religious culture people could feel lifted up over against the self-righteous moralism, arrogant technology, and exploitative capitalism of northern Yankees, all of which also belonged to the causes of the Vietnam War. The technically-driven war strategies of Secretary of Defense Robert McNamara sacrificed the soldiers formed in the traditional military culture of personal virtues into the fire of a nationalistic egoism. Questions about American militarism awakened once again the deep historic division between the cultures and experiences of North and South. The search for the cultural basis of the war threatened an important part of the compromise between North and South that had been reached earlier in trying to come to terms with the Civil War and its consequences.

The history of the encounter of European settlers with the original inhabitants of the land had also deeply formed the struggle for memory and understanding. The Vietnamese appeared to many as a projection of the Indians, the original Americans, as portrayed in white America's mythos of its history.[9] The American fighter appeared as a Pioneer-Scout; Vietnam, like the moon, the "Last Frontier." Therefore, the Vietnamese appeared either as the "noble savage" or the inhuman enemy. The task of the American was to civilize them, that is to preserve them from Communism. If that wasn't possible, they must annihilate them. So emerged the famous sentence out of the My Lai incident: "We had to destroy the village to save it."

In light of the Vietnam debacle, many tried to save this mythic image. Thus arose the myth of the Green Berets, the special unit that was to infiltrate the local civilian population. And so also emerged the Vietnam Memorial. Most notably, Jan Scruggs was motivated to begin work on this project by the film "Deerhunter." This film is—in Vietnamese costume—a reproduction of the famous nineteenth-century novel by James Fenimore Cooper, "The Deerslayer," that took up the conflict between noble settlers and noble Indians. In the figure of Natty Bumpo, the deer slayer, we see a "civilized" man who is fully at home in the culture and environment of the Indian. He is a prototype of the Green Berets. Once again we are led to the honoring of the figures of this mythic American drama. America could remember its horrible Vietnam nightmare as a re-enactment of its founding myth of invasion and settlement. The memory and public conception of the Vietnam era could be comprehended, but at the cost of actual reality.

One reality of American life had to be expressed by the memorial in any case—the plurality of its population. In the selection of figures to stand at the entrance to the memorial, with their anxious faces turned to an uncertain environment, one had to be African American. Appearing next to him are the faces of a European American and a Hispanic American. Here there are no heroic generals on horses. In this portrayal the long oppressed black soldier finds his rightful place. He is now partner, not servant or slave, in America's struggle with its mission in the world. The Latin-American soldier also shows that those invaded by the wars of

---

9. Hellman, *American Myth*, 78.

westward expansion of the United States now find their new participation in a pluralistic nation. Here we find elements of a gradual reconciliation of the fundamental forms of alienation in American history. It is striking that such a recognition would arise precisely in light of this tragic devastation.

These three repressed memories—of the Civil War, of genocide, and slavery—always lie behind the deep fracture in American cultural history manifested in the Vietnam Era. This fracture was manifold. John Wheeler saw it as a turning point in the understanding of the relationship between men and women and in the definition of masculinity. The powerful commander is replaced by the participatory partner in life. Robert Bellah saw it as a time of testing for the American Civil Religion.[10] He asked whether America could rework its historically broken covenant with God with respect to its relationship to the Native Americans, to enslaved Africans and to the poor. America had to take on in its Civil Religion a new partnership role in the world. Its fundamental belief in Manifest Destiny must be transformed in order to avoid future Vietnams. Vietnam was the fracture through which this change could become possible.

Such transformations required remorse and conversion. For many this conversion was impossible. They fought strenuously against the Memorial Committee for a "heroic" memorial that would reflect Vietnam as a victory. They had gotten their flags and figures, but nothing more. Was Vietnam, as many claimed, the last war against Communism, indeed the beginning of its downfall? But the end of global Communism came fifteen years later, and even then not in China. At the construction of the memorial it wasn't even in sight. All sides were in agreement that the memorial would offer no interpretation of the war, whether as victory, as defeat, as lesson or as turning point. One could only provide a public reckoning of the sheer fact of the 58,000 who had lost their lives in service to their country. Only these individual sacrifices could be acknowledged. Now every visitor had to develop his or her own interpretation, or just touch the names, find the room to grieve, and sympathize beyond all political opinions and oppositions. Only under these conditions could the memorial be erected. This was its success but also the limit of its effectiveness as a means of reconciliation.

---

10. Bellah, *Broken Covenant*.

THEOLOGICAL REFLECTIONS

The memorial was from the beginning envisioned as a part of a process of reconciliation. It was to be constructed as a means for the healing of the nation. Almost two million people come annually to remember, to grieve, and find a kind of relationship with the missing and unknown who sacrificed their lives to respond to the nation's call. In the words of John Wheeler, "Who among us was not touched, or even wounded, in some way by the Vietnam War? The walls shine like mirrors. So we begin to see hurts inside us, too, when we see our own reflections in the walls." Veterans and all others "now know that healing begins only when you look deeply into yourself and when you honor those who have suffered on your behalf."[11]

Now we confront the question: Is this grief process to be understood as a process of reconciliation? Is reconciliation a kind of healing? And what does healing mean for a people and for institutions? Since reconciliation is so central for Christian faith, we always stand in danger of portraying all kinds of reunion after a conflict as "reconciliation." In terms of Biblical understanding, reconciliation is not to be understood as the dissolution of conflict within a juridical system, but as a new creation over against the threatening, fundamental divisions among human beings that undermine the possibility of social life altogether. That is, such divisions are not simply violations of contracts but, seen biblically, of covenants. They are divisions that utterly block the possibility of communication. In face of such deep collapse of the very bonds of humanity, one cannot simply retreat into silence and isolation. That would be to give up the possibility of public life. One would have no voice before others, no shared histories and no common history. People lose their capacity for participation. Reconciliation means the overcoming of division in this sense. Here we see why reconciliation is a fundamental theological concept and why it stands at the crossroads between social life and theological grounding.

Although the biblical concept of reconciliation is packed with various levels of meaning, it seems to me very important to draw several leading elements out of this complex of meanings in order to illuminate specific theological dimensions and identify their consequences for the church's engagement with processes of reconciliation. Three elements in biblical and church tradition appear to be central. In light of the overlapping relationship between social-political and theological dimensions

---

11. Scruggs and Swerdlow, *To Heal a Nation,* 160.

I identify these as *public*, *covenant*, and *participation*. Put concisely, a genuine public sphere needs relationships of trust grounded in covenants in order to make possible widespread participation in processes of consensus formation. In biblical history covenant-making was the way out of the constricted network of family and clan into wider relationships. Relationships grounded in covenants led out into the whole history of federalism, which is indispensable for the just coexistence of peoples in our own time. The public life that can emerge in the encounter of people who are foreign to each other was already envisioned in the *kahel* YHWH (congregation of God), in the synagogue, and above all in the *ekklesia* (public assembly) of Christ. This interrelationship of covenant and public opens up an orientation to important aspects of the present search to develop processes of reconciliation.

Processes of reconciliation surface all over the world as struggles for public life. The practical problem is how people can make their witness public where previously silence, lies, mistrust, and darkness reigned. This search for public openness is not only the precondition for judicial process but also the foundation of ever comprehensive process of reconciliation. There was no judicial process in the case of Vietnam. Nevertheless, people needed something they could only comprehend with the word reconciliation. Similarly, the Truth and Reconciliation Commission in South Africa held publicly established truth as more important than reparative justice. Even if an "objective" truth cannot be reached, people need a comprehensive public space in order to make possible a common search for a new way to live together. Only in such a public can people develop a common life grounded in mutual understanding.

The memorial represented only a beginning for the development of such a public reality. It recognized that there was no public consensus about the war, but it recognized also the barest facts of the war—the dead and their names. Beyond that, it let the stones, the surrounding grass, the topography, context, gifts, and behavior of the visitors express the elements of a new commonality.

The memorial has the function of encouraging a wider public discussion. Visitors themselves are the meaning-givers. Each makes out his or her own meaning as well as their own grief, memories, and special relationships and estrangements. Here is where we find the ambiguity of the memorial. It is simultaneously fully public and completely private. Visitors are thrown back on themselves in the process. It opens up a way to individualization and privacy. But it also opens up a way to a pluralistic and differentiated "democratic" truth. In any case the memorial offers the opportunity to proclaim respect for the fallen, for the tragedy of the war, and for the sheer humanity of the fallen.

This democratic element in the effect of the memorial is also deeply bound up with its source. It was established on a voluntary basis by ordinary people themselves. It arose out of a public life that stood outside of the state, the administration or large corporations. It is "of the people." It sets forth no official meaning. It is to be a public space where "the People" can find their own unofficial meanings. Here we also find a possible characteristic of reconciliation. It must arise directly out of the life of people. The problem with the search for truth and common meanings is always that it can be steered by institutional interests or that this or that marginalized group sees the "truth" of the official institutions as one-sided or ideological. It is not "their" truth and therefore no basis for reconciliation. The memorial and the struggle about its abstract character reminds us of the nature of public life itself, whose dynamics echo those of reconciliation as well.

Reconciliation exercises and demands a public sphere, indeed a dynamic of publicity, that must vitally undergird the institutions of a society. In the Vietnam Memorial this dynamic emerged to make possible a new underlying public reality. It stands at the boundary between traumatized silence and the verbal search for a new mutual understanding. It is this reality that is meant in the biblical concept of covenant. The public exists in order to make new covenants possible.

John Wheeler begins his reflections on Vietnam in *Touched with Fire* with the concept of promise. For him this concept is central for understanding the war and for a way out of the war to a positive task for the Vietnam generation. The soldiers and the protest movement, he claims, knew what "promise" means. It means the duty of maintaining the fundamental principles of our life. People answer to "the call." They had completely different understandings of the "call" of the situation—on one side to support the war, on the other to resist the war policies. Both were "promise keepers." From the standpoint of a biblical concept (which Wheeler, as a former theology student knows), there were people in both camps who were in the situation of making covenant commitments. (The ironic but also apposite juxtaposition of "Bund" [covenant] and "zum 'Bund" gehen" [to enlist in the Army] in Germany I simply mention here in passing.)

The question for the process of reconciliation is: Which covenant do we choose? Perhaps the memorial speaks of a duty and of a covenant relationship, out of which arises a duty. But then the question is posed: Into which covenant are we called? In a biblical perspective reconciliation arises as the rebuilding or institution of a covenant between God and the people, God and the creation. It arises out of the salvation history of God with humankind—a history of wounds and also of grace. The paradigm of this history lies in the Exodus and is repeated for Christians in a new way in the crucifixion and resurrection of Christ. This framework of reconciliation as entrance into a new covenant offers people like John Wheeler a language by which the problem of reconciliation in the context of America's war in Vietnam can be understood. For him it is a problem of promise and duty, especially at the individual level. Moreover, the concept of covenant can lead us further, since biblical covenant always takes its full form with groups, peoples, and nations. In light of the covenantal idea, reconciliation is not only conversion but also a new relationship. It is a way into the future.

Wheeler himself, like the other supporters of the memorial, began with the individuals. This heightened acceptance of the project so that it could succeed. But beyond that he speaks of the many other new relationships that nourish the healing of wounds of the war. He speaks of the overcoming of a dualistic perspective in order to achieve a perspective that can embrace a pluralistic reality. He speaks above all of the new relationships between men and women that came out of the war experience. We can also cite other relationship possibilities. The wall itself represents a kind of reconciliation between people and nature, also destroyed in the war. Robert Bellah speaks of a new covenant relationship between America and other nations of the world—a new covenant that is not yet concluded. Out of such a new covenant comes the demand that America itself should develop away from national egoism to a capacity for genuine partnership.

Such questions lead us finally to the obvious question about the place of the Vietnamese in the process of reconciliation that the memorial should serve. They don't appear there. They are not mentioned in the literature. They don't exist as the chief partner in reconciliation. Why not? It appears to me that their absence is a sign that America has not overcome its original myth of the noble savage and the inhuman enemy. The reconciliation between Americans and Vietnamese comes only gradually through individual relationships like marriage and economic ties. For most Americans what really matters about Vietnam is the homecoming of the lost bodies or certain knowledge about their fate. The list of names will be continually updated. As always reconciliation remains in the public consciousness a matter of individuals. Reconciliation as a new covenant between alienated peoples remains in the future.

Here we see how reconciliation always contains an eschatological element. Reconciliation is not the rebuilding of an original position. It is never a return to the "Original Position" of social contract thought. Should reconciliation in America be understood as a return to a time before the European invasion? Can reconciliation between Vietnam and America be considered as a return to some kind of historical position? Obviously, such "original position" concepts of reconciliation play an important role in ethical thought, since they present elements of an effort for some kind of reparations. And such reparation is a very important sign and symbol of a new relationship as reciprocal partners in the common undertaking of building up a new future.

Such an effort always stands in the light of a vision of a new future. Without such a vision one cannot enter the promises of a new covenant. One can almost say that reconciliation is the experience of being caught up in a new common future. Theologically speaking, reconciliation arrives as the new creation given by God. It is participation in this new creation that reconciles people with each other and with nature. It is this dimension of reconciliation with which Americans traumatized by the Vietnam War struggle. Successive Presidents have difficulties with the "Vision Thing" (like George W. Bush) because they haven't come to terms with the realities and challenges of the Vietnam War. The Vietnam Memorial has enabled a homecoming for veterans, but has not created a vision for the future. It is a step. The journey goes on.

19

Journey Images *and the* Search *for* Reconciliation

Our search for justice and reconciliation is shaped by many forces and conditions. Some of these are economic, military, ecological, and political. Others are religious and cultural. Among these religious and cultural factors are our fundamental images of how we relate to other people and the wider world. One of the most powerful and pervasive such governing images is that of the journey.

In this essay I will first sort out what we mean by a journey and how these journey images generally function in our cultural experiences. I will do this by examining three powerful journey traditions—the Cherokee "Trail of Tears" in North America, the South African "Great Trek", and the "Journey to the West" of the Chinese pilgrim Xuanxang, with its reverberations in "The Long March" of Mao Zhedong's Red Army. Secondly, I will turn to the ways these journey images shape our ethical aspirations and actions, especially within the perspective of the search for reconciliation.

This connection has been made most vividly by the banner that presided over deliberations of South Africa's Truth and Reconciliation Commission—"Truth, the Road to Reconciliation." It was echoed in the title of Nelson Mandela's autobiography, *Long Walk to Freedom* and the South African Broadcasting Corporation's documentary, "Long Night's Journey into Day." Achieving reconciliation was not a quick "solution" to a "problem," but a long, difficult, and uncharted journey.

The journey image is powerfully connected with the struggle for reconciliation and just relationships, but exactly how does this connection work? Classical cultures were shaped by the story of the Odyssey, the Biblical cultures of Judaism and Christianity by the Exodus and Conquest, and Islam by the flight from and return to Mecca. They contain rich evidence for our inquiry. Rather than begin with these already mythicized and culturally burdened stories, let us first turn to some other powerful human experiences to begin to gather our insights.

THE TRAIL OF TEARS, 1835-39

The Cherokee ("Tsa-la-gi") people probably arose among the Iroquois group and migrated south to occupy the southern Appalachian Mountains by 1000 C.E.[1] This migration was not part of their mythic orientation. Rather, they knew themselves as the people of the hunt and the cultivation of corn. In the seventeenth and eighteenth centuries they quickly adapted to Spanish, French, and English technologies. In spite of frequent clashes with Europeans, primarily over land, they intermarried frequently with white tradesmen, soldiers, and settlers. By the beginning of the nineteenth century, they rapidly took on settler farming techniques, slaveholding, and, by 1830, Christian faith. While many learned English, in the 1820s the majority learned to write and read with the Cherokee syllabary developed by Sequoyah. At the same time, they adopted a republican form of government patterned after the US Constitution and assumed an overall prosperity equal to any settler. They had become paramount among the "civilized" aboriginal nations.

Though the fledgling US government solemnly guaranteed Cherokee national sovereignty in the Treaty of Hopewell (1785), it also promised the state of Georgia in 1802 that it would arrange removal of all native peoples from Georgia in return for Georgia's relinquishing its western lands to form the states of Alabama and Mississippi. With the discovery of gold on Cherokee land in north Georgia in 1828, Georgia began to press its claims in earnest and with increasing coercion. Creeks, Choctaws, Seminoles, and Chickasaw were already in the process of being forced out.[2] The Cherokee turned to the courts to defend themselves. In

---

1. Still the indispensable introduction to Cherokee history and culture is Mooney, *Myths of the Cherokee*.

2. For this history see Foreman, *Indian Removal*. For the Trail of Tears see Ehle,

spite of Supreme Court decisions in the Cherokee's favor (*Worcester v. Georgia*, 1832), they found themselves under constant harassment from white prospectors, settlers, and lawless rabble.

In December of 1835 some of the more influential Cherokee leaders signed, at the Cherokee capital of New Echota, a treaty with federal representatives selling their eastern lands in exchange for assisted passage to the Indian territory of western Arkansas and the future state of Oklahoma. They did this knowing that such an act was a capital crime under Cherokee law, yet they felt this was the only way out of an impossible situation for their people. Their action stirred an enormous protest from the some 16,000 Cherokees governed by Chief John Ross, who continued to resist, by all legal means possible, the enormous pressures to leave their ancestral lands.

Andrew Jackson, whose life had been saved by prominent Cherokee warriors in the battle of Horseshoe Bend in 1812, was determined to evict the Cherokee now that he was President and pushed through the Removal Act of 1830 against fierce political opposition. Thus, the irresistible pressure of law and force gathered first to remove the Cherokee "voluntarily," and then, in 1838, under the terms of the rump treaty of New Echota, to force them out.

After General John Wool resigned from what he found to be an odious assignment, General Winfield Scott took up the task. Thirteen stockades were built in the area of north Georgia, eastern Tennessee, and western North Carolina. In late May and early June individuals and families, including their slaves, were driven to the stockades with only a few minutes notice. James Mooney, the foremost early ethnographer of Cherokee life, recounts that one Georgia volunteer said in his later years, "I fought through the civil war and have seen men shot to pieces and slaughtered by thousands, but the Cherokee removal was the cruelest work I ever knew." John G. Burnett, a Tennessee soldier who had grown up with Cherokee and spoke their language, later recounted in a famous passage "Men working in their fields were arrested and driven to the stockades. Women were dragged from their homes by soldiers whose language they could not understand. Children were often separated from their parents and driven into the stockades with the sky for a blanket and

---

Trail of Tears; and Anderson, *Cherokee Removal*.

the earth for a pillow. And often the old and infirm were prodded with bayonets to hasten them to the stockades. . . ."[3]

About 1000 Cherokee from the mountainous villages held legal title, though tenuous, to their land. Others hid out in the mountains and avoided the military dragnet. In one famous incident, a Cherokee named Tsali killed two soldiers while resisting their abuse of his wife. He was hunted down with help from Cherokee men (and their white companion, Will Thomas) and executed near Bryson City, North Carolina, with the tacit understanding that the rest could stay. Tsali's sacrificial death is memorialized to this day.[4]

The first groups were forced onto flatboats near Chattanooga to go down the Tennessee, Ohio, and Mississippi rivers and then up the Arkansas to the western territories. So many died in the heat and summer diseases that Chief Ross persuaded General Scott to let Cherokee leaders take the remainder in the fall. With this arrangement, 10 groups, of about 1000 each, left between September 1 and the end of October to walk overland through Tennessee, crossing the Ohio to southern Illinois and then across the Mississippi at Cape Girardeau before heading across Missouri. It was an unusually harsh winter and the ice clogging the Mississippi held up many groups for weeks. As a result of illnesses and exhaustion approximately 4000 people—almost one-fourth of the eastern tribe—died during the removal and shortly afterwards. The Cherokee came to call it "The Trail Where We Cried" ("Nunna-da-ul-tsun-yi"), or simply "The Trail of Tears."

The bitter conflict and near warfare between the New Echota treaty party and those who suffered the Trail of Tears under Chief Ross persisted for years. It was not until 1984 that the Western Band in Oklahoma and the Eastern Band in North Carolina had a reunion council.

Making Meaning of the Trail of Tears

The traumas of removal and national reconstruction caused many Cherokee to repress the memories of the Trail, but over the years the stories persisted and congealed into an almost mythic form. Within this mythicized, selective memory Cherokees began to construct a meaning that could sustain them as a unique people within the American aggregate.

3. Both passages cited in Thornton, "Demography," 79.
4. King and Evans, "Tsali."

After World War II, both Western and Eastern Bands produced outdoor dramas to rehearse this memory and begin to generate sustaining meanings.

What was the meaning of the Trail? Many Cherokee Christians and their missionary friends tried to see it as an Exodus with Oklahoma as a Promised Land, but it was not. It was more like an Exile, though this interpretation does not figure strongly in the literature. However, like ancient Israel, some traditionalist Cherokee searched for the sin that had justified their expulsion. The Trail, they held, occurred because they had adopted the ways of the settlers.[5]

The advocates of Exodus meanings searched for ways to turn Oklahoma into a promised land, to prove once again that Cherokees had a unique gift for survival, adaptation, and creative innovation. The Exile interpreters sought to pursue a narrower course of cultural retrieval and, later, ecological spirituality.

Coursing through these two images of Exile and Exodus, with their quite different cultural implications, was the theme of betrayal. The Trail was the outcome of multiple betrayals of treaty obligations by the British, the US government, and the states. At the end, the majority were betrayed by a minority faction, the bitterest pill of all. The memory of betrayal, however, only reinforced commitment to the honoring of promises. The suffering of the Trail was the supreme evidence of Cherokee commitment to promise-keeping rather than violent reprisal or devious escapes and communal disintegration. Thus, the very humiliation on the Trail was the sign of a higher virtue—a call, if you will, to all Americans to a higher level of civilized negotiation and the faithful honoring of agreements.[6]

At this point, the memory of the Trail begins to take on features resonant with Christian images of the Way of the Cross. The Trail of Tears was a collective crucifixion for the sake of a new order of peace. It could thus yield a sense of vocation that redeems the Cherokee people from resentment and revenge as well as the rest of America from arrogant self-delusion. In the midst of casino blandishments and global American hegemony, this is a lofty calling indeed.

The second sin producing the Trail, as many commentators reiterate, was greed. First of all, it was the greed of settlers bent on gold and

---

5. The faction led by Whitepath promulgated this view in the 1820s and it persisted in various "Kitowah" societies into the present day. See Mooney, *Myths of the Cherokee*, 113–14; and Leeds, *United Keetowah Band*, ch. 1.

6. This is certainly the message of the Eastern Band's drama, "Unto These Hills."

land. It was this expression of greed that overran all rule of law. However, as some of the Cherokee conservatives claimed, the Cherokee desire to make money off the land lured them into intensive farming, led them to leave the Spartan life of the mountains, and adopt slavery and plantation life.[7] This was a form of greed that seduced them into believing they could assimilate into white society, forgetting that finally white society would spit them out onto a journey of suffering.

The peculiar form of cultural greed we call racism swallowed up the myth of cultural equality. However, the pervasiveness of actual intermarriage with whites as well as Cherokee complicity in slavery have made it very difficult to pursue this strand of interpretation and ethical consequence very far. This is the strand that emerges to the fore in the journey myth found in South Africa's Great Trek.

THE GREAT TREK, 1835-39

The migrations that came to be called the Great Trek were quite different, though chronologically simultaneous, products of the conflicts of a frontier society. Both journey narratives have deeply shaped personal and collective consciousness. The Trail story is a subversive memory that seeks alignment with wider, dominant American myths, the Trek story is a memory of ethnic victory that became subversive of contemporary South African civil aspirations.

"The Great Trek" as a single journey narrative represents the coalescence of a wave of migration by Boer farmers (descendants of Dutch immigrants) from British-controlled areas of the Western Cape to the lands of various Bantu language peoples—primarily Xhosa, Sotho, and Zulu.[8] Though voluntary in itself, it arose from various pressures experienced by the farmers; namely, drought, increasing scarcity of grazing land, vacillating British policies toward the Xhosa, and, finally, the abolition of slavery in 1834 (with a transition period to 1838), which threatened not only their economic base but their sense of cultural superiority.

Farmer dissatisfaction with the Dutch East India Company's administration of affairs in South Africa led not only to abortive efforts

---

7. For this shadow side of the history see Halliburton, Jr., *Red Over Black*; and Perdue, *Slavery*. Halliburton estimates that 125 to 175 African-American slaves died on the Trail (Halliburton, *Red Over Black*, 61).

8. For two representative histories see Walker, *Great Trek*; and Meintjes, *Voortrekkers*. For a contemporary revisionist effort see Etherington, *Great Treks*.

to form independent republics in Graaff-Reinet and Swellendam at the end of the eighteenth century but also to a general acceptance of British rule in 1806. However, on the eastern frontier, where farmers clashed frequently with Xhosa groups, the British oscillated between a policy of firm separation of the clashing populations and efforts at expansion through the introduction of English settlers. Sometimes memories of the Trek's origins focus on the tensions between new settlers and the old farmers, sometimes on their common anger at British failure to protect them against Xhosa invasions and reprisals. Thus, the motives for the migration include escape from persecution, the desire to form independent republics far from the British crown's grasp, and desperation born of economic need and desire for personal security.

While independent farmers and hunters had roamed as Trek Boers far beyond the official colonial boundaries for decades, the actual efforts to settle north of the Orange (Gariep) River began with exploratory treks in 1835 by Louis Trichardt (or Tregart), Hans van Rensburg, and Andries Potgieter. Trichardt tried to find a way north of the Drakensberg Mountains (modern Lesotho) and over to Delagoa Bay in present-day Mozambique. Such an outlet would make it possible to bypass the British controlled port of Durban. Potgieter and van Rensburg confined their search to the open high veldt north of the Vaal River up to the Limpopo. While Trichardt's venture ended in ruin and death for most of his party, the others established the possibility of settlement, even in the face of stout resistance by the forces of Mzilikazi, the displaced Zulu leader. With his defeat at the battle of Vegkop in October 1836, the way was cleared for settlement by the Boers. These "Kommissie Treks" became part of the mythic framework of the Trek narrative.

The organized treks by families began in earnest with the "Manifesto" by Piet Retief in Grahamstown in 1837, declaring the reasons that drove the farmers to emigrate. Unlike the Trail of Tears, the Trek was a virtual military campaign in which muskets defeated spears, cannons defeated assegai, and the determined laagers of the Boers held off the reprisals of Zulu troops. At the same time, the Trek narrative overwhelms us with the seemingly impossible rigors of crossing mountains, fording rivers, and maneuvering oxen and carts up and down steep cliffsides. All of this is retold among references to the ever-present leatherbound Bible, the evening songs around the campfire, the nearly obsessive ministrations of Erasmus Smit, their irregular minister, and the women's heroic care of young children and their husbands. Punctuating these stories of perseverance and stoic dignity are the horrors of the defensive massacres by Zulu forces at Bloukrans, Bushman's River, and at Umgungundhlovo, the kraal of the Zulu King Dingane, who put Piet Retief and seventy others to death while negotiating over the cession of land to the Boers.

The turning point for the Voortrekker's invasion of Zulu country occurs at the battle of Blood River, December 16, 1838, where a determined party of Boer commandos held off the army of Dingane, with terrible loss to the Zulus and minor casualties for the Boers. This victory, so the story goes, was anticipated with a "vow" (*Gelofte*) to God that the Boers would always remember this moment with thanks and the construction of a church (in Pietermaritzburg as it turned out) in memory of God's providential support. English speakers translated this vow as a "covenant," an interpretation which later reinforced images of the Afrikaner as a chosen people, similar to those of English Puritans and their American progeny.[9]

The Great Trek narrative ends with the expulsion of Dingane's forces, even though the white Afrikaans-speaking people turned to sixty years of struggle with the British, ending with the Anglo-Boer War of 1899–1902. It was actually in the events leading up to this war that the Trek narrative was first constructed as a mythic template for the Afrikaner people and their struggle for nationhood.

The Meaning of the Great Trek

The stories of the original migration and conquest have gone through several reconstructions to serve successive purposes in the struggles of Afrikaners to shape and preserve their way of life.[10] In the struggle against the British, the themes of escape, liberation, and even expulsion shape the narrative, and the Voortrekkers are characterized with firm resolve, a thirst for freedom, the preservation of white civilization (which included master-slave paradigms of paternal authority), and maternal self-sacrifice. The work of Gustav Preller and the artist W. H. Coetzee

---

9. James Michener then used this idea as the inspiration for his own novelized history in Michener, *Covenant*. Donald Harman Akenson works through this Biblical template in Akenson, *God's Peoples*. This perspective on South African history is sharply questioned by Andre DuToit in DuToit, "No Chosen People"; and Dirk J. Smit, in Smit, "Covenant and Ethics?"

10. The most extensive history of the civil religion grounded in the Trek mythos is Moodie, *Rise of Afrikanerdom*.

sank deep into people's consciousness.[11] The personal virtues of perseverance, courage, and sheer strength fused with an emergent national self-understanding focused on freedom, republican government, and the cultivation of Christian piety.

In the later struggle leading up to the Apartheid era (1948–1994) the Trek highlighted the collision between European civilization and noble but primitive native African peoples. In 1938 Afrikaner labor leaders organized a "Memorial Trek" to commemorate and rehearse the now mythicized journey of the Voortrekkers. Ox wagons set out from major cities and historic sites, stopping along the old routes for pictures and weddings in traditional costume, and ending up at Blood River and at the future site of the Voortrekker monument. It was a deeply gripping event further cementing a new national consciousness. In this narrative the people of the Trek had a call to create a certain kind of civilization in an awesomely spectacular but often violent land. "Die Stem [Call] van Suid Afrika," still sung as part of the national anthem, reflected this self-understanding. The frieze of the Voortrekker Monument in Pretoria, conceived in the 1930s, exemplifies this Trek narrative. This version of the Trek then served to legitimate white hegemony within a kind of purported racial-ethnic federalism (the "homeland" system) and a white-controlled industrial economy dependent on Black labor. This crystallization of the Trek mythos produced a kind of neo-fascist strongman culture and architecture, still visible in statuary in Pretoria.

In the post-Apartheid era, most of the Trek mythology has been abandoned or closeted. However, the Trek mentality is too deeply embedded in the stories of all the South African ethnic groups to be escaped. Whether it is Mandela's "Long Walk to Freedom" (in a prison cell!), or Piet Meiring's use of the Trek metaphor to chronicle the TRC, or the pervasive use of Trek metaphors to talk about the journey into personal self-transformation, it remains a powerful symbol for people's orientation to the new South African world.[12]

However, there is no shortcut to this transformation of cultural symbols. In 1998 a museum of Zulu culture was built across the Ncome River at the site of the Blood River battle and plans were made to have a joint celebration with the Voortrekker Memorial across the river. This ceremony included declaration of a new national covenant of reconciliation

11. Preller, *Voortrekkermense*. W. H. Coetzee's paintings were used extensively to illustrate works relating to the Trek.

12. Meiring, *Chronicle*.

to supplant the so-called covenant of the Blood River battle. While the form of this occasion was enacted, there was no bridge across the river, so only a few committed whites were able to ford or wade the river to join in the ceremonies. The people, despite good intentions, remained tragically divided. The new trek is only beginning.

In the present transformation, people must take an internal trek to follow Mandela's walk from violence to patient negotiation and inclusion. The freedom sought in the trek now has to be sought in cooperation rather than conquest, while the conquest of nature contained in Trek stories now has to be a conquest of our greed for the sake of ecological sustainability. The earlier Trek conception of honor as defense of personal autonomy has to be transmuted into an honor based in honest negotiation and promise keeping. The new civil society requires a move from interpersonal honor governed by a sense of status to common citizenship under the impersonal rule of law. The Trek to freedom has to be a trek toward the truth, as displayed in the TRC's own banner. Finally, the valor of women exemplified in the Trek has to be translated into the virtue of equal citizenship in all walks of life. Only with the clothing of these kinds of interpretations can the Trek narrative re-emerge as a public myth to guide people in their search for a new way of life together.

XUANZANG'S JOURNEY TO THE WEST AND MAO'S LONG MARCH

The Trail and the Trek stories occurred in frontier situations between different but intermingled cultures competing for land and goods. In China we come to another classic journey story that incorporates themes of boundaries, struggle, and self-discipline, first in an ancient form and then a contemporary one. Both have provided mythic templates deeply shaping Chinese life, both in its traditional Confucian and modern Communist forms.

The Buddhist monk and scholar Xuanzang was born in central China in 596 CE.[13] Early in the Tang dynasty the emperor Taizong (626–49) became interested in Buddhism as a way to unify the country culturally and morally. Presumably he saw in Buddhism a means to subordinate local kinship ties for the sake of a universal outlook while at the same time

---

13. The most thorough history is Wriggins, *Xuanzang*. I am indebted to Nancy Boulton LeGates for guiding me into Xuanzang's story and significance.

orienting people away from a quest for power that would destabilize the social order. In the words of Nancy Boulton, a prominent scholar on Xuanzang, "The traditional Chinese preoccupation with the destiny of the family and the state was replaced by Buddhists with concern for the salvation of the individual and the promotion of universal enlightenment."[14]

Xuanzang decided to seek out the most ancient and authentic Buddhist scriptures and practices as they were being lived in the land of Buddha's birth so that he could promote a "pure" form of Buddhism in his own country. His purpose, under the authority of the Emperor, was no less than the religious reform of his own country.

Though others had preceded him on this trip to India, it was Xuanzang who became the archetypal mythic pilgrim in Chinese history. The general route and the approximate location of the sacred places in India were already known. He set out on his journey in 629, probably not knowing it would take him sixteen years and 10,000 miles, returning to the imperial capital of Chang'an (Xian) in 645. The terra cotta warriors and horses from the early Qing and Han emperors had already been in the ground over 700 years when Xuanzang set out from Chang'an.

He then proceeded around the edges of the Taklamakan Desert ("The place from which no one returns"). There were a number of Buddhist shrines and centers along this way as well as oasis towns and centers fed by the melting glaciers on the Chingling Mountains and other ranges. Many of these survive to this day, at least as sources for archaeology. This part of the journey itself took several years, not only to examine the Buddhist relics, writings, and religious artifacts of these places, but also to recuperate and wait out scorching summers and brutal winters.

He then made his way through the passes at the far western end of modern China and eastern Afghanistan, surely passing by the caves of Bamiyan, where the recently destroyed giant Buddha stood. The extreme weather and hazards, including bandit gangs, entailed the loss of about one-third of his entourage.

He finally emerged down through the Khyber Pass into the Indus plain and made his way to the Ganges basin, where he stayed for many years of study and gathering of materials at Ayodhya and Bodh-gaya. While there he became a favorite of King Harsha, one of the early great unifiers of much of India. His reign provided a modicum of security for Xuanzang, making possible his continued journey.

---

14. Boulton, "Early Chinese Buddhist Travel Records," 168.

He then turned south through tropical forests to Kancipurum and sacred sites near Madras (Chennai) before heading back inland to the region near Pune, and then up to Bombay (Mumbai). After taking some ten years to secure or copy all the ancient texts he could, he then had to return over these same treacherous mountain passes, crossing once again the Taklamakan desert, and returning to Chang'an, where he presented the record of his journey and findings to the Emperor Taizong, written up as "Record of the Western Regions." He then devoted the rest of his life to the translation and promulgation of these writings and practices in order to spread "true Buddhism" in China.

The journey was first of all a courageous act of opening China up to its distant neighboring civilizations. It also provided the basis for the growth of a unifying religious tradition in China. Equally important on a religious dimension, was that the journey, while first of all an effort to get the scriptures and relics of true Buddhism, constituted a testimony to many personal acts of self-purification. It provided both collective identity for China and also a program of personal moral formation for its citizens. This basic story formed the core narrative for its subsequent iterations as national saga and revolutionary myth.

"Journey to the West"

Xuanzang's disciple Shaman Huili (Hwui-li) wrote the first biography of Xuanzang soon after his death. His hagiographic testimony begins the legend of Xuanzang as more than a scholar and devoted pilgrim. He begins to take on the shamanistic qualities that characterize the later depictions of him. Li Yongshi, his second biographer, begins his development into a folk hero by the end of the seventh century. It takes 800 years, however, before Xuanzang the monk metamorphoses into a virtually mythical figure in the tales gathered together as *Journey to the West* by the Ming dynasty author Wu Cheng'an (ca. 1570).[15]

Wu's saga of clearly mythical events reads more like a Kung Fu novel or scenes from "Crouching Tigers, Hidden Dragons" than the original sober report of a monk scholar and pilgrim. In this mythic story the monk, here named Sangzan, is accompanied by a mischievous, magical Monkey (Sun Wukong), a blithering Pig (Zhu Bajie), Friar Sand (Sha Wujing), and a cast of demons, princes, princesses, and monsters. Monkey, Pig, and

15. Cheng'an, *Journey to the West*.

Friar accompany him to conquer mythological foes and reach Thunder Monastery in the Western Heaven to recover the sacred Sutras.

A story that looked like Pilgrim's Progress now becomes a very fanciful saga of Don Quixote, or perhaps the Wizard of Oz, in which Xuanzang appears as a muddled and bumbling incompetent who is constantly being saved by the wiles of Monkey, an otherwise unpredictable and troublesome braggart, though utterly devoted to him and gifted with magical powers to protect the endearing but confused monk. As a children's story and an adult fable, the journey of Xuanzang burrowed deep into Chinese culture and consciousness.

For all its fabulous content, *Journey to the West* still preserves a core of popular, Buddhist-influenced religion. In *Journey*, Xuanzang receives the epithet "Tripitaka," after the "Three Baskets" of Buddhist wisdom. Like any brave monk he is beguiled by a queen who almost ensnares him, but he overcomes this sensual allure for the sake of his quest. Near Ayodhya, the Indian city now made famous for the destruction of the Muslim shrine there by Hindu zealots, he is seized by pirates for sacrifice to Durga. A typhoon routs the pirates, who convert to Buddhist monkhood. In a bow to ancient Hinduism, the mischievous Monkey is clearly patterned after Hanuman, the rescuer of King Rama in the Ramayana, who is extolled in endless stories and venerated with countless statues throughout southern India. In the last century Monkey became a mythic figure in comic books, murals, TV shows, plays, and movies. Mao himself wrote several poems about Monkey and made elaborate use of contemporary versions of the story to advance his political aims.[16]

Journey to the West became the sugar-coated pill that introduced Buddhist values of self-denial, compassion, self-transcending humor, and detachment into millions of Chinese for centuries. W. J. F. Jenner, one translator of *Journey*, says of the book's symbolic message: "The quest could be any quest, any long and difficult undertaking in which patience, ingenuity and courage, together with an excellent understanding of the ways of the world, are essential for success."[17] So here we have the journey as pilgrimage and as quest. It is a formula or road map for cultivating personal discipline in service of a higher goal for the benefit of the wider society. Added to that, it conveys a sense of support by cosmic and magical

---

16. Wagner, "Monkey King," ch. 3.
17. Jenner, "Afterword," 645.

forces to enable a person or a people to persevere against all earthly odds for the sake of a higher and purer life.[18]

Journey of National Transformation: The Long March of Mao Zhedong

It may seem like a very improbable leap to go from a story that became a mythical children's saga to a very real historical experience in twentieth-century military and political history, the so-called Long March of 1934–35. I would like to argue that it is actually of one cloth with that cultural heritage. Just as Xuanzang's journey became the symbol of Buddhism as a national unifier, so the Long March became the symbolic source for the Communist unification of China.

After the death of Sun Yat-sen, the founder of modern China, in 1925, division immediately emerged not only within his followers in the Koumintang ("People's Party"), which at that time included Zhou Enlai (Chou En-lai), but also between those who would take a conventional republican path (Chiang Kai-shek) and those who would take a more revolutionary, socialist path (Mao and others in the Chinese Communist Party). This led to bitter battles in the late twenties in Canton and Shanghai between the Koumintang and the Communists. Nearly annihilated, the Communists established a proto-socialist "soviet"—Chingkangshan—on the rugged and remote border between Hunan and Jiangxi province. However, Chiang Kai-shek's forces proved too much for the fledgling socialist experiment and Mao's forces were routed. This is where the long march, which began as a retreat, commences.[19]

The way was led by Mao and Chu The ("Red Virtue"), his indispensable military leader, along with Lin Piao, general of the first army. Chu's wife, Kang Ke-ching (his fourth) also accompanied them. Amazingly, Mao's second wife, Ho Tzu-chen (He Zhizhen), accompanied him on this march while pregnant. Not only did she give birth on the march but saved another person from a bomb with her own body, suffering seventeen shrapnel wounds. She survived the march and went for medical

---

18. In Bernstein, *Ultimate Journey*, we find a western appropriation of Xuanzang's pilgrimage.

19. For this account I am relying mainly on Fairbank and Goldman, *China*; Salisbury, *Long March*; Wilson, *Long March 1935*; and Yang, *From Revolution to Politics*.

treatment for some years in Moscow. On her return to China, Mao had already divorced her to marry his third wife.

Despite this egregious rejection of He Zhizen, women were always politically very important to Mao and the Communists, in opposition to Chinese traditions of subordination of women. Thirty-five were on the march, several of whom engaged in combat. All survived the march, though bearing enormous physical and psychological scars.

The full march wound from Ruijin (Jiangxi Province) to Yenan (Shaanxi province). Starting with up to 10,000 soldiers, they recruited thousands of additional fighters along their march. They would end up traveling some 6,000 miles in one year, averaging seventeen miles a day, mostly on foot. They once marched twenty-seven consecutive days without rest, and sometimes they walked all day and night, occasionally making their way in the dark, each with his hand on the shoulder ahead. People were carried on litters, supplies on poles. Mao rode a horse, but most walked. Much apparatus and most medical cases had to be discarded along the way. Almost all of their historical documentation and records were lost in rivers and swamps.

They crossed eighteen mountain ranges, five of them snow-capped, and twenty-four rivers. Mao had malaria, others had tuberculosis, dysentery, typhoid, and the flu. In the midst of these rigors they averaged a skirmish a day, with fifteen pitched battles. From this long campaign emerged some key events that became the core of the Long March story.

The march began on October 16, 1934, with the crossing of the Yudu river. After battles in which they lost 50,000 men they crossed the Hsiang river. They then entered the Lushan Pass, seized it and went on to take Zunyi and Tongzi in early January, a crucial, morale-boosting victory.

In the wake of this battle they promulgated a code of ethics for the soldiers:

> Obey orders in all your actions.
> Don't take a needle or a piece of thread from the people.
> Turn in everything you capture.
> Speak politely.
> Pay fairly for what you buy.
> Return everything you borrow.
> Pay for any damage.
> Don't strike or swear at people.
> Don't damage the crops.
> Don't take liberties with women.

Don't mistreat captives.[20]

These were set to music and became fundamental for Mao's whole peasant-based strategy, quite in contrast to Chiang Kai-shek's army and, at the level of Russian Communist theory, a sharp departure from a focus on the industrial proletariat. More immediately, this code of conduct made possible their very survival on the long march. In the end, with the virtual elimination of other, urban approaches by other Communists, the Long March, with its peasant strategy, brought Mao to the head of the Communist Revolution with his final proclamation of victory in Beijing on October 1, 1949.

The ragged column of troops then crossed the upper Yangtze under heavy fire, May 1–9. Crossing the mountains of Sikang, including the particularly rugged Fire Mountain, a party elder encouraged the soldiers by reminding them that Monkey had burned off his buttocks hair there but still managed to cross. They could do the same!

At the Tatu river they crossed the Luting chain bridge by braving flaming boards at the end. Crossing over the great Snow Mountains and four other ranges at 16,000 feet claimed many more lives through exposure, disease, and stomach disorders. Finally reaching Maoergai, they rested to gather the abundant wheat of the region. Here they confronted the infamous and deadly grasslands. Locked between the Yangtze River basin to the south and the Yellow River basin to the north, the grasslands were actually hundreds of miles of swamps at 5,000 feet elevation, providing no shelter or food, and frequently pounded with freezing rain and snow. They could swallow a whole regiment in their deadly quicksand muck. In addition, Mao's forces suffered deadly encounters with hostile, non-Han, tribals, including Tibetans, always resistant to control from the Han. They finally made it across as other army units were trying to break through Nationalist lines to get to the safety of the high plateaus and mountain areas beyond the Yellow River.

Their final obstacle was the Latzu Pass, with its narrow and almost impassable river. Through some daring maneuvers they finally reached Senshi, near the Great Wall, in October and secured Yenan, the base from which they waged war first against the Japanese, and then against Chiang after the Japanese defeat. All in all, only about 5,000 made the whole trip, with about 70,000 deaths.

---

20. Cited in Salisbury, *Long March*, 117. See also Wilson, *Long March* 1935, 71.

The Meaning of the Long March

As John Fairbank and Merle Goldman observe, "The Long March has always seemed like a miracle, more documented than Moses leading his Chosen People through the Red Sea."[21] However, as Harrison Salisbury notes, "It was not a 'march' in the conventional sense, not a military campaign, not a victory. It was a triumph of human survival, a deadly, endless retreat from the claws of Chiang Kai-shek; a battle that again and again came within a hair's breadth of defeat and disaster. It was fought without a plan."[22]

On the way west and northwest Mao regained the leadership of the Chinese Communist Party in early 1935 and thereafter never relinquished it. Zhou Enlai, his former superior, became his chief supporter from then on. In the process he solidified the Maoist doctrine of peasant-based warfare. More importantly he forged the symbolic resources for a Maoist personality cult and the centralized leadership of his Communist party. The survivors of the March became the revered elders of the Communist leadership.

In order to construct this mythic memory the tortuous campaign first of all had to become a single march. Because there was a relative paucity of documentation, due to the hazards of the journey and intentional destruction of records, the Party was able to control the initial years of this myth-making process. The resulting story became a template to legitimate the regime and its reconstruction of Chinese character by instilling the virtues of self-sacrifice, struggle, perseverance, and party loyalty.

In 1966 the Red Guards took a "Long March" to discipline themselves for the revolution and re-ignite revolutionary zeal.[23] The study of Mao's words would help them "wage revolution in their souls." In the words of one commentary, "the process of the Long March becomes the process of the transformation of one's own soul through Mao Tse-tung thought."

In one oft-repeated story, a Captain told his protégé soldier about an instructor on the Long March, who, in his dying moments in the snow, had given his cap and shoes to a young recruit. The recruit went on to leadership in the army. In a battle against the Japanese, he led a squad to

---

21. Fairbank and Goldman, *China*, 305.
22. Salisbury, *Long March*, 1.
23. Wagner, "Reading the Chairman Mao Memorial Hall," 380.

secure a post. Mortally wounded in the process, he gave the same cap to his own junior soldier in order to carry on the chain of sacrifice.[24]

The story of the March, as with the Red Guards, undercut stable hierarchies of place, a key impact of journeys in a pre-electronic age.[25] Mao's reforms of the party and Red Guard attacks on revisionism appealed back to the solidarity and single-headed leadership that emerged in the long march. Party leadership had to emulate the values and practices instilled in the March in order to hold onto its legitimacy. Loyalty to Mao, as the leader of the Long March, had to override all other loyalties.

As Rudolf G. Wagner has observed,

> The entire revolutionary advance to the distant goal of communism is structurally so closely linked to pilgrimage in China that the most famous pilgrimage novel of China, *Journey to the West*, has served since 1949 as the standard metaphor for the long pilgrimage to communism. The purpose of an actual pilgrimage—to bridge the painful chasm between the distant goal and present-day triviality through short-term exultation—fully anticipated Communist individual behavior and social relations for a short time, and was to energize further advance."[26]

Indeed, Wagner points out, the memorial hall to Mao in Beijing was built to re-enact the pilgrimage of revolution in the process of visiting the hall.[27]

While there are obvious differences between these two journey stories (Mao and Xuanzang are radically different personalities!), certain features have persisted to shape a Chinese ethos. Certainly, the stereotyping and demonization of enemies intensifies problems of inter-group reconciliation, whether between east and west in China or between Taiwan and the mainland. In particular the March intensified the alienation of the western, non-Han, peoples, including Tibetan groups, who have been relentlessly subordinated to the Han leadership ever since. *The Journey to the West* became a call to empire as much as a pilgrimage to the source of pure doctrine.

As in Xuanzang's journey, the geography was more an enemy than a friend. The toll it took reinforced a traditional reverence for and fear of

---

24. Recounted in Wilson, *Long March 1935*, 280–83.
25. Wagner, "Reading the Chairman Mao Memorial Hall," 381.
26. Wagner, "Reading the Chairman Mao Memorial Hall," 379.
27. Wagner, "Reading the Chairman Mao Memorial Hall," 401.

mountains and rivers but it may also have reinforced the desire to tame them (as with the Three Gorges Dam) for the sake of the people.

Above all, both journeys are stories of the sacrifices necessary to find and to spread a One Pure Doctrine to unify China. This, we can say, is the ancient vocation of the pilgrim. Both Xuanzang and Mao were founding pilgrims of two very different efforts. Whether the order of Mao's China will survive as long as the Tang dynasty is yet to be told.

Both journey stories contain the seeds of a subversive undermining of political order for the sake of the pure doctrine. Also, both have carried profound messages of social conformity. This ambiguity of radical dissent and obedient, indeed heroic, conformity and self-denial was carried forth in the stories of their journeys. It lies at the heart of the challenge of cultivating a democratic citizenry in a republican China that no longer kowtows to either emperor or chairman.

RECONSTRUCTING JOURNEY NARRATIVES

As illustrated in the Chinese case, journey stories are constantly reinterpreted and reconfigured by other, classic journey myths. The Biblical image of the Exodus and Conquest, for instance, has been an American generative journey narrative that shaped the meaning of European immigration to North America. The resulting expulsions and "removals" of the original inhabitants drew in turn on stories of exile as well as of exodus, both of which revolve around covenantal conditions for keeping the land. The Cherokee, for instance, have tried to make sense of their traumatic experience of "removal," first as punishment for betraying the old ways, but then, through the annual outdoor drama, "Unto These Hills," as a journey to a new Promised Land that carries forth the wider American hope for democracy and peace. Buried in this conjunction of the Trail and the American dream, however, is the subversive claim that this goal must be reached through voluntary agreement, suffering, and respect for the land. The journey as exile reshapes the journey as conquest.

The effort to strike a new covenant of national reconciliation at Blood River in 1998 was a similar attempt to reconstruct the meaning of that destination of the Voortrekker's journey as the beginning of a new journey toward a just sharing of the land. While the journey rituals of the tourism industry may do more for the economy, the new journey toward cultural, ethnic, and political reconciliation will be much harder. It will have to be a reconstructed great trek. Such a trek has to be both intensely personal as well as collective. A powerful journey myth can do both.

THE FUSION OF PERSONAL AND COLLECTIVE JOURNEYS

An effective journey story fuses personal and collective stories to forge new identities and characters. This was perhaps clearest in the Chinese journeys, which explicitly challenged people to make the Journey to the West or the Long March, a personal pilgrimage that continually reclaims the values of those journeys for the sake of personal moral transformation.

In reconstructing the story of the Great Trek, we not only reclaim the fact that there were many collective treks, but also, that each individual needs to engage in a personal inner trek. Each of these inner treks may be quite different, since people in South Africa have come to a common land from very different origins. With the advent of a political constitution based primarily on individual rights and relations, these trek narratives have to become personalized while at the same time the land in which these personal journeys take place has to become a shared land, a shared ecology.

THE SIGNIFICANCE OF JOURNEY STORIES FOR JUSTICE AND RECONCILIATION

Journey myths, whether they recount an intentional search or pilgrimage, a gradual, unplanned migration, a coerced expulsion, or a hopeful exploration, shape what we might expect about our future, for they tell us not only where we were but where we are going. They are guides in changing and uncertain times, providing a script to follow in an otherwise unfamiliar world. They illuminate our search for a way to do the next right thing, to take the right course of action, to find our way into an uncertain future. They generate our sense of identity, our orientations toward strangers, and our concepts of the land we occupy or seek to enter. They are generative in the sense of providing an inter-generational thread of identity as well as of generating our orientation to the world. They generate the perceptions that shape our capacity to be reconciled to other groups and to the planet we share.

Such generative journey images have decisively shaped the way particular peoples establish or re-establish their relationships with other people, with animals, and with the land and its plants. In seeking to establish just relationships with these others we also continually reintegrate our own identities as migrants, exiles, settlers, and refugees. These narratives thus constitute bedrock cultural images that define the contours within which we can seek a path to a just peace among competing peoples. In the suffering and hope expressed in such narratives we uncover the texture of loyalties, faith, and fundamental assumptions that can promote or undermine the arduous work of reconciliation. How we envision those we have left behind, either in hatred or regret, those whom we encounter in fear, love, or anger, and the land we traverse sets up the expectations with which we approach the task of establishing right relationships with them.

While journey stories share some common characteristics—separation of traveler from land and home; the crossing of boundaries; struggle and suffering in the face of dangers; a sense of origin, itinerary, destination, and purposefulness—each type of journey has some typical peculiar characteristics that affect the key themes of alienation, personal and collective formation, and reconciliation.

When considering other classic journey stories in addition to these four, we can identify at least eight types: (1) Exodus and liberation; (2) Removal or exile; (3) Wandering; (4) Exploration; (5) Quest; (6) Pilgrimage; (7) Settlement (and conquest); and (8) Return or home-coming.[28] Each of these yields different meanings for alienation, personal and collective virtues, and reconciliation. We can only consider a few connections related to our cases and other prominent examples here.

In the exodus pattern people reject their past associations with people, land, and economy. They cultivate the virtues of gratitude for liberation, courage in the face of the unknown, and hope for a better life. Collectively they pursue values of unity, mutual defense, and self-sacrifice for the sake of survival for the collective. The challenge facing people in exodus is how to reconcile with those who are rejected or left behind. This was particularly poignant for the Cherokee, who were separated into two estranged bands for almost 150 years before their formal efforts at reconciliation.

---

28. For another list and an in-depth examination of the American context see Stout, *Journey Narrative*.

Those in removal or exile suffer alienation from their home, their identity, and their ancestral bonds. This theme courses through the Cherokee experience and stands at the center of the experience of Africans wrenched from their homes and robbed even of ancestral and cultural memory in the Middle Passage of the slave trade. Like exodus, exile is one of the core journey myths in the Bible. Together they form a dominant lens by which many people have searched for meaning in the many other exiles of human history.

Exiles typically are called to practice the virtues of perseverance, mutual care, hope, and remembrance. Hope of return is inextricably linked to preservation of collective identity. Exile as a journey helps preserve a sense that return is not only desirable but possible. For the Cherokee the hope of return, and with it their sense of exile, has tended to be replaced by an exodus-like interpretation that orients them to the development of a new promised land.

The challenge for those in exile is how to be justly reconciled with those who expelled them. For the Cherokee it was the task of reconciliation not only with white settlers and governments, but also with the Cherokees of "the treaty group" who paved the way for their removal. African-Americans have faced the challenge of reconciliation with whites as well as with the Africans that sold them into slavery.

With the quest and the pilgrimage we begin with a sense of estrangement from a lofty or transcendent source of meaning. In the quest we seek to obtain the symbol of this source of meaning. In the pilgrimage we seek to have contact with places along a somewhat known itinerary in order to secure this connection to the transcendent source of meaning. In both cases we nourish virtues of curiosity (found most explicitly in the journey of exploration), self-denial, perseverance, and faith. When this formation of virtue takes a collective form, as with Mao's march, it emphasizes unity of devotion and faith—the Pure Doctrine.

The challenge for the quester and pilgrim is reconciliation with the source of ultimate power and meaning. This entails also reconciliation with the true self within. One purifies one's self in the journey in order to be reconciled to who one really is or ought to be. The same can be said for the collective that undertakes this journey.

The journey of settlement and conquest, while it may contain elements of exodus and even exile, has its own typical contours. In this journey people are first of all estranged from their land of origin. For whatever reason they cut themselves off from the ecology of their homes.

They also tend to be estranged from the people they encounter in their new land. In the case of the Voortrekkers, as with the American pioneers, they either had to claim the land was unpopulated or they sought, as with Piet Retief or the US government, to alienate the inhabitants from the land through a treaty or cash purchase.

The journey of settlement typically cultivates virtues of strength, endurance, courage, ingenuity, adaptability, and self-discipline. It also tends to yield collective virtues of cooperation, mutual aid, and ecological accommodation. For the settler, hope is more important than memory, novel adaptation more important than tradition.

The challenge facing settlers is reconciliation with the people they displace or conquer as well as with the land they occupy. One of the complexities of the Cherokee story is that the land to which they were exiled was already inhabited by several other tribes, so that they were immediately absorbed in hostilities not only among Cherokee factions but also with the Osage and others in their new land. Moreover, they had to adapt to a very different agricultural environment in order to survive. The Voortrekkers, however, sought land that was familiar to them and hospitable to their established practices, even as they continued a life of friction or warfare with indigenous peoples—a warfare which only now is opening into new possibilities of reconciliation. The problem of reconciliation with the land takes us to another dimension of the journey pattern—stories of place.

PLACES AND JOURNEYS: THE CHALLENGE TO RECONCILING JUSTICE

Journey stories are often counterpoised to myths and stories of place. We are also powerfully shaped by deep memories and myths of place—of a primordial land and household where "we have always been." In this case identity is not shaped by a journey but by an eternal participation in a particular geography. Indeed, many powerful journey myths presume a place from which or to which one travels. Motifs of wholeness, return, and conciliation arise in great part from this image of place. The clash between peoples with an intense journey myth and those with an intense mythos of place constitutes one of the most difficult and often tragic conflicts in human history.

These two contrasting narratives deeply shape contests over land. People of the journey claim land as a promise and destination. This claim, however, has to be rooted in some authority beyond the simple connection of people and land found in narratives of place. This kind of land claim rests in some act of will and election, whether human or divine. This is very different from the bond with the land found among people who "have always been here," for their very being is tied to the land. Thus, they cannot "possess" it in the way that people of the journey seek to do. These are very different paradigms of connection between people and land. Reconciliation in this situation requires that we find some way of moving beyond the notion that the land is a prize, a gift, or possession at the end of the journey. At the same time, we have to honor the fact of our distinction from the land in our journeys. Some notion of trusteeship and shared use may be able to join these two demands to reconcile journey and place as defining our human identity.

This tension between journey and place challenges us to ask how we can affirm our being as creatures of journeys as well as of place in a way that helps us seek patterns of justice not only with other journeyers on this planet but also with the earth which is our place in the universe. Journeys imply a break between people and land. We become nomads, pilgrims, or refugees without any claim to a particular land. Our identity is not tied to a place. We therefore tend to overlook the need for a genuine ecological ethic. The ecological ravages of South Africa's Transkei, inextricably tied to the pattern of removal, of inappropriate ecological practices, and economic deprivation, are just one of many images of this alienation.

Alienation from the land can take many forms. The land that sustained the settler journeys, like the journeys of exiles, could be seen as godforsaken, as "promised," as open space, or hostile wilderness. Refracted through the generative journey narratives of the Bible the land of America continually oscillated between being the harsh wilderness of Sinai and the recovered garden of a promised land. When natural geography was exhausted, Americans turned in the twentieth century to mythic journeys to outer space to define their mission.

Journey narratives also shape the way we conceive of our planet. While European settlers and Asian immigrants conceive of the planet in terms of the east-west movement of their journeys, African Americans and Hispanic Americans see their world in terms of north and south. While the north exerts the power of a hoped-for better life economically, south still exerts the pull of "home," somewhat the way people of the Transkei might feel in the Western Cape or Gauteng.

For many people the notion of "home," rooted in a story of place, is no longer the origin or destination of a journey. Their journey becomes an oscillation of commuting between family and work rather than a single journey to a holistic ecological goal—a new creation. The family's primary residence is no longer a whole ecology of work, land, family, and community, but simply a temporary place of personal nurture and, perhaps, longing.

THE CONFLICT OF JOURNEY NARRATIVES

Many of the conflicts of our time occur among groups with very different journey narratives. In these conflicts people come from different places to struggle for control of a single land. South African as well as American history is shaped by these clashing narratives of migration. Sometimes a journey narrative of exile, rubbed raw with the rehearsal of past wrongs, collides with the optimistic, hopeful, and sometimes arrogant journey narrative of exodus, settlement, and conquest, as in the American or Middle Eastern stories.

The original Exodus story, coupled with that of the Babylonian exile, has been paradigmatic for Jewish efforts to understand what Jews ought to do within the cataclysms of uprooting and genocide in the twentieth century. What for many people would have been a time of cultural destruction has become a time of "return" (*shivat Tzion*) as well as of settlement. Many American Jews live in the tension between the call to settle in the land of Israel and the call to develop Jewish life within American geography as the focus of this paradigmatic journey. Many American Christians, oriented by the same stories, have great trouble distinguishing this American geography from the promised land itself.

For Muslims, the journey from Mecca to Medina (*Hejira*) has played a similar foundational role. Indeed, the Muslim calendar begins with this event. Even now, in the midst of the traumatic conflict of recent months, the means by which Mohammed reclaimed the city of Mecca from the "pagans" has become a benchmark for debates over violent and non-violent means to secure justice for Muslims.

JOURNEY STORIES AND THE PROCESS OF RECONCILIATION

From the standpoint of journey narratives, reconciliation is the act of entering into each other's journeys in order to construct a common journey. Doing this demands the discovery of the deep ambiguities and dualities within each of our lives and stories. It also means that no reconciliation occurs without a long, painful, and uncertain process. It is not a "problem" to be "solved." We have to remember the journey we have been on in the past and what it means to start a new journey with new partners. A deep recovery of our generative journey myths can help us do that.

To do this, we have to claim our generative journey narratives and then find how they bring us into conflict with others, who may be steered by quite different narratives of journey or of place. These re-claimed journey stories then have to be shared in dramatic ways so that people can enter into the shoes, the moccasins, that walked those other trails, treks, journeys, and marches.

Reconciling these narratives, whether of journey or of place, so that people can move into a more peaceful future is extremely difficult. The American project has generally tried to assimilate everyone into a story of migration from every corner of the earth into the construction of an entirely new multi-ethnic people. This migration narrative includes the earliest migrants, whether from across the Bering Straits or by Polynesian boat, as well as the most recent from Laos, Nigeria, or Afghanistan. Lately, there have also been efforts, especially with the spiritualist and naturalist movements, to reconstruct the journey images of these migrant peoples around participation in the natural cycles of the land, using a reconstruction of Native American myths as their generative structure.

This ecological moment then raises a global problem: How can people pursue all their little journeys by car, plane, bus, ship, and train, without polluting and defacing the land, the air, and the water? How can the land sustain our journeying in pursuit of our individualized promised lands?

To meet these challenges, we have to imagine how these disparate narratives might be knit into a broader journey toward a new future. How can we move from a memory of our past to a scenario for the future? How can the story of a journey through time and space become a story about our own psychological as well as political transformation? Remembering that an effective journey myth enables us to move back and forth between our collective history and personal biography, we have to find a resonance between our own little journey and that of the people with whom we identify. Beyond that, however, we need to find a rapprochement between this generative journey myth, the journey myths of others, and the claims of the planet that binds us inextricably together within the vastness of God's galaxies.

Bibliography

Achtemeier, Elizabeth. *The Committed Marriage*. Philadelphia: Westminster, 1976.
Ackerman, Joseph, and Marshall Harris. *Family Farm Policy*. Chicago: University of Chicago Press, 1947.
Adams, Charles. "Some Aspects of Black Worship." *Andover Newton Quarterly* XI (1971) 124–38.
Ahrons, Constance. "Joint Custody Arrangements in the Post-divorce Family." *Journal of Divorce and Remarriage* 3 (1980) 189–205.
Aiken, William, and Lafollette, Hugh, eds. *Whose Child? Children's Rights, Parental Authority, and State Power*. Lanham, MD: Rowman and Littlefield, 1980.
Akenson, Donald Harman. *God's Peoples: Covenant and Land in South Africa, Israel and Ulster*. Ithaca and London: Cornell University Press, 1992.
Aldous, Joan, ed. *Two Paychecks: Life in Dual-Earner Families*. Beverly Hills, CA: Sage, 1982.
Alexander, Neville. *Nation Building in the New South Africa*. Series B. Applied and Interdisciplinary Papers 257. Duisburg: L.A.U.D., 1994.
Alyea, Paul E., and Blanche R. *Fairhope, 1894–1954: The Story of a Single Tax Colony*. University, AL: University of Alabama Press, 1956.
Ammerman, Nancy. *Baptist Battles*. New Brunswick, NJ: Rutgers University Press, 1990.
Anderson, William L., ed. *Cherokee Removal: Before and After*. Athens: University of Georgia Press, 1991.
Apter, David. "Political Religion in the New Nations." In *Old Societies and New States: The Quest for Modernity in Asia and Africa*, edited by Clifford Geertz, 57–104. New York: Free Press, 1963.
Arendt, Hannah. "Thinking and Moral Considerations." In *Social Research* 38 (Autumn 1971) 417–47.
———. "What Is Authority?" In *Between Past and Future: Six Exercises in Political Thought*, 91–142. Cleveland: World Publishing Company, 1963.
———. *Between Past and Future*. Cleveland: World Publishing Company, 1963.
———. *On Revolution*. New York: Viking, 1965.
———. *The Human Condition*. Garden City, NY: Doubleday, 1959.
Ashcraft, Morris. "Southeastern Seminary in Crisis: 1986–87." *Faith and Mission* 6:1 (Fall 1988) 47–61.
Austin, Granville. *The Indian Constitution: A Cornerstone of the Nation*. London: Oxford University Press, 1966.

Austin, Richard C. "Three Axioms for Land Use." *The Christian Century* 94/32 (October 12, 1977) 910–11, 915.

Baden, Clifford. *Work and Family: An Annotated Bibliography, 1978–80.* Boston: Wheelock College Center for Parenting Studies, 1981.

Bank of America Corporation. *Bibliography of Corporate Social Responsibility Programs and Policies.* New York Bank of America, 1977.

Barber, Bernard. "Some Problems in the Sociology of the Professions." In *The Professions in America*, edited by Kenneth S. Lynn, 15–34. Boston: Houghton Mifflin, 1967.

Barnes, Peter. "Buying Back the Land." In *Working Papers for a New Society* 1:2 (Summer 1973) 43–50.

Barnes, Peter, ed. *The People's Land: A Reader on Land Reform in the United States.* Emmaus, PA: Rodale, 1975.

Barnes, Peter, and Larry Casalino. *Who Owns the Land? A Primer on Land Reform in the USA.* San Francisco: Center for Rural Studies, 1972.

Barnett, Frank and Sharan. *Working Together: Entrepreneurial Couples.* Berkeley: Ten Speed, 1988.

Barnett, Richard, and Ronald Mueller. *Global Reach.* New York: Simon and Schuster, 1975.

Barr v. United Methodist Church, 90 Cal. App. 3d 259 (1979), and 153 Cal. Rep. 328 (1979).

Barry, Vincent. *Moral Issues in Business.* Belmont, CA: Wadsworth, 1979.

Barth, Karl. *Church Dogmatics*, 3/4. Edited by G. Bromiley and T. Torrance. Edinburgh: T & T Clark, 1961.

Bartolome, Fernando. "The Work Alibi: When It's Harder to Go Home." *Harvard Business Review* 61 (March–April, 1983) 67–74.

Battin, Margaret P. *Ethics in the Sanctuary: Examining the Practices of Organized Religion.* New Haven: Yale University Press, 1990.

Baum, Gregory. *The Priority of Labor: A Commentary on "Laborem Exercens".* New York: Paulist, 1982.

Beauchamp, Thomas L., and Norman Bowie, eds. *Ethical Theory and Business.* Englewood Cliffs, NJ: Prentice-Hall, 1979.

Beck v. Beck, 432 A.2d 63 (N.J. 1981).

Beisser, Arnold R. *The Madness in Sports: Psycho-social Observations on Sports.* New York: Appleton-Century Crofts, 1967.

Bellah, Robert. *The Broken Covenant: American Civil Religion in Time of Trial.* New York: Seabury, 1974.

———. "Civil Religion in America." *Daedalus* (Winter, 1967) 1–21. Also in Donald Cutler, ed., *The Religious Situation* 1968, 331–356. Boston: Beacon, 1968.

———. "Religious Evolution." *American Sociological Review* 29 (1964) 358–74.

Bellah, Robert, et al. *Habits of the Heart: Individualism and Commitment in American Life.* Berkeley: University of California Press, 1985.

Berchin, Sondra E. "Regulation of Land Use: From Magna Charta to a Just Formulation." *UCLA Law Review* 23 (June, 1976) 904–35.

Berger, Peter, and Thomas Luckmann. *The Social Construction of Reality.* Garden City, NY: Doubleday, 1966.

Bernstein, Richard. *Ultimate Journey: Retracing the Path of an Ancient Buddhist Monk who Crossed China in Search of Enlightenment.* New York: Vintage, 2001.

Berry, Wendell. *The Unsettling of America: Culture and Agriculture*. San Francisco: Sierra Club, 1978.
Bertalanffy, Ludwig von. *Robots, Men, and Minds*. New York: George Braziller, 1967.
Best, Fred. *Flexible Life Scheduling: Breaking the Education-Work-Retirement Lockstep*. New York: Praeger, 1980.
———. *Work Sharing: Issues, Policy Options and Prospects*. Kalamazoo: W. E. Upjohn, 1981.
Black, John. *The Dominion of Man: The Search for Ecological Responsibility*. Chicago: Aldine, 1970.
Blackmore, John. "Community Trusts Offer a Hopeful Way Back to the Land." *Smithsonian* 9:3 (June 1978) 97–109.
Blackstone, William T., ed. *Philosophy and Environmental Crisis*. Athens: University of Georgia Press, 1974.
Blenkinsopp, Joseph. "The Structure of P." *Catholic Biblical Quarterly* 38 (July 1976) 275–99.
Blotnick, Srully. *The Corporate Steeplechase: Predictable Crises in a Business Career*. New York: Facts on File, 1984.
Blustein, Jeffry. *Parents and Children: The Ethics of the Family*. New York: Oxford University Press, 1982.
Bobo, Benjamin F., et al. *No Land is an Island: Individual Rights and Government Control of Land Use*. San Francisco: Institute for Contemporary Studies, 1975.
Bock, Wolfgang. *Das für all geltende Gesetz und christliche Selbstbestimmung in Dienstrecht*. PhD diss., J. W. Goethe Universität, Frankfurt am Main, 1992.
Book of Discipline of The United Methodist Church 1992. Nashville: The United Methodist Publishing House, 1992.
Boraine, Alex, and Janet Levy, eds. *The Healing of a Nation?* Cape Town: Justice in Transition, 1995.
Boraine, Alex, Janet Levy, and Ronel Scheffer, eds. *Dealing with the Past: Truth and Reconciliation in South Africa*. Cape Town: IDASA, 1994.
Borsodi, Ralph. *Flight from the City*. New York: Harper and Row, 1972.
Bosselman, Fred, and David Callies. *The Quiet Revolution in Land Use Control*. Washington: US Council on Environmental Quality, 1971.
Bosselman, Fred, David Callies, and John Banta. *The Taking Issue*. Washington: Council on Environmental Quality, 1973.
Botman, H. Russel, and Robin M. Petersen, eds. *To Remember and to Heal: Theological and Psychological Reflections on Truth and Reconciliation*. Cape Town, Human & Rousseau, 1996.
Boulding, Kenneth. *The Organizational Revolution*. Chicago: Quadrangle, 1968.
Boulton, Nancy Elizabeth. "Early Chinese Buddhist Travel Records as a Literary Genre." PhD diss., Georgetown University, 1982; Ann Arbor: UMI, 1983.
Bowles, Samuel, and Herbert Gintis. *Schooling in Capitalist America: Educational Reform and the Contradictions of Capitalism*. New York: Basic Books, 1976.
Bowman, James S. "Altering the Fabric of Work: Beyond the Behavioral Sciences." *Business Horizons* 27 (September–October 1984) 42–53.
Bradley, Martin B., et al., eds. *Churches and Church Membership 1990*. Atlanta: Glenmary Research Center, 1992.

Bradshaw, Thornton, and David Vogel, eds. *Corporations and their Critics: Issues and Answers to the Problem of Corporate Social Responsibility.* New York: McGraw-Hill, 1980.

Bratt, C. S. "Joint Custody." *Kentucky Law Journal* 71 (1982-83) 271-308.

Brewster, John M. "The Relevance of the Jeffersonian Dream Today." In *Land Use Policy and Problems in the United States,* edited by Howard W. Ottoson, 86-136. Lincoln: University of Nebraska Press, 1963.

Brueggeman, Walter. *The Land.* Philadelphia: Fortress Press, 1977.

Bryson, R. B. and J. B., and M. H. and B. G. Licht. "The Professional Pair: Husband and Wife Psychologists." *American Psychologist* 31 (1976) 10-17.

Buckley, James M. *Constitutional and Parliamentary History of the Methodist Episcopal Church.* New York: Methodist Book Concern, 1912.

Bureau of National Affairs, *Work and Family: A Changing Dynamic.* Washington: BNA, 1986.

Burgsmüller, Alfred, ed. *Kirche als "Gemeinde von Brüdern" (Barmen III),* vol. 1. Gütersloh: Gütersoher Verlagshaus, 1980.

Byron, William J. "The Ethics of Stewardship." In *The Earth is the Lord's: Essays on Stewardship,* edited by Mary E. Jegen and Bruno V. Hanno, 44-50. New York: Paulist Press, 1978.

Canacakos, Ellen. "Joint Custody as a Fundamental Right." *Arizona Law Review* 23 (1981) 785-800.

Canadian Conference of Catholic Bishops. "Ethical Reflections on the Economic Crisis." *Origins* 12 (January 27, 1983) 521-27.

Canon Law Society of America. *The Code of Canon Law: A Text and Commentary.* Mahwah, NJ: 1985.

Capps, Walter. *The Unfinished War.* Boston: Beacon, 1982.

Carnes v. Smith. 236 [1976] Ga. 30.

Cassetty, Judith. *Child Support and Public Policy.* Lexington, MA: Lexington, 1978.

Cheng'an, Wu. *The Journey to the West* (ca. 1570). 4 vols. Trans. and ed. Anthony C. Yu. Chicago: University of Chicago Press, 1977.

Cherlin, Andrew. "Remarriage as an Incomplete Institution." *American Journal of Sociology* 84:3 (1978) 634-650.

Cherry, Conrad. *God's New Israel.* New York: Prentice-Hall, 1970.

Chickering, A. Laurence. "Land Use Controls and Low Income Groups: Why Are There No Poor People in the Sierra Club?" In *No Land is an Island: Individual Rights and Government Control of Land Use,* Benjamin F. Bobo et al., 87-92. San Francisco: Institute for Contemporary Studies, 1975.

"Children of Divorce." *Journal of Social Issues* 35:4 (Fall 1979).

Clark, Donald C., Jr. "Sexual Abuse in the Church: The Law Steps In." *The Christian Century* 110 (April 14, 1993) 396-98.

Clark, Henry. "The Calling of Christian Ethics." (Manuscript available from author.)

———. *Ministries of Dialogue.* New York: Association Press, 1971.

Clawson, Marion. *The Land System of the United States: An Introduction to the History and Practice or Land Use and Land Tenure.* Lincoln: University of Nebraska Press, 1968.

———. *America's Land and Its Uses.* Baltimore: Johns Hopkins University Press, 1972.

Cloward, Richard A. "Illegitimate Means, Anomie, and Deviant Behavior." *American Sociological Review* 24 (1959) 164-76.

Cobb, John B., Jr. *Is It Too Late?* Beverly Hills: Bruce, 1972.
Cochrane, Arthur C. *The Church's Confession Under Hitler.* Philadelphia: Westminster, 1962.
Coe, George Albert. *A Social Theory of Religious Education.* New York: Arno, [1917] 1979.
Coleman, John A. "Civil Religion." *Sociological Analysis* 31 (1970) 67–77.
Commoner, Barry. *The Closing Circle: Nature, Man, and Technology.* New York: Alfred Knopf, 1971.
Connolly, William E., ed. *The Bias of Pluralism.* New York: Lieber-Atherton, 1971.
Conservation Foundation, The. *National Parks for the Future.* Washington: The Conservation Foundation, 1972.
Coriden, James. "Human Rights in the Church: A Matter of Credibility and Authenticity." In *The Church and the Rights of Man*, edited by Alois Muller and Norbert Greinacher, 67–76. New York: Seabury, 1979.
———. *We, The People of God.* Huntington, IN: Canon Law Society of America, 1969.
Coriden, James A., ed. *The Case for Freedom: Human Rights in the Church.* Washington: Corpus, 1969.
———. ed. *Sexism and Church Law: Equal Rights and Affirmative Action.* New York: Paulist, 1977.
Costonis, John, et al. "Development Rights Transfer." *Urban Land* 34 (January 1975) 5–15, 28–34.
Cox, Harvey. *The Feast of Fools: A Theological Essay on Festivity and Fantasy.* Cambridge, MA: Harvard University Press, 1969.
Coyne, Thomas A. "Who Will Speak for the Child?" In *The Rights of Children: Emergent Concepts in Law and Society*, edited by Albert E. Wilkerson, 193–211. Philadelphia: Temple University Press, 1973.
Cronin, John F. *Social Principles and Economic Life.* Milwaukee: Bruce, 1966.
Crouch, Richard E. "An Essay on the Critical and Judicial Reception of Beyond the Best Interests of the Child." *Family Law Quarterly* 13:1 (Spring 1979) 49–103.
Curran, Charles E. "The Gospel and Culture: Christian Marriage and Divorce Today." In *Ministering to the Divorced Catholic*, edited by James Young, 15–36. New York: Paulist, 1979.
Curtis, James E., and John W. Petras, eds. *The Sociology of Knowledge: A Reader.* New York: Praeger, 1970.
D'Antonio, William V., and Joan Aldous, eds. *Families and Religions: Conflict and Change in Modern Society.* Beverly Hills: Sage, 1983.
Danco, Katy. *From the Other Side of the Bed: A Woman Looks at Life in the Family Business.* Englewood Cliffs, NJ: Prentice-Hall, 1982.
Danco, Leon. *Inside the Family Business.* Englewood Cliffs, NJ: Prentice-Hall, 1982.
Darin-Drabkin, Haim. *Land Policy and Urban Growth.* New York: Pergamon, 1977.
Davidson, John M. *Concerning Four Precursors of Henry George and the Single Tax as also the Land Gospel according to Winstanley "The Digger."* Port Washington, NY: Kennikat [1899] 1971.
Davies, W. D. *The Gospel and the Land.* Berkeley: University of California Press, 1974.
Davis, Keith, and William C. Frederick. *Business and Society: Management, Public Policy, Ethics.* 5th ed. New York: McGraw-Hill, 1984.
de Klerk, Willem A. *The Puritans in Africa: A Story of Afrikanerdom.* London: Bok Books International, 1988.

de Villiers, Simon A. *Robben Island: Out of Reach, Out of Mind.* Cape Town: C. Struik, 1971.

Deacon, Harriet, ed. *The Island: A History of Robben Island, 1488–1990.* Mayibuye History and Literature Series. Cape Town: Mayibuye Books and David Philip, 1997.

Deal, Terrence E., and Allan A. Kennedy. *Corporate Cultures: The Rites and Rituals of Corporate Life.* Reading, MA: Addison-Wesley, 1982.

DeGeorge, Richard T. *Business Ethics.* New York: Macmillan, 1982.

Degler, Carl N. *At Odds: Women and the Family in America from the Revolution to the Present.* New York: Oxford University Press, 1980.

Delafons, John. *Land Use Controls in the United States.* 2d ed. Cambridge, MA: MIT Press, 1969.

Demerath, Nicholas J., and Richard A. Peterson, eds. *System, Change and Conflict.* New York: Free Press, 1967.

Derr, Thomas S. *Ecology and Human Need.* Philadelphia: Westminster, 1975.

Derrett, J. Duncan M. *The Death of a Marriage Law: Epitaph for the Rishis.* New Delhi: Vikas, 1978.

Deutsch, Karl. *The Nerves of Government.* New York: Free Press, 1966.

Devadason, Edwin D. *Christian Law in India: Law Applicable to Christians in India.* Madras: DSI, 1974.

Dibble, Vernon. "Occupations and Ideologies." In *The Sociology of Knowledge: A Reader,* edited by James E. Curtis and John W. Petras, 434–51. New York: Praeger, 1970.

Dillistone, F. W. *The Structure of the Divine Society.* Philadelphia: Westminster Press, 1951.

Doctrines and Discipline of The Methodist Church 1939. Edited by John W. Langdale et al. New York: Methodist Publishing House, 1939.

Dombois, Hans, ed. *Das Recht der Gnade: Ökumenisches Kirchenrecht.* Bd. 1. Witten: Luther Verlag, 1961.

Domhoff, G.W., and H.B. Ballard, comps. *C. Wright Mills and the Power Elite.* Boston: Beacon, 1968.

Douglas, Ann. *The Feminization of American Culture.* New York: Knopf, 1977.

Draheim, Philip. "The Lutheran Approach." In *Ascending Liability in Religious and Other Non-profit Organizations,* edited by Edward McGlynn Gaffney Jr., Philip C. Sorenson. and Howard R. Griffin, 137–142. Macon, GA: Mercer University Press, 1984.

du Toit, Andre. "No Chosen People: The Myth of the Calvinist Origins of Afrikaner Nationalism and Racial Ideology." *American Historical Review* 88 (October 1983) 920–52.

Dubos, Rene. "Franciscan Conservation versus Benedictine Stewardship." In *Ecology and Religion in History,* edited by David and Eileen Spring, 114–36. New York: Harper and Row, 1974.

Dudden, Arthur P. *Joseph Fels and the Single-Tax Movement.* Philadelphia: Temple University Press, 1971.

Duerksen, Christopher J. "England's Community Land Act: A Yankee View." *Urban Law Annual* 12 (1976) 49–76.

Dulles, Avery. *Models of the Church: A Critical Assessment of the Church in all its Aspects.* Rev. ed. Garden City, NY: Image Books, 1987.

Duncan, Hugh. *Communication and Social Order.* New York: Bedminster, 1962.

———. *Symbols in Society*. New York: Oxford University Press, 1968.
Dunlap, Riley E., and Kent D. Van Liere. "Land Ethic or Golden Rule: Comment on 'Land Ethic Realized' by Thomas A. Heberlein." *Journal of Social Issues* 33:3 (Summer 1977) 200–207.
Dunn, John. *The Political Thought of John Locke*. Cambridge: Cambridge University Press, 1969.
Durkheim, Emile. *The Elementary Forms of the Religious Life*. Trans. J. W. Swain. New York: Free Press, 1965.
———. *Professional Ethics and Civic Morals*. London: Routledge and Paul, 1957.
Duvall, Evelyn Mills. *Marriage and Family Development*. 5th ed. Philadelphia: J. B. Lippincott, 1977.
Ehle, John. *Trail of Tears: The Rise and Fall of the Cherokee Nation*. New York: Doubleday, 1988.
Ehrenreich, Barbara. *The Hearts of Men: American Dreams and the Flight from Commitment*. Garden City, NY: Doubleday, 1983.
Eisler, Riane. *Dissolution: No-Fault Divorce, Marriage, and the Future of Women*. New York: McGraw-Hill, 1977.
Elazar, Daniel J. *Covenant and Polity in Biblical Israel*. Vol. 1 in *The Covenant Tradition in Politics*. New Brunswick, NJ: Transaction, 1995.
———. *The Covenant Tradition in Politics*. 4 vols. New Brunswick, NJ: Transaction, 1994–98.
Engel, J. Ronald, ed. *Voluntary Associations: Socio-Cultural Analyses and Theological Interpretation*. Chicago: Exposition, 1986.
Engineer, Ali Asghar, ed. *The Shah Bano Controversy*. Bombay: Orient Longman, 1987.
Epstein, Cynthia F. "Law Partners and Marital Partners." *Human Relations* 24 (1971) 549–64.
Erikson, Erik. *Childhood and Society*. 2d. ed. New York: W. W. Norton, 1963.
Etherington, Norman. *The Great Treks: The Transformation of Southern Africa, 1815–1854*. New York: Longman, 2001.
Ethics Resource Center. *Codes of Ethics in Corporations and Trade Associations and the Teaching of Ethics in Graduate Business Schools*. Princeton: Opinion Research Corporation, 1979.
Eurich, Nell P. *Corporate Classroom: The Learning Business*. Princeton: Carnegie Foundation for the Advancement of Teaching, 1985.
Everett, William Johnson. *Blessed Be the Bond: Theological Perspectives on Marriage and Family*. Philadelphia: Fortress, 1985.
———. *God's Federal Republic: Reconstructing our Governing Symbol*. New York: Paulist, 1988.
———. *Gottes Bund und menschliche Öffentlichkeit*, Trans. Gerd Decke. Munich: Christian Kaiser, 1991.
———. *Making My Way in Ethics, Worship, and Wood: An Expository Memoir*. Eugene, OR: Resource Publications, 2021.
———. "*OIKOS*: Convergence in Business Ethics." *Journal of Business Ethics* 5 (1986) 313–25.
———. *The Politics of Worship*. Cleveland: United Church Press, 1999.
———. "Religion and Constitutional Development in India." *Journal of Church and State* 37:1 (1995) 61–85.

Everett, William Johnson, and Thomas Edward Frank. "Constitutional Order in United Methodism and American Culture." In *Connectionalism: Ecclesiology, Mission, and Identity*, edited by Russell E. Richey and Dennis Campbell, 41–73. Nashville: Abingdon, 1997.

Everett, William W. "Body Thinking in Ecclesiology and Cybernetics." PhD diss., Harvard University, 1970.

———. "Liturgy and American Society: An Invocation to Ethical Analysis." *Anglican Theological Review* 56 (January 1974) 16–34.

Fairbank, John K., and Merle Goldman. *China: A New History*. Cambridge, MA: Harvard University Press, 1998.

Farmer, Richard N., and W. Dickerson Hogue. *Corporate Social Responsibility*. 2d ed. Lexington, MA: Lexington, 1985.

Federalist Papers, The. (1788). New York: Penguin, 1961.

Feinberg, Joel. "The Rights of Animals and Unborn Generations." In *Philosophy and Environmental Crisis*, edited by William T. Blackstone, 43–68. Athens: University of Georgia Press, 1974.

Finke, Roger, and Rodney Stark. *The Churching of America 1776–1990. Winners and Losers in our Religious Economy*. New Brunswick, NJ: Rutgers University Press, 1992.

Fisher, Miles Mark. *Negro Slave Songs in the United States*. New York: Citadel, [1953] 1969.

Folberg, H. Jay, ed. *Joint Custody and Shared Parenting*. Washington: Bureau of National Affairs, 1984.

Folberg, H. Jay, and Marva Graham. "Joint Custody of Children Following Divorce." *University of California at Davis Law Review* 12 (1979) 523–81.

Foley, Vincent. "Family Therapy." In *Current Psychotherapies*, 2d ed., edited by Raymond Corsini. Itasca, IL: F. E. Peacock, 1979.

Foreman, Grant. *Indian Removal: The Emigration of the Five Civilized Tribes of Indians*. New ed. Norman, OK: University of Oklahoma Press, 1953.

Fortune, Marie M. "How the Church Should Imitate the Navy." *The Christian Century* 109 (August 26/September 2, 1992) 765–66.

Frank, Thomas Edward. *Polity, Practice, and the Mission of the United Methodist Church*. Rev. ed. Nashville: Abingdon, 2006.

Freed, Doris Jonas, and Henry H. Foster Jr. "Divorce in the Fifty States: An Overview as of 1978." *Family Law Quarterly* 13:1 (Spring 1979) 105–28.

Freire, Paulo. *Pedagogy of the Oppressed*. Trans. Myra B. Ramos. New York: Herder and Herder, 1972.

Friedman, Milton. *Capitalism and Freedom*. Chicago: University of Chicago Press, 1962.

Friedrich, Carl. *The Philosophy of Law in Historical Perspective*. Chicago: University of Chicago Press, 1963.

Friesenhahn, Ernst, Ulrich Scheuner, and Joseph Listl, ed. *Handbuch des Staatskirchenrechts der Bundesrepublik Deutschland*. Berlin: Duncker und Humblot, 1974–1975.

Fustel de Coulanges, Numa Denis. *The Ancient City*. Garden City, NY: Doubleday-Anchor, 1955 [1856]).

———. *The Origin of Property in Land*. Trans. M. Ashley. London: Swan Sonnenschein, 1892. Reprint, New York: B. Franklin, n.d.

Gaffney, Edward McGlynn, Jr., Philip C. Sorenson, and Howard R. Griffin. *Ascending Liability in Religious and Other Non-profit Organizations.* Macon, GA: Mercer University Press, 1984.
Geertz, Clifford. "Ritual and Social Change: A Javanese Example." *American Anthropologist* 59 (1957) 32–54.
Geiger, George R. *The Theory of the Land Question.* New York: Macmillan, 1936.
General Assembly of the Presbyterian Church in the United States. "International Economic Justice." 1980.
George, Henry. *Progress and Poverty.* New York: Robert Schalkenbach Foundation, 1955.
Gilbert, James B. *Work Without Salvation: America's Intellectuals and Industrial Alienation, 1880–1910.* Baltimore: Johns Hopkins University Press, 1977.
Gillett, Richard. *The Human Enterprise: A Christian Perspective on Work.* Kansas City, MO: Leaven, 1985.
Ginzberg, Eli. *Good Jobs, Bad Jobs, No Jobs.* Cambridge, MA: Harvard University Press, 1979.
Goldstein, Joseph, Anna Freud, and Albert Solnit. *Beyond the Best Interests of the Child.* New York: Free Press, 1973.
Goldthorpe, J. E. *Family Life in Western Societies: A Historical Sociology of Family Relationships in Britain and North America.* Cambridge: Cambridge University Press, 1987.
Goode, William J. "The Theoretical Limits of Professionalization." In William J. Goode, *Explorations in Social Theory,* 341–382. New York: Oxford University Press, 1973.
Goudzwaard, Bob, and Harry de Lang. *Beyond Poverty and Affluence: Toward an Economy of Care.* Trans. Mark R. Vander Vennen. Grand Rapids: Eerdmans, 1995.
Gouge, Marcy A. "Joint Custody: A Revolution in Child Custody Law? *Washburn Law Journal* 20 (1980–1981) 326–43.
Gouldner, Alvin W. *The Coming Crisis of Western Sociology.* New York: Basic Books, 1970.
Gramm, W. Philip, and Robert B. Ekelund Jr. "Land Use Planning: The Market Alternative." In *No Land is an Island: Individual Rights and Government Control of Land Use,* Benjamin F. Bobo et al., 127–40. San Francisco: Institute for Contemporary Studies, 1975.
Grant, Balcom. "The Community Land Act: An Overview." *Journal of Planning and Environmental Law* (October 1976) 614–26; (November 1976) 675–90; (December 1976) 732–48.
Greeley, Andrew. *The Denominational Society: A Sociological Approach to Religion in America.* Glenview, IL: Foresman, 1972.
Green, Robert W., ed. *Protestantism, Capitalism and Social Science: The Weber Thesis Controversy.* 2d ed. Lexington, MA: D. C. Heath, 1973.
Greenleaf, Robert K. *Servant Leadership: A Journey into the Nature of Legitimate Power and Greatness.* New York: Paulist, 1977.
Griffiths, Edward. *One Team, One Country: The Greatest Year of Springbok Rugby.* London and New York: Viking Penguin, 1996.
Griswold, A. Whitney. *Farming and Democracy.* New York: Harcourt, Brace, 1948.
Groome, Thomas. *Christian Religious Education.* San Francisco: Harper & Row, 1980.
Grubb, W. Morton, and Lazerson, Marvin. *Broken Promises: How Americans Fail Their Children.* New York: Basic Books, 1982.

Grundlingh, Albert, Andre Odendaal, and Burridge Spies. *Beyond the Tryline: Rugby and South African Society*. Johannesburg: Ravan, 1995.

Gustafson, James. "The Voluntary Church: A Moral Appraisal." In *Voluntary Associations: A Study of Groups in Free Societies*, edited by D. B. Robertson, 299-322. Richmond: John Knox, 1966.

———. "The Burden of the Ethical: Reflections on Disinterestedness and Involvement." *The Foundation* (Gammon Theological Seminary) 66 (Winter 1970) 8-15.

Gutman, Herbert. *Work, Culture and Society in Industrializing America*. New York: Knopf, 1976.

Hall, Francine S. and Douglas E. *The Two-Career Couple*. Reading, MA: Addison-Wesley, 1979.

Hall, Douglas J. *Imaging God: Dominion as Stewardship*. Grand Rapids: Eerdmans, 1986.

Halliburton, Rudi, Jr. *Red Over Black: Black Slavery Among the Cherokee Indians*. Westport, CT: Greenwood, 1977.

Hamilton, Marshall L. *Father's Influence on Children*. Chicago: Nelson-Hall, 1977.

Harper, Charles, ed. *Impunity: An Ethical Perspective. Six Case Studies from Latin America*. Geneva: WCC Publications, 1996.

Harriman, Ann. *The Work/Leisure Trade Off: Reduced Work Time for Managers and Professionals*. New York: Praeger, 1982.

Harrington, James. *Oceana*. In *The Political Works of James Harrington*, edited by J.G.A. Pocock. New York: Cambridge University Press, 1977.

Harris, Marshall. "Private Interest in Private Lands: Intra- and Inter-private." In *Land Use Policy and Problems in the United States*, edited by Howard W. Ottoson, 307-35. Lincoln: University of Nebraska Press, 1963.

Harris, Marshall, and Joseph Ackerman, eds. *Agrarian Reform and Moral Responsibility*. New York: Agricultural Mission, 1949.

Harrison, Paul. *Power and Authority in the Free Church Tradition*. Princeton: Princeton University Press, 1959.

Harriss, C. Lowell, ed. *The Good Earth of America: Planning Our Land Use*. Englewood Cliffs, NJ: Prentice-Hall, 1976.

Hart, Marie M., ed. *Sport in the Socio-cultural Process*. Dubuque: William C. Brown, 1972.

Hatch, Nathan O. "The Puzzle of American Methodism." *Church History* 63:2 (June 1994) 175-89.

Hauerwas, Stanley. *A Community of Character: Toward a Constructive Christian Social Ethic*. Notre Dame: University of Notre Dame, 1981.

Healy, Robert G. *Land Use and the States*. Baltimore: Johns Hopkins University Press, 1976.

Heckman, Norma A., and Rebecca and Jeff Bryson. "Problems of Professional Couples: A Content Analysis." *Journal of Marriage and the Family* 39 (1977) 323-30.

Heim, S. Mark. *Grounds for Understanding: Ecumenical Resources for Responses to Religious Pluralism*. Grand Rapids: Eerdmans, 1998.

Heimert, Alan. *Religion and the American Mind: From the Great Awakening to the Revolution*. Cambridge, MA: Harvard University Press, 1966.

Hellman, John. *American Myth and the Legacy of Vietnam*. New York: Columbia University Press, 1986.

Herman, Jeanne, and Karen K. Gyllstrom. "Working Men and Women: Inter- and Intra-role Conflict." *Psychology of Women Quarterly* 1 (1977) 319-30.

Hertz, Rosanna. *More Equal than Others: Women and Men in Dual-Career Marriages.* Berkeley: University of California Press, 1986.

Hess, Robert D., and Kathleen A. Camara. "Post-Divorce Family Relationships as Mediating Factors in the Consequences of Divorce for Children." *Journal of Social Issues* 35:4 (Fall 1979) 79-96.

Hesse, Konrad. "Die Entwicklung des Staatskirchenrechts seit 1945." *Jahrbuch des Öffentlichen Rechts der Gegenwart, Neue Folge* 10 (1988) 1-80.

Hexham, Irving. *The Irony of Apartheid: The Struggle for National Independence of Afrikaner Calvinism Against British Imperialism.* Toronto: Edwin Mellen Press, 1981.

Heyne, Paul T. *Private Keepers of the Public Interest.* New York: McGraw-Hill, 1968.

Hinds, William A. *American Communities and Co-operative Colonies.* 2d rev. ed. Philadelphia: Porcupine, [1908] 1975.

Hochschild, Arlie. *Second Shift.* New York: Viking Press, 1989.

Hodges, Donald. "Class Analysis and its Presuppositions." *American Journal of Economics and Sociology* 20 (October 1960) 23-38.

―――. "The Class Significance of Ethical Traditions." *American Journal of Economics and Sociology* 20 (April 1961) 241-52.

Hoffmann, Lenore, and Gloria DeSole, eds. *Career and Couples: An Academic Question.* New York; Modern Language Association, 1976.

Holloway, Mark. *Heavens on Earth: Utopian Communities in America, 1680-1880.* 2d ed. rev. New York: Dover, 1966.

Hood, Jane C. *Becoming a Two-Job Family.* New York: Praeger, 1983.

Horton, John. "The Dehumanization of Anomie and Alienation." In *The Sociology of Knowledge: A Reader,* edited by James E. Curtis and John W. Petras, 586-604. New York: Praeger, 1970.

―――. "Order and Conflict Theories of Social Problems as Competing Ideologies." *American Journal of Sociology* 71 (May 1966) 701-13. Reprinted in *Sociology of Sociology,* edited by Larry T. and Janice M. Reynolds, 152-71. New York: McKay, 1970.

Hough, Joseph C., Jr. "Christian Social Ethics as Advocacy." *Journal of Religious Ethics* 5:1 (Spring 1977) 115-33.

Huber, Wolfgang, and Heinz Eduard Tödt. *Menschenrechte: Perspektiven einer menschlichen Welt.* Stuttgart: Kreuz, 1977.

Hunt, Janet G. and Larry L. "Dilemmas and Contradictions of Status: The Case of the Dual-Career Family." *Social Problems* 24 (1977) 407-16.

Hyams, Edward. *Soil and Civilization.* New York: Harper Colophon, 1976.

Illich, Ivan. *De-Schooling Society.* New York: Harper & Row, 1972.

―――. *Gender.* New York: Pantheon, 1982.

Institute of the Church and Urban Industrial Society. *Land Use Issues and the Quality of Urban and Rural Life: An ICUIS Working Bibliography.* Chicago: Institute of the Church and Urban-Industrial Society, 1976.

International Independence Institute. *The Community Land Trust: A Guide to a New Model for Land Tenure in America.* Cambridge, MA: Center for Community Economic Development, 1972.

Irving, Howard, Michael Benjamin, and Nicholas Trocme. "Shared Parenting: An Empirical Analysis Utilizing a Large Data Base." *Family Process* 23 (1984) 561–69.
Jack, Homer. "SANE as a Voluntary Organization." In *Voluntary Associations: A Study of Groups in Free Societies*, edited by D. B. Robertson, 243–54. Richmond: John Knox, 1966.
Jackson, Barbara Ward, and Rene Dubos. *Only One Earth: The Care and Maintenance of a Small Planet*. New York: W. W. Norton, 1972.
Jackson, John A. *Professions and Professionalization*. "Sociological Studies" 3. New York: Cambridge University Press, 1970.
Jacobs, J. W. "Effect of Divorce on Fathers: A Review of the Literature." *American Journal of Psychiatry* 139 (1982) 1235–41.
Jaeger, Werner. "The Moral Value of the Contemplative Life." In *Moral Principles of Action: Man's Ethical Imperative*, edited by R.N. Anshen, 77–93. New York: Harper, 1952.
Jegen, Mary E., and Bruno V. Hanno, eds. *The Earth is the Lord's: Essays on Stewardship*. New York: Paulist Press, 1978.
Jenner, W. J. F. "Afterword." In *Journey to the West*, vol. 3. Trans. W. J. F. Jenner. Beijing: Foreign Languages Press, 1997.
Johnpoll, Bernard K. *Pacifist's Progress: Norman Thomas and the Decline of American Socialism*. Chicago: Quadrangle, 1970.
Johnson v. Johnson, 564 P.2d (Alaska 1977), 71.
Jones, Cathy J. "The Tender Years Doctrine: Survey and Analysis." *Journal of Family Law* 16 (1977–78) 695–738.
Jones, Donald. *Doing Ethics in Business: New Ventures in Management Development*. Cambridge, MA: Oelgeschlager, Gunn and Hain, 1982.
———. "Teaching Business Ethics: State of the Art and Normative Critique." In *Annual of the Society of Christian Ethics*, 1981, 185–215. Waterloo, Ontario: Council on the Study of Religion, 1981.
Jones, Donald, ed. *Business, Religion, and Ethics: Inquiry and Encounter*. Cambridge, MA: Oelgeschlager, Gunn and Hain, 1982.
Jones, Ezra Earl. *Quest for Quality in the Church. A New Paradigm*. Nashville: Discipleship Resources, 1993.
Kamerman, Sheila B. *Parenting in an Unresponsive Society: Managing Work and Family Life*. New York: The Free Press, 1980.
Kamerman, Sheila, and Alfred Kahn. *The Responsive Workplace*. New York: Columbia University Press, 1987.
Kanter, Rosabeth Moss. *Men and Women of the Corporation*. New York: Basic Books, 1977.
———. "The Organization Child: Experience Management in a Nursery School." *Sociology of Education* 45 (Spring 1972) 186–212.
———. *Work and Family in the United States: A Critical Review and Agenda for Research and Policy*. New York: Russell Sage Foundation, 1977.
Kanter, Rosabeth Moss, and Barry Stein. *Life in Organizations: Workplaces as People Experience Them*. New York: Basic Books, 1979.
Kathrada, Ahmed, et al. *The Robben Island Exhibition Esiqithini*. Cape Town: South African Museum and Mayibuye Books, 1996.
Kathrada, Ahmed, Nelson Mandela, and Mxolisi Mgxashe. *Robben Island: The Reunion*. Bellville: Mayibuye Books, 1996.

Kaufmann, Ludwig. *Ein ungelöster Kirchenkonflikt: Der Fall Pfürtner, Dokumente und zeitgeschichtlicher Analysen*. Freiburg/Schweiz: Ed. Exodus, 1987.
Kedroff v. St. Nicholas Cathedral, 344 U.S. 94 (1952)
Keen, Sam. *To a Dancing God*. New York: Harper & Row, 1973.
Kelley, Dean, ed. *Government Intervention in Religious Affairs*. New York: Pilgrim, 1986.
Kelly, Joan B. "Further Observations on Joint Custody." *University of California at Davis Law Review* 16:3 (1983) 762-70.
Kelso, H. H. 1963. "Resolving Land Use Conflicts." In *Land Use Policy and Problems in the United States*, edited by Howard W. Ottoson, 282-303. Lincoln: University of Nebraska Press, 1963.
Khodie, Narmada, ed., *Readings in Uniform Civil Code*. Bombay: Thacker, 1975.
Khory, Kavita R. "The Shah Bano Case: Some Political Implications." In *Religion and Law in Independent India*, edited by Robert D. Baird, 121-38. New Delhi: Manohar, 1981.
King, Duane, and E. Raymond Evans. "Tsali: The Man Behind the Legend." *Journal of Cherokee Studies* 4:4 (Fall 1979) 194-99.
King, William McGuire. "Denominational Modernization and Religious Identity: The Case of the Methodist Episcopal Church." In *Perspectives on American Methodism: Interpretive Essays*, edited by Russell Richey, Kenneth Rowe, Jean M. Schmidt, 343-55. Nashville: Abingdon, 1993.
Kirby, Donald J. *Prophecy vs. Profits: An Ethical Investment Dilemma for Churches*. Maryknoll, NY: Orbis-Probe, 1980.
Kiser, W. David. "Termination of Parental Rights—Suggested Reforms and Responses," *Journal of Family Law* 16 (1977-78) 239.
Kishwar, Madhu, and Ruth Vanita. "Inheritance Rights for Women: A Response to Some Commonly Expressed Fears." *Manushi* 57 (March/April 1990) 3-14.
Kittel, Gerhard. "Leitourgen." In *Theological Dictionary of the New Testament*, trans. G. W. Bromiley, vol. 4, 215-31. Grand Rapids: Eerdmans, 1967.
Koch, James V. "The Economics of 'Big-Time' Intercollegiate Athletics." *Social Science Quarterly* 52 (1971) 248-60.
Kochen, Manfred, and Karl W. Deutsch. "Toward a Rational Theory of Decentralization: Some Implications of a Mathematical Approach." *American Political Science Review* 63 (1969) 734-49.
Krause, Elliot A. *The Sociology of Occupations*. Boston: Little, Brown, 1971.
Krikorian, Robert V. "The Time for Self-Regulation is Now." In *Self Regulation: Conference Proceedings, November 16, 1982*, 61-70. Washington: Ethics Resource Center, 1982.
Krueger, David, and Bruce Grelle, eds. *Christianity and Capitalism: Perspective on Religion, Liberalism and the Economy*. Chicago: Center for the Scientific Study of Religion, 1986.
Krupa, Arlene, with Chris Kirk-Kuwaye. *Couple-Power: How to be Partners in Love and Business*. New York: Dodd-Mead, 1987.
Kubie, Lawrence S. "Provisions for the Care of Children of Divorced Parents: A New Legal Instrument." In *The Rights of Children: Emergent Concepts in Law and Society*, edited by Albert E. Wilkerson, 212-17. Philadelphia: Temple University Press, 1973.
Lang, Martin. *Acquiring our Image of God: The Emotional Basis for Religious Education*. New York: Paulist, 1983.

Langer, Susanne. *Philosophy in a New Key.* New York: New American Library, 1951.
Lantz, Herman, Martin Schultz, and Mary O'Hara. "The Changing American Family from Preindustrial to the Industrial Period: A Final Report." *American Sociological Review* 42 (1977) 406–21.
Lasch, Christopher. *Haven in a Heartless World: The Family Besieged.* New York: Basic Books, 1977.
Lawrence, Bruce. "Woman as Subject/Woman as Symbol: Islamic Fundamentalism and the Status or Women." *Journal of Religious Ethics* 22:1 (Spring 1994) 163–85.
Laycock, Douglas. "The Right to Church Autonomy as Part of Free Exercise of Religion." In *Government Intervention in Religious Affairs*, edited by Dean Kelley, 28–39. New York: Pilgrim, 1986.
Leach, Edmund. "Ritual." In *International Encyclopedia of the Social Sciences*, edited by D. Sills, 13: 520–26. New York: Macmillan, 1968.
Leeds, Georgia Rae. *The United Keetowah Band of Cherokee Indians in Oklahoma.* New York: Peter Lang, 2000.
Leiffer, Murray H. "Affadavit to California Superior Court, San Diego." In *Barr v. United Methodist Church*, January 6, 1978.
———. "The Episcopacy in the Present Day." In *The Study of the General Superintendency of the Methodist Church. A Report to the General Conference of 1964.* The Co-ordinating Council of the Methodist Church, January 1964, 169.
Leopold, Aldo. *A Sand County Almanac, with other Essays on Conservation from Round River.* New York: Oxford University Press, 1966.
Levine, David. "Child Custody: Iowa Corn and the Avant Garde." In *The Rights of Children: Emergent Concepts in Law and Society*, edited by Albert E. Wilkerson, 193–200. Philadelphia: Temple University Press, 1973.
Levinger, George, and Oliver Moles, eds. *Divorce and Separation: Context, Causes, and Consequences.* New York: Basic Books, 1979.
Levinson, Daniel, et al., *Seasons of a Man's Life.* New York: Ballantine, 1979.
Levinson, Harry. *Executive Stress* New York: Harper & Row, 1970.
Lindner, Eileen, Carol Mattis, and June Rogers. *When Churches Mind the Children: A Study of Day Care in Local Parishes.* Ypsilanti, MI: High Scope Press, 1983.
Lippmann, Walter. *The Method of Freedom.* New York: Macmillan, 1934.
Listl, Joseph, ed. *Grundriss des nachkonziliaren Kirchenrechts.* Regensburg: F. Pustel, 1980.
Locke, John. *A Letter Concerning Toleration* [1689]. Ed. Mario Montuori. The Hague: Martinus Nijhoff, 1963.
———. *Two Treatises of Government* [1689]. Ed. Peter Laslett. Cambridge: Cambridge University Press, 1960.
Long, Edward L., Jr. "Christian Ethics and the Problem of Credibility." *Selected Papers of the Annual Meeting of the American Society of Christian Ethics*, 1973.
Lutheran Church in America. "Stewardship of Creation in Human Community." New York: LCA-Division on Mission in North America, 1980.
Lutz, Donald S. *Documents of Political Foundation Written by Colonial Americans: From Covenant to Constitution.* Philadelphia: Institute for the Study of Human Issues, 1986.
Lyles, Jean Caffey. "Methodist Litigations and Public Relations." *The Christian Century* 96:42 (December 19, 1979) 1256–57.
Lynn, Kenneth S., ed. *The Professions in America.* Boston: Houghton Mifflin, 1967.

Lyson, Thomas A. "Husband and Wife Work Roles and the Organization and Operation of Family Farms." *Journal of Marriage and the Family* 47:3 (August 1985) 759-64.
Maccoby, Michael. *The Gamesman: The New Corporate Leaders*. New York: Simon and Schuster, 1976.
Macpherson, C. B. *Hobbes to Locke*. New York: Oxford University Press, 1962.
———. *The Political Theory of Possessive Individualism: Hobbes to Locke*. London: Oxford University Press, 1962.
Maltz, Maxwell. *Psychocybernetics*. New York: Essandess Special Editions, 1967.
Mandela, Nelson. *Long Walk to Freedom: The Autobiography of Nelson Mandela*. Randburg: Macdonald Purnell, 1994.
Mannheim, Karl. *Essays on the Sociology of Knowledge*. New York: Oxford University Press, 1952.
———. *Ideology and Utopia*. Trans. Louis Wirth. New York: Harcourt, Brace, 1936.
Mansfield, John H. "The Personal Laws or a Uniform Civil Code?" In *Religion and Law in Independent India*, edited by Robert D. Baird, 139-78. New Delhi, Manohar, 1981.
Margolis, D. R. *The Managers: Corporate Life in America*. Clifton, NJ: William Morrow, 1979.
Marshall, Gordon. *In Search of the Spirit of Capitalism: An Essay on Max Weber's Protestant Ethic Thesis*. New York: Columbia University Press, 1982.
Martin, Karl, ed. *Frieden staat Sicherheit: Von der Militärseelsorge zum Dienst der Kirche unter den Soldaten*. Gütersloh: Gütersloher Verlagshaus, 1989.
Mary Roy v. State of Kerala. 1986 Supreme Court [India] 1011.
Masi, Dale A. *Human Services in Industry*. Lexington, MA: D. C. Heath, 1982.
Mathew, Babu, et al. *Cases and Materials on Family Law I*. Bangalore: National Law School of India University, 1990.
Mathews, James K. *Set Apart to Serve. The Meaning and Role of Episcopacy in the Wesleyan Tradition*. Nashville: Abingdon, 1985.
May, William W., ed. *Vatican Authority and American Catholic Dissent*. New York: Crossroad, 1987.
McCann, Dennis. *New Experiment in Democracy: The Challenge for American Catholicism*. Kansas City: Sheed & Ward, 1987.
McCarthy, Maureen, and Gail Rosenberg. *Work Sharing Case Studies*. Kalamazoo: W.E. Upjohn, 1981.
McClaughry, John. "The Future of Private Property and its Distribution." *Ripon Quarterly* 2 (Fall 1974).
———. "A Model State Land Trust Act." *Harvard Journal on Legislation* 12:4 (June 1975) 563-609.
McClellan, Grant S., ed. *Land Use in the United States: Exploitation or Conservation*. New York: H. W. Wilson, 1971.
McCoy, Charles S., and J. Wayne Baker. *The Fountainhead of Federalism: Heinrich Bullinger and the Covenantal Tradition*. Philadelphia: Westminster Press, 1991.
McKiernan-Allen, Linda, and Ronald J. Allen. "Colleagues in Marriage and Ministry." In *Women Ministers*, new and expanded edition, edited by Judith L. Weidman, 207-20. San Francisco: Harper & Row, 1985.
Mead, Sidney E. *The Lively Experiment: The Shaping of Christianity in America*. New York: Harper & Row, 1963.

———. "The Nation with the Soul of a Church." *Church History* 36 (September 1967) 262–83.
Meeks, M. Douglas. *God the Economist*. Philadelphia: Fortress, 1987.
Meier, Gretl S. *Job-Sharing: A New Pattern for Work and Life*. Kalamazoo: W.E. Upjohn Institute for Employment Research, 1979.
Meintjes, Johannes. *The Voortrekkers: The Story of the Great Trek and the Making of South Africa*. London: Cassell, 1973.
Meiring, Piet. *Chronicle of the Truth Commission: A Journey through the Past and Present into the Future of South Africa*. Vanderbiljpark: Carpe Diem, 1999.
Merrill, Richard, ed. *Radical Agriculture*. New York: Harper Colophon, 1976.
Merton, Robert. "Social Problems and Sociological Theory." In *Contemporary Social Problems*, edited by Robert K. Merton and Robert Nisbet, 697–737. New York: Harcourt, Brace and World, 1961.
Methodist Episcopal Church, General Conference. *Journal*, 1896.
Meyer, Donald. *The Protestant Search for Political Realism*. Berkeley: University of California Press, 1960.
Michaelsen, Robert. *Piety in the Public Schools*. New York: Macmillan, 1970.
Michener, James. *The Covenant*. New York: Ballantine, 1980.
Miller, Allen O., ed. *A Christian Declaration on Human Rights*. Grand Rapids: Eerdmans, 1977.
Miller, David J. "Joint Custody." *Family Law Quarterly* 13:3 (1979) 345–412.
Miller, David L. *Gods and Games: Toward a Theology of Play*. New York: Harper & Row, 1973.
Miller, John W. "The Contemporary Fathering Crisis: The Bible and Research Psychology." *Conrad Grebel Review* (1983) 21–37.
Miller, Perry. *The New England Mind: The Seventeenth Century*. Boston: Beacon, 1961.
Miller, Sherod, Daniel Wachman et al. *Working Together: Improving Communication on the Job*. Minneapolis: Interpersonal Communications Programs, 1980.
Milton, John. "The Readie and Easie Way to Establish a Free Commonwealth and the Excellence Thereof Compared with the Inconveniences and Dangers of Readmitting Kingship in This Nation." In *Complete Prose Works of John Milton*, vol. 7, 1659–60, rev. ed., 407–63. New Haven, Yale University Press, 1980.
Minuchin, Salvador. *Families and Family Therapy*. Cambridge, MA: Harvard University Press, 1974.
Mitchell, Henry. *Black Preaching*. Philadelphia: J. P. Lippincott, 1970.
Moede, Gerald F. "Bishops in the Methodist Tradition: Historical Perspectives." In *Episcopacy. Lutheran—United Methodist Dialogue II*, edited by Jack M. Tuell and Roger W. Field, 52–69. Minneapolis: Augsburg, 1991.
Moede, Gerald F., ed. *The COCU Consensus. In Quest of a Church of Christ Uniting*. Princeton: Consultation on Church Union, 1985.
Mohammed Ahmed Khan v. Shah Bano Begum and others. 1985 Supreme Court Cases 556 [India].
Moltmann, Jürgen. *A Theology of Hope: On the Ground and the Implications of a Christian Eschatology*. New York: Harper & Row, 1967.
———. *The Coming of God: Christian Eschatology*. Trans. Margaret Kohl. Minneapolis: Fortress, 1996.
———. *Theology of Play*. Trans. Reinhard Ulrich. New York: Harper and Row, 1972.

Moodie, T. Dunbar. *The Rise of Afrikanerdom: Power, Apartheid, and the Afrikaner Civil Religion*. Berkeley: University of California Press, 1975.
Mooney, James. *Myths of the Cherokee and Sacred Formulas of the Cherokees*. Nashville: Charles and Randy Elder, 1982.
Moore, John M. *The Long Road to Methodist Union*. New York: Abingdon-Cokesbury, 1943.
Moore, Leroy, Jr. "From Profane to Sacred America: Religion and the Cultural Revolution in the United States." *Journal of the American Academy of Religion* 39 (1971) 321–38.
Morgan, Gareth. *Images of Organization*. Newbury Park and London: Sage, 1986.
Morgenbesser, Mel, and Nadine Nehls. *Joint Custody: An Alternative for Divorcing Families*. Chicago: Nelson-Hall, 1981.
Mount, Eric, Jr. *Professional Ethics in Context: Institutions, Images, and Empathy*. Louisville: Westminster/John Knox, 1990.
Mudge, Lewis S. *The Church as Moral Community: Ecclesiology and Ethics in Ecumenical Debate*. New York: Continuum, and Geneva: WCC Publications, 1998.
Naidoo, Indres. *Island in Chains: Ten Years on Robben Island by Prisoner 885/63*. Harmondsworth: Penguin, 1982.
Nally v. Grace Community Church of the Valley. 47 Cal. 3d 278 (1988).
National Conference of Catholic Bishops. "U.S. Bishops' Pastoral Letter on Catholic Social Teaching and the U.S. Economy." Washington: US Catholic Conference, 1984.
National Conference of State Legislatures. "50 State Overview of Joint Custody Statutes as of Legislative Year 1982," Dec. 15, 1983.
National Labor Relations Board v. The Catholic Bishop of Chicago, et al. 440 US 490 (1979).
Nauright, John. "Sustaining Masculine Hegemony: Rugby and the Nostalgia of Masculinity." In *Making Men: Rugby and Masculine Identity*, edited by John Nauright and Timothy J. L. Chandler, 227–44. London: Frank Cass, 1996.
Nauright, John, and David Black. "'Hitting Them Where It Hurts': Springbok-All Black Rugby, Masculine National Identity and Counter-Hegemonic Struggle, 1959–1992." In *Making Men: Rugby and Masculine Identity*, edited by John Nauright and Timothy J. L. Chandler, 205–26. London: Frank Cass, 1996.
Neale, Robert E. *In Praise of Play: Toward a Psychology of Religion*. New York: Harper and Row, 1969.
Neilson, Francis. *The Cultural Tradition and Other Essays*. Freeport, NY: Books for Libraries Press, 1969.
Neitz, Mary Jo. *Charisma and Community: A Study of Religious Commitment within the Charismatic Renewal*. New Brunswick, NJ: Transaction, 1987.
Nelton, Sharon. *In Love and in Business: How Entrepreneurial Couples are Changing the Rules of Business and Marriage*. New York: John Wiley & Sons, 1986.
Netzer, Dick. *Economics of the Property Tax*. Washington: Brookings Institute, 1965.
Neuhaus, Richard John. *In Defense of People*. New York: Macmillan, 1971.
———. *The Naked Public Square: Religion and Democracy in America*. Grand Rapids: Eerdmans, 1984.
———. "The War, the Churches, and Civil Religion." *The Annals* 387 (January 1970) 128–40.
Niebuhr, H. Richard. *Christ and Culture*. New York: Harper & Row, 1956 [1951].

———. *The Kingdom of God in America*. New York: Harper & Row, 1937.
O'Brien, Sharon. *American Indian Tribal Governments*. Norman, OK: University of Oklahoma Press, 1989.
O'Neill, Onora, and William Ruddick, eds. *Having Children: Philosophical and Legal Reflections on Parenthood*. New York: Oxford University Press, 1979.
Olmstead, Barney, and Suzanne Smith. *The Job Sharing Handbook*. Berkeley: Ten Speed Press, 1983.
Open Space Action Committee. *Stewardship: The Land, the Landowner, the Metropolis*. New York: Open Space Action Committee, 1965.
Orna, Mary Virginia. *Cybernetics, Society, and the Church*. Dayton, OH: Pflaum, 1969.
Ossowska, Maria. *Social Determinants of Ideas*. Philadelphia: University of Pennsylvania Press, 1969.
Ottoson, Howard W., ed. *Land Use Policy and Problems in the United States*. Lincoln: University of Nebraska Press, 1963.
Palmer, Parker. *To Know as We Are Known: A Spirituality of Education*. San Francisco: Harper & Row, 1983.
Papanek, Hannah. "Men, Women and Work: Reflections on the Two-Person Career." *American Journal of Sociology* 78:4 (January 1973) 853–72.
Parsons, Talcott. "Christianity and Modern Industrial Society." In *Sociological Theory and Modern Society*, 385–421. New York: The Free Press, 1967.
Passmore, John. *Man's Responsibility for Nature: Ecological Problems and Western Traditions*. New York: Charles Scribner's Sons, 1974.
Pennock, J. Roland, and John W. Chapman, eds. *Voluntary Associations*. Nomos XI. New York: Lieber-Atherton, 1969.
Pepitone-Rockwell, Fran, ed. *Dual Career Couples*. Beverly Hills: Sage, 1980.
Perdue, Theda. *Slavery and the Evolution of Cherokee Society, 1540–1866*. Knoxville: University of Tennessee Press, 1979.
Peters, Thomas J., and Robert H. Waterman Jr. *In Search of Excellence: Lessons from America's Best-Run Companies*. San Francisco: Harper and Row, 1982.
Pignone, Mary Margaret. "Concentrated Ownership of Land." In *The Earth is the Lord's: Essays on Stewardship*, edited by Mary E. Jegen and Bruno V. Hanno, 112–29. New York: Paulist, 1978.
Pingree, Suzanne et al. "Anti-Nepotism's Ghost: Attitudes of Administrators Toward Hiring Professional Couples." *Psychology of Women Quarterly* 3:1 (Fall 1978) 22–29.
Pinto, William E. *Law of Marriage and Matrimonial Reliefs for Christians in India*. Bangalore: Theological Publications in India, 1991.
Pleck, Joseph H. *Working Wives/Working Husbands*. Beverly Hills: Sage, 1985.
Portner, Joyce, with Larry A. Etkin. *Work and Family: Friends or Foes*. Minneapolis: Minnesota Council on Family Relations, 1978.
Poster, Mark. *Critical Theory of the Family*. New York: Seabury, 1978.
Powers, Charles W. *Ethics in the Education of Business Managers*. Hastings-on-Hudson, NY: Hastings Institute of Society, Ethics, and the Life Sciences, 1980.
Preller, Gustav S. *Voortrekkermense*. 6 vols. Kaapstad: National Pers, 1918–1938.
Provost, James, and Knut Walf, eds. *Canon Law—Church Reality*. Edinburgh: T&T Clark, 1986.
Provost, James, and Knut Walf, eds. *The Tabu of Democracy Within the Church*. Concilium 5. London: SCM Press, 1992.

Quaritsch, Helmut, and Hermann Weber, eds. *Staat und Kirchen in der Bundesrepublik: Staats-kirchliche Aufsätze*. Bad Homburg: Gehlen, 1967.
Railings, E. M., and David J. Pratte. *Two-Clergy Marriages: A Special Case of Dual Careers*. Lanham, MD: University Press of America, 1984.
Raines, John, and Donna Day-Lower. *Modern Work and Human Meaning*. Philadelphia: Westminster, 1986.
Raiser, Konrad. *Ecumenism in Transition*. Trans. Tony Coates. Geneva: WCC Publications, 1991.
Rapoport, Rhona and Robert. *Fathers, Mothers, and Society: Towards New Alliances*. New York: Basic Books, 1977.
Rapoport, Rhona and Robert N., eds. *Working Couples*. New York: Harper & Row, 1978.
Rauschenbusch, Walter. *Christianity and the Social Crisis*. New York: Harper, 1964.
———. *A Theology for the Social Gospel*. New York: Abingdon, 1945.
Rawls, John. *A Theory of Justice*. Cambridge, MA: Harvard University Press, 1971.
Reuther, Rosemary. *Gaia and God: An Ecofeminist Theology of Earth Healing*. San Francisco: Harper, 1992.
Reynolds, Larry T. and Janice M., eds. *Sociology of Sociology*. New York: McKay, 1970.
Richardson, Elliott. *Work in America: Report of a Special Task Force to the Secretary of Health, Education, and Welfare*. Cambridge, MA: MIT Press, 1978.
Richey, Russell, and Donald Jones, ed. *American Civil Religion*. New York: Harper & Row, 1974.
Rightor, Henry Haskell. *Pastoral Counseling in Work Crises: An Introduction for both Lay and Ordained Ministers*. Valley Forge, PA: Judson, 1979.
Ringeling, Hermann. "Der Fall Stephan Pfürtner." *Zeitschrift für evangelische Ethik* 32:4 (1988) 292–96.
Robertson, D. B., ed. *Voluntary Associations: A Study of Groups in Free Societies*. Richmond: John Knox Press, 1966.
Rodgers, Daniel T. *The Work Ethic in Industrial America, 1850–1920*. Chicago: University of Chicago Press, 1978.
Rose, Arnold. *The Power Structure: Political Process in American Society*. New York: Oxford University Press, 1967.
Roth, Allan. "The Tender Years Presumption in Child Custody Disputes." *Journal of Family Law* 15:3 (1976–77) 423–62.
Rowatt, G. Wade and Mary Jo. *The Two-Career Marriage*. Philadelphia: Westminster, 1980.
Rubington, Earl, and Martin Weinberg. *The Study of Social Problems: Five Perspectives*. New York: Oxford University Press, 1971.
Rudge, Peter F. *Ministry and Management*. London: Tavistock Publications, 1968.
Russell, Letty M. *Household of Freedom: Authority in Feminist Theology*. Philadelphia: Westminster, 1987.
Sakolski, Aaron M. *Land Tenure and Land Taxation in America*. New York: Robert Schalkenbach Foundation, 1957.
Salisbury, Harrison. *The Long March: The Untold Story*. New York: Macmillan, 1985.
Santmire, Paul H. *Brother Earth: Nature, God and Ecology in Time of Crisis*. New York: Thomas Nelson, 1970.
Santosky v. Kramer. 102 S. Cl. 1388 (1982).
Schadeberg, Jürgen, comp. *Voices from Robben Island*. Randburg: Ravan, 1994.
Schall v. Schall. 251 Pennsylvania Superior Court 262.380 A.2d (1977), 478.

Schmidt, Warren H., and Barry Z. Posner. *Managerial Values and Expectations: The Silent Power in Personal and Organizational Life*. New York: American Management Association, 1982.

Schmidt-Eichstaedt, Gerd. *Kirchen als Körperschaften des Öffentlichen Rechts*. Köln: Carl Heymans, 1975.

Schulman, Joanne, and Valerie Pitt. "Second Thoughts on Joint Custody Analysis of Legislation and its Impact for Women and Children." *Golden Gate University Law Review* 12:3 (1982) 539-77.

Schumacher, E. F. *Small is Beautiful: Economics as if People Mattered*. New York: Harper and Row, 1973.

Schüssler-Fiorenza, Elisabeth. *In Memory of Her: A Feminist Theological Reconstruction of Christian Origins*. New York: Crossroad, 1984.

Scruggs, Jan C. and Joel L. Swerdlow. *To Heal a Nation: The Vietnam Veterans Memorial*. New York: Harper & Row, 1985.

Segalen, Martine. *Historical Anthropology of the Family*. Trans. J. C. Whitehouse and S. Matthews. Cambridge: Cambridge University Press, 1986.

Senge, Peter. *The Fifth Discipline. The Art and Practice of the Learning Organization*. New York: Doubleday Business, 1990.

Serbian Orthodox Diocese v. Milivojevic. 426 U.S. 696 (1976).

Seymour, Jack L., Robert T. O'Gorman, and Charles R. Foster. *The Church in the Education of the Public: Refocusing the Task of Religious Education*. Nashville: Abingdon, 1984.

Sheldon, Charles. *In His Steps: "What Would Jesus Do?"* New York: Hurst, n.d.

Shelton v. Tucker. 364 U.S. 479. 488 (1960).

Short, Roy H. *Chosen to be Consecrated. The Bishops of The Methodist Church, 1784-1968*. Lake Junaluska, NC: Commission on Archives and History, 1976.

Sider, Ronald J. 1978. "A Biblical Perspective on Stewardship." In *The Earth is the Lord's: Essays on Stewardship*, edited by Mary E. Jegen and Bruno V. Hanno, 1-21. New York: Paulist Press, 1978.

Siegan, Bernard H. "No Zoning is the Best Zoning." In *No Land is an Island: Individual Rights and Government Control of Land Use*, Benjamin F. Bobo et al., 157-67. San Francisco: Institute for Contemporary Studies, 1975.

Simon, Yves. *Work, Society and Culture*. Trans. V. Kuic. New York: Fordham University Press, 1971.

Smit, Dirk J. "Covenant and Ethics? Comments from a South African Perspective." In *Annual of the Society of Christian Ethics*, 265-82. Washington: Society of Christian Ethics, 1996.

Smith, Charles L., ed. *A Bibliography on Land Reform in America*. San Francisco: Center for Rural Studies, 1974.

Smith, David H. "The Importance of Formal Voluntary Organizations for Society." *Sociology and Social Research* 1 (July 1966) 483-96.

Smith, Elwyn A., ed. *The Religion of the Republic*. Philadelphia: Fortress, 1971.

Smith, Harmon L. *Where Two or Three Are Gathered*. Cleveland: Pilgrim, 1995.

Smock, Audrey Chapman, ed. *Christian Faith and Economic Life*. New York: United Church of Christ Board for World Ministries, 1987.

Solomon, Robert C., and Kristine R. Hanson. *Above the Bottom Line: An Introduction to Business Ethics*. New York: Harcourt, Brace, Jovanovich, 1983.

Sommer, Elyse and Mike. *The Two-Boss Business: The Joys and Pitfalls of Working and Living Together—And Still Remaining Friends*. New York: Butterick, 1980.
Soulen, Richard. "Black Worship and Hermeneutic." *The Christian Century*, Feb. 11, 1970, 168–71.
Spanier, Graham, and Paul Glick. "Marital Instability in the United States. Some Correlates and Recent Changes." *Family Relations* 30:3 (1981) 329–38.
Sparks, Allister. *The Mind of South Africa*. New York: Alfred A. Knopf, 1990.
Spector, Malcolm, and John Kitsuse. "Social Problems: A Reformulation." *Social Problems* 21 (Fall 1973) 145–58.
Speltz, George H. "Restrict Ownership for the Common Good? Here's Where the Church Stands." *Rural Life* 25:3 (March 1976) 3–5.
Spring, David and Eileen, eds. *Ecology and Religion in History*. New York: Harper & Row, 1974.
Stackhouse, Max L. *Creeds, Societies and Human Rights*. Grand Rapids: Eerdmans, 1986.
———. *Ethics and the Urban Ethos*. Boston: Beacon, 1972.
———. *Public Theology and Political Economy: Christian Stewardship in Modern Society*. Grand Rapids: Eerdmans, 1987.
Staines, Graham L., and Joseph H. Pleck. *The Impact of Work Schedules on the Family*. Ann Arbor, MI: Institute of Social Research, 1983.
Stassen, Glen H. *Just Peacemaking: Transforming Initiatives for Justice and Peace*. Louisville: Westminster/John Knox, 1992.
Stein, Barry. "The Company Family: The Early Years of the Andrews Winery." In *Life in Organizations: Workplaces as People Experience Them*, edited by Rosabeth Moss Kanter and Barry Stein, 290–301. New York: Basic Books, 1979.
Steinman, Susan. "Joint Custody: 'What We Know, What We Have Yet to Learn, and the Judicial and Legislative Implications'." *University of California at Davis Law Review* 16 (1983) 739–62.
Stevens, Edward. *Business Ethics*. New York: Paulist, 1979.
Stone, Christopher D. *Should Trees Have Standing? Toward Legal Rights for Natural Objects*. Los Altos, CA: William Kaufman, 1974.
Stout, Janis P. *The Journey Narrative in American Literature: Patterns and Departures*. Westport, CT: Greenwood, 1983.
Straughn, James H. *Inside Methodist Union*. Nashville: The Methodist Publishing House, 1958.
Superior Court of California, County of San Diego. "Minute Order," No. 40-4611 (1978).
Swartly, Willard M. "Biblical Sources of Stewardship." In *The Earth is the Lord's: Essays on Stewardship*, edited by Mary E. Jegen and Bruno V. Hanno, 22–43. New York: Paulist, 1978.
Sweet, Leonard. *The Minister's Wife: Her Role in Nineteenth-Century American Evangelicalism*. Philadelphia: Temple University Press, 1982.
Sweet, William Warren. *Methodism in American History*. Nashville: Abingdon, 1954.
Swidler, Leonard. "Demokratia, The Rule of the People of God, or Consensus Fidelium." In *Authority in the Church and the Schillebeeckx Case*, edited by Leonard Swidler and Piet F. Fransen, 226–43. New York: Crossroad, 1982.
Swidler, Leonard W., ed. *Küng in Conflict*. Garden City, NY: Doubleday, 1981.

Swidler, Leonard W. and Piet F. Fransen, eds. *Authority in the Church and the Schillebeeckx Case.* New York: Crossroad, 1982.

Taylor, Mary G., and Shirley F. Hartley. "The Two-Person Career: A Classic Example." *Sociology of Work and Occupations* 2:4 (November 1975) 354-72.

TeSelle, Sallie, ed. *The Family, Communes, and Utopian Societies.* San Francisco: Harper & Row, 1973.

Thangaraj, M. Thomas. *Relating to People of Other Religions: What Every Christian Needs to Know.* Nashville: Abingdon, 1997.

Thornton, Russell. "Demography of the Trail of Tears." In *Cherokee Removal*, edited by William L. Anderson, 75-93. Athens, GA: University of Georgia Press, 1991.

Tigert, John J. *A Constitutional History of American Episcopal Methodism.* 6th ed., revised and enlarged. Nashville: Publishing House of the M. E. Church, South, 1916.

Tillich, Paul. "The Religious Symbol." Trans. James L. Adams. *Journal of Liberal Religion* 2 (1940) 13-14.

Timmons, John F., and William G. Murray, eds. *Land Problems and Policies.* New York; Arno Press, 1972 [1950].

Tobin, Richard J. "Some Observations on the Use of State Constitutions to Protect the Environment." *Environmental Affairs* 3:3 (1974) 474-93.

Tocqueville, Alexis de. *Democracy in America.* Trans. Henry Reeve. Cambridge: Sever and Francis, 1863.

Trigg, et al, vs. Pacific Methodist Investment Fund, et al. US District Court for the Southern District of California, No. 78-0198-S (1978).

Troeltsch, Ernst. *The Social Teachings of the Christian Churches.* Trans. Olive Wyon. New York: Harper, 1960.

Tufte, Virginia, and Barhara Myerhoff, eds. *Changing Images of the Family.* New Haven: Yale University Press, 1979.

Turner, C., and M.N. Hodge. "Occupations and Professions." In *Professions and Professionalization*, "Sociological Studies" 3, edited by John A. Jackson, 17-50. New York: Cambridge University Press, 1970.

Turner, Louis, ed. *Multinational Companies and the Third World.* New York: Hill and Wang, 1974.

Ulrich, David N., and Harry P. Dunne Jr. *To Love and Work: A Systemic Interlocking of Family, Workplace and Career.* New York: Brunner/Mazel, 1986.

United Methodist Church. "Report to the 1980 General Conference Regarding the Pacific Homes Litigation." In *General Conference Proceedings*, 1980, Exhibit C, No. 17, 42-48.

United States Catholic Conference. *The Küng Dialogue: Facts and Documents.* Washington: United States Catholic Conference, 1980.

van der Bent, Ans T. "Ideology and Ideologies in Current Ecumenical Thinking and Understanding." 1972. Available from the Library, World Council of Churches, Geneva, Switzerland.

Vernon, Raymond. *Storm Over the Multinationals: The Real Issues.* Cambridge, MA: Harvard University Press, 1977.

Victor, Ira, and Winkler, Win Ann. *Fathers and Custody.* New York: Hawthorn, 1977.

von Lackum, John P., and Nancy Jo Kemper. *Clergy Couples.* New York: National Council of Churches, 1979.

von Rad, Gerhard. "The Promised Land and Yahweh's Land in the Hexateuch." In *The Problem of the Hexateuch and Other Essays*, trans. E. W. T. Dicken, 79–93. New York: McGraw-Hill, 1966.
Voydanoff, Patricia, ed. *Work and Family: Changing Roles of Men and Women*. Mountain View, CA: Mayfield, 1984.
Wagner, Philip L. *Environments and Peoples*. Englewood Cliffs, NJ: Prentice-Hall, 1972.
Wagner, Rudolf G. "Monkey King Subdues the White Bone Demon: A Study in PRC Mythology." In *Contemporary Chinese Historical Drama: Four Studies*, ch. 3. Berkeley: University of California Press, 1990.
———. "Reading the Chairman Mao Memorial Hall in Peking: The Tribulations of the Implied Pilgrim." In *Pilgrims and Sacred Sites in China*, edited by Susan Naquin and Chuen-fang Yue, 378–423. Berkeley: University of California Press, 1996.
Wald, Michael S. "Children's Rights: A Framework for Analysis." *University of California at Davis Law Review* 12 (1979) 255–82.
Walker, Eric A. *The Great Trek*. 4th ed. London, A. & C. Black, 1960.
Wallerstein, Judith S., and Joan B. Kelly. *Surviving the Breakup: How Children and Parents Cope with Divorce*. New York: Basic Books, 1980.
Walton, Clarence C. *Ethos and the Executive: Values in Managerial Decision-Making*. Englewood Cliffs, NJ: Prentice-Hall, 1969.
Ware, Ciji. *Sharing Parenthood after Divorce: An Enlightened Custody Guide for Mothers, Fathers and Kids*. New York: Viking, 1982.
Watson v. Jones. U.S. (13 Wall) 679 (1872).
Weber, Max. "Proposal for the Study of Voluntary Associations." A lecture delivered in Frankfurt, Germany, in 1911. Ms. trans. Everett C. Hughes. Cited in James Luther Adams, "The Protestant Ethic with Fewer Tears." In *Voluntary Associations: Socio-Cultural Analyses and Theological Interpretation*, edited by J. Ronald Engel, 117. Chicago: Exposition Press, 1986.
———. *From Max Weber: Essays in Sociology*. Trans. and ed. by H. H. Gerth and C. Wright Mills. New York: Oxford/Galaxy, 1958.
———. *The Protestant Ethic and the Spirit of Capitalism*. Trans. Talcott Parsons. New York: Scribner's, 1985.
Weeks, Kent M. "The Methodist Approach." In *Ascending Liability in Religious and Other Non-profit Organizations*. edited by Edward McGlynn Gaffney Jr., Philip C. Sorenson, and Howard R. Griffin, 133–37. Macon, GA: Mercer University Press, 1984.
Weiss, Paul. *Sport: A Philosophic Inquiry*. Carbondale, IL: Southern Illinois University Press, 1969.
Wheatley, Margaret. *Leadership and the New Science*. Oakland, CA: Berrett-Koehler, 1992.
Wheeler, John. *Touched with Fire: The Future of the Vietnam Generation*. New York: Franklin Watts, 1984.
Wheeler, John, and A. D. Horne. *The Wounded Generation: America after Vietnam*. Englewood Cliffs, NJ: Prentice-Hall, 1981.
White, Kendall O. "Constitutional Norms and the Formal Organization of American Churches." *Journal of Sociological Analysis* 33:2 (1972) 95–109.
Whitehead, James and Evelyn. *Marrying Well: Possibilities in Christian Marriage Today*. New York: Doubleday, 1981.
Wiener, Norbert. *Cybernetics*. 2d ed. Cambridge, MA: M.I.T. Press, 1965.

———. *The Human Use of Human Beings: Cybernetics and Society.* 2d rev. ed. Garden City, NY: Doubleday, 1954.

Wilber, Charles K. "The Role of Property in an Economic System." In *The Earth is the Lord's: Essays on Stewardship,* edited by Mary E. Jegen and Bruno V. Manno, 70-80. New York: Paulist Press, 1978.

Wilkerson, Albert E., ed. *The Rights of Children: Emergent Concepts in Law and Society.* Philadelphia: Temple University Press, 1973.

Williams, Oliver. "Who Cast the First Stone?" *Harvard Business Review* 84 (September-October, 1984) 151-60.

Williams, Oliver, and John Houck. *Full Value: Cases in Christian Business Ethics.* San Francisco: Harper and Row, 1978.

Wilson, Dick. *The Long March 1935: The Epic of Chinese Communism's Survival.* New York: Viking, 1971.

Winter, J. Alan. "Elective Affinities between Religious Beliefs and Ideologies of Management in Two Eras." *American Journal of Sociology* 79 (March 1974) 1134-50.

Wogaman, J. Philip. *The Great Economics Debate: An Ethical Analysis.* Philadelphia: Westminster, 1977.

Wolcott, Roger T. "The Church and Social Action: Steelworkers and Bishops in Youngstown." *Journal for the Scientific Study of Religion* 21:1 (March 1982) 71-79.

Wolf, William J. *The Religion of Abraham Lincoln.* Rev. ed. New York: Seabury, 1963.

Woodhouse, A.S.P., and William Clark, eds. *Puritanism and Liberty.* 2d ed. Chicago: University of Chicago Press, 1965.

Woodruff, A. M. "A Comparison between Henry George and Karl Marx in their Approach to Land Reform." In *Readings in Land Reform,* edited by Sein Lin, 1-31. Taipei: Good Friend Press, 1970.

World Council of Churches Central Committee. "Statement on the New International Economic Order." *Church Alert* (Geneva) 16 (September-October 1977) 16-17.

Wrenn, Lawrence, ed. *Divorce and Remarriage in the Catholic Church.* New York: Paulist, 1973.

Wrich, James T. *The Employee Assistance Program Updated for the 1980s.* Minneapolis: Hazelden Foundation, 1980.

Wriggins, Sally Hovey. *Xuanzang: A Buddhist Pilgrim on the Silk Road.* Boulder, CO: Westview, 1996.

Wuthnow, Robert. "The Moral Crisis in American Capitalism." *Harvard Business Review* 60 (March-April, 1982) 76-84.

Yang, Benjamin. *From Revolution to Politics: Chinese Communists on the Long March.* Boulder, CO: Westview, 1990.

Young, James. *Growing Through Divorce.* New York: Paulist, 1979.

Young, Michael, and Peter Willmott. *The Symmetrical Family.* New York: Pantheon, 1973.

www.ingramcontent.com/pod-product-compliance
Lightning Source LLC
Chambersburg PA
CBHW050612300426
44112CB00012B/1477